Advances in Intelligent Systems and Computing

Volume 401

Series editor

Janusz Kacprzyk, Polish Academy of Sciences, Warsaw, Poland
e-mail: kacprzyk@ibspan.waw.pl

About this Series

The series "Advances in Intelligent Systems and Computing" contains publications on theory, applications, and design methods of Intelligent Systems and Intelligent Computing. Virtually all disciplines such as engineering, natural sciences, computer and information science, ICT, economics, business, e-commerce, environment, healthcare, life science are covered. The list of topics spans all the areas of modern intelligent systems and computing.

The publications within "Advances in Intelligent Systems and Computing" are primarily textbooks and proceedings of important conferences, symposia and congresses. They cover significant recent developments in the field, both of a foundational and applicable character. An important characteristic feature of the series is the short publication time and world-wide distribution. This permits a rapid and broad dissemination of research results.

More information about this series at http://www.springer.com/series/11156

Krassimir T. Atanassov · Oscar Castillo
Janusz Kacprzyk · Maciej Krawczak
Patricia Melin · Sotir Sotirov
Evdokia Sotirova · Eulalia Szmidt
Guy De Tré · Sławomir Zadrożny
Editors

Novel Developments in Uncertainty Representation and Processing

Advances in Intuitionistic Fuzzy Sets and Generalized Nets – Proceedings of 14th International Conference on Intuitionistic Fuzzy Sets and Generalized Nets

Editors
Krassimir T. Atanassov
Institute of Biophysics and Biomedical Engineering
Bulgarian Academy of Sciences
Sofia
Bulgaria

Oscar Castillo
Division of Graduate Studies
Tijuana Institute of Technology
Tijuana, Baja California
Mexico

Janusz Kacprzyk
Systems Research Institute
Polish Academy of Sciences
Warsaw
Poland

Maciej Krawczak
Warsaw School of Information Technology
Systems Research Institute
Polish Academy of Sciences
Warsaw
Poland

Patricia Melin
Division of Graduate Studies and Research
Tijuana Institute of Technology
Tijuana, Baja California
Mexico

Sotir Sotirov
Intelligent Systems Laboratory
University Professor Dr. Assen Zlatarov Burgas
Burgas
Bulgaria

Evdokia Sotirova
Intelligent Systems Laboratory
University Professor Dr. Assen Zlatarov Burgas
Burgas
Bulgaria

Eulalia Szmidt
Systems Research Institute
Polish Academy of Sciences
Warsaw
Poland

Guy De Tré
Department of Telecommunications
and Information Processing
Ghent University
Gent
Belgium

Sławomir Zadrożny
Systems Research Institute
Polish Academy of Sciences
Warsaw
Poland

ISSN 2194-5357 ISSN 2194-5365 (electronic)
Advances in Intelligent Systems and Computing
ISBN 978-3-319-26210-9 ISBN 978-3-319-26211-6 (eBook)
DOI 10.1007/978-3-319-26211-6

Library of Congress Control Number: 2015953272

Springer Cham Heidelberg New York Dordrecht London

Printed on acid-free paper

Springer International Publishing AG Switzerland is part of Springer Science+Business Media (www.springer.com)

Preface

This volume contains, first of all, the papers presented at the Fourteenth International Workshop on Intuitionistic Fuzzy Sets and Generalized Nets (IWIFSGN-2015) organized, and collocated with the Flexible Query Answering Systems 2015 (FQAS-2015) held on October 26–28, 2015 in Cracow, Poland. Moreover, the volume contains some papers of a particular relevance not presented at the Workshop.

This Workshop is a next edition of a series of the IWIFSGN Workshops organized for years by the Systems Research Institute, Polish Academy of Sciences, Warsaw, Poland, Institute of Biophysics and Biomedical Engineering, Bulgarian Academy of Sciences, Sofia, Bulgaria, and WIT—Warsaw School of Information Technology, Warsaw, Poland, and co-organized by: Matej Bel University, Banska Bystrica, Slovakia, Universidad Publica de Navarra, Pamplona, Spain, Universidade de Tras-Os-Montes e Alto Douro, Vila Real, Portugal, Prof. Asen Zlatarov University, Burgas, Bulgaria, Complutense University, Madrid, Spain, and the University of Westminster, Harrow, UK.

The workshop is mainly devoted to the presentation of recent research results in the broadly perceived fields of intuitionistic fuzzy sets and generalized nets initiated by Professor Krassimir T. Atanassov whose constant inspiration and support is crucial for such a widespread growing popularity and recognition of these areas.

An important contribution of the Workshop is also the fact that it greatly facilitates, and often makes it possible, a deeper discussion on papers presented which as a rule results then in new collaborative works and a further progress in the areas.

The workshop has been partially supported, financially and technically, by many organizations, notably: Systems Research Institute, Polish Academy of Sciences; Department IV of Engineering Sciences, Polish Academy of Sciences; Cracow Branch, Polish Academy of Sciences; Academia Europaea—The Hubert Curien Initiative Fund; Ghent University; Polish Association of Artificial Intelligence, and Polish Operational and Systems Research Society. Their support is acknowledged and highly appreciated.

We hope that the collection of main contributions presented at the workshop, completed with papers by leading experts who have not been able to participate, will provide a source of much needed information and inspiration on recent trends in the topics considered.

We wish to thank all the authors for their excellent contributions and their collaboration during the editing process of the volume. We are looking forward to the same fruitful collaboration during the next workshops of this series that are planned for the years to come. Special thanks are due to the peer reviewers whose excellent and timely work has significantly contributed to the quality of the volume.

And last but not least, we wish to thank Dr. Tom Ditzinger, Dr. Leontina di Cecco, and Mr. Holger Schaepe from Springer for their dedication and help to implement and finish this large publication project on time maintaining the highest publication standards.

August 2015 The Editors

Contents

Part II Intuitionistic Fuzzy Sets: Foundations, Tools and Techniques

Part III Intuitionistic Fuzzy Sets: Applications

Part I
General Issues in the Representation and Processing of Uncertainty and Imprecision

Paired Structures, Imprecision Types and Two-Level Knowledge Representation by Means of Opposites

J. Tinguaro Rodríguez, Camilo Franco, Daniel Gómez and Javier Montero

Abstract Opposition-based models are a current hot-topic in knowledge representation. The point of this paper is to suggest that opposition can be in fact introduced at two different levels, those of the predicates of interest being represented (as *short/tall*) and of the logical references (*true/false*) used to evaluate the verification of the former. We study this issue by means of the consideration of different paired structures at each level. We also pay attention at how different types of fuzziness may be introduced in these paired structures to model imprecision and lack of knowledge. As a consequence, we obtain a unifying framework for studying the relationships between different knowledge representation models and different kinds of uncertainty.

Keywords Intuitionistic fuzzy sets · Bipolar fuzzy sets · Paired fuzzy sets

1 Introduction

Knowledge acquisition and representation is a complex task where many scientific fields interact. Let simply point out here the relevance of concept representation in our knowledge process (see, e.g., [13, 25]). Our brain is able to produce concepts

J. Tinguaro Rodríguez (✉) · J. Montero
Faculty of Mathematics, Complutense University, 28040 Madrid, Spain
e-mail: jtrodrig@mat.ucm.es

J. Montero
e-mail: javier_montero@mat.ucm.es

C. Franco
IFRO, Faculty of Science, University of Copenhagen, Copenhagen, Denmark
e-mail: cf@ifro.ku.dk

D. Gómez
Faculty of Statistics, Complutense University, 28040 Madrid, Spain
e-mail: dagomez@estad.ucm.es

K.T. Atanassov et al. (eds.), *Novel Developments in Uncertainty Representation and Processing*, Advances in Intelligent Systems and Computing 401,
DOI 10.1007/978-3-319-26211-6_1

that mean a compact, reliable and flexible representation of reality, and this representation is the basis for an efficient decision making process, and perhaps more important, the basis for an efficient communication language, when put into words.

But most concepts we use are complex in nature, far from being binary (hold/do not hold), and most probably with no associated objective measure. Most of the time, in order to understand a concept we need to explore its related concepts. Only taking into account surrounding concepts we can capture the borders of the concept we are considering, as well as the transition zones from one concept to the other.

Of course, concepts in a complex reality do not have a unique related concept. But indeed our knowledge use to start by putting a single concept in front of the concept under study. In Psychology, for example, the importance of bipolar reasoning in human activity has been well stated (see [19], but also [6–8]). But it is relevant to observe that particularly in this context the semantic bipolar scale positive/negative comes with a neutral valuation stage.

Within fuzzy sets theory we can find several models that fit into the above approach. For example, Atanassov's fuzzy sets (see [3, 4]) were originally presented in terms of a concept and its negation, allowing some indeterminacy state. And Dubois and Prade offered a unifying view of three kinds of bipolarities (see [8–10]).

This paper continues the work in [21] by analyzing the role of opposition and imprecision in both logic and knowledge representation models through the recently proposed notion of paired structures (see [12, 18, 20, 22]). Particularly, here we focus on how opposition (and thus paired structures) can simultaneously operate at two different levels, the logical and the representational ones. Our point is that this approach enables a unifying view of different knowledge representation models, as well as a comprehensive view of the different kinds of uncertainty and neutrality they allow to express.

2 Opposite L-Fuzzy Sets

Let $L = (L, \leq, 0_L, 1_L)$ be a lattice. In this work we will usually assume that L is linearly ordered by $\leq$. In this context, a negation function has been traditionally defined (see [26, 27]) as a non-increasing function

$$n: L \rightarrow L$$

such that $n(0_L) = 1_L$ and $n(1_L) = 0_L$. A negation function will be called a strong negation if it is in addition a strictly decreasing, continuous negation being also involutive (i.e. such that $n(n(v)) = v$ for all v in L). In this paper we shall consider only strong negations. Then, if we denote with $F_L(X)$ the set of all L-fuzzy sets (i.e. predicates) over a given universe X, then any strong negation n determines a negation operator

$$N: F_L(X) \rightarrow F_L(X)$$

such that $N(\mu)(x) = n(\mu(x))$ for any predicate $\mu \in F_L(X)$ and any object $x \in X$.

Definition 1 A function $O: F_L(X) \rightarrow F_L(X)$ will be called an *opposition* operator if the following two properties hold:

(A1) $O^2 = Id$ (i.e. O is involutive);
(A2) $\mu(x) \leq \mu(y) \Rightarrow O(\mu)(y) \leq O(\mu)(x)$ for all $\mu \in F_L(X)$ and $x, y \in X$;

The above definition generalizes the following definition given in [28], where a particular negation operator $N: F_L(X) \rightarrow F_L(X)$ is being assumed:

Definition 2 An *antonym* operator is a mapping $A: F_L(X) \rightarrow F_L(X)$ verifying the following properties:

(A1) $A^2 = Id$;
(A2) $\mu(x) \leq \mu(y) \Rightarrow A(\mu)(y) \leq A(\mu)(x)$ for all $\mu \in F_L(X)$ and $x, y \in X$;
(A3) $A \leq N$.

Hence, the following definition seems also natural, as another family of relevant opposites:

Definition 3 An *antagonism* operator is a mapping $A: F_L(X) \rightarrow F_L(X)$ fulfilling the following properties:

(A1) $A^2 = Id$;
(A2) $\mu(x) \leq \mu(y) \Rightarrow A(\mu)(y) \leq A(\mu)(x)$ for any $\mu \in F_L(X)$ and $x, y \in X$.
(A3) $A \geq N$.

In this way, any chosen negation N is an opposite, and it determines two main families of opposites, antonym and antagonism, names that should be usually assigned to any antonym or antagonism different than N. But notice that there are opposites not being antonym or antagonism.

3 Paired Fuzzy Sets and Paired Fuzzy Structures

Previous definitions offer in our opinion an appropriate range for defining opposites, depending on a previous negation that acts as a reference. In this way we can represent both the case in which two opposite fuzzy sets overlap and the case in which two opposite fuzzy sets do not cover the whole reality under study.

Definition 4 Two predicates (or fuzzy sets) P, Q are *paired* if and only if $P = O(Q)$, and thus also $Q = O(P)$, holds for a certain semantic *opposition* operator O.

In other words, a paired fuzzy set is a couple of opposite fuzzy sets. Our point is that neutral predicates will naturally emerge from opposites: as in classification context (see, e.g., [1, 2]), two opposite predicates (e.g., *tall/short*) that refer to the same notion (*height*) and, depending on their semantics, can generate different neutral concepts. When opposites overlap (e.g., *more or less tall/more or less short*), both opposites are perhaps too wide and *ambivalence* appears as a neutral predicate (to some extent both opposite predicates hold). But if opposites do not overlap (e.g., *very tall/very short*), we find that both opposites are perhaps too strict, and *indeterminacy* appears. And of course both situations might hold, depending on the object under consideration. *Indeterminacy* and *ambivalence* therefore appear as two main neutral valuation concepts. Alternatively, specific intermediate predicates might appear, in some cases leading to a non-paired structure perhaps by modifying the definition of the two basic opposite predicates and/or introducing new non-neutral intermediate predicates, defining perhaps a linear scale (see, e.g., [14]).

But a third main neutrality can appear when opposites are *complex* predicates (e.g., *good/bad*), mainly due to the underlying multidimensional nature of the problem. In this context it is usually suggested a decomposition in terms of simpler predicates, and then we can easily be faced to a *conflict* between different criteria.

Hence, from a basic predicate and its negation we can define an opposite that might imply, to some extent, the existence of indeterminacy (antonym) and/or ambivalence (antagonism), and also conflict. Hence, we have reached to a qualitative scale S = {*concept, opposite, indeterminacy, ambivalence, conflict*}, which contains the relevant representational situations into consideration and constitute the basis of any paired structure. The main idea of a paired structure is that the verification (for a given object) of each of these primary, qualitative situations is then measured in a secondary, quantitative scale L (usually a lattice).

Definition 5 Given an appropriate lattice L, paired structures are represented through a multidimensional L-fuzzy set A_L given by

$$A_L = \{\langle x; (\mu_s(x))_{s \in S}\rangle | x \in X\},$$

where X is our universe of discourse and each object $x \in X$ is assigned up to a degree $\mu_s(x) \in L$ to each one of the above five predicates $s \in S$, S = {*concept, opposite, indeterminacy, ambivalence, conflict*}.

Although more details will be found in [18], we should stress that, consistently with [17], an appropriate logic (or logics if [15, 16] are taken into account), should be then associated to this structure (X, S, L, A_L). It is important to point out that, consistently with [1, 2], a condition similar to Ruspini's partiton [24]

$$\sum_{s \in S} \mu_s(x) = 1_L \text{ for all } x \text{ in } X$$

is not being a priori imposed, although certain circumstances, constraints or generalizations, might suggest specific adaptations, particularly in the management of predicates (see, e.g., [5]).

4 The Imprecision Issue

Now we follow [21] to remark that in addition to the above types of neutrality, there are other representational situations that can be identified as a kind of neutrality, since they produce in the decision maker some kind of symmetry among opposites. However, this symmetry does not actually originate from the semantic tension between reference opposites, but as a result of allowing both such references and the formal tools we use to represent them to be imprecise. Let us discuss here these two related types of imprecision and the way they are related to and introduced in our paired structures.

On one hand, if we regard the reference predicates P, Q to be imprecise concepts (such as e.g., *young/old* or *short/tall*), we need to introduce a certain kind of gradualness in our quantitative scale L in order to enable our representation model to express such imprecision. For example, departing from $\{0, 1\}$ and iteratively allowing intermediate valuations between each pair of previous valuations, we may reach a continuous linear scale as $L = [0, 1]$. In this way, we may obtain in the limit a fuzzy paired structure, in which our references are represented through fuzzy sets, as well as the other representational neutralities in S. We refer to this kind of imprecision, leading to gradual secondary scales L in which a precise degree $\mu_s \in L$ represents the verification of the primary (imprecise) categories $s \in S$, as type-1 imprecision.

In this context, whenever the considered negation $n{:}L \rightarrow L$ has an equilibrium point e (i.e. an $e \in L$ such that $n(e) = e$), some objects $x \in X$ may for instance verify that $\mu_P(x) = \mu_Q(x) = e$. If $L = [0, 1]$ and n is the standard negation, we would obtain $\mu_P(x) = \mu_Q(x) = 0.5$. In this kind of situation, decision makers find difficulties in choosing among both opposites. But this representational symmetry does not constitute a different (neutral) concept providing a new primary category in S as a result of a semantic tension between references. Instead, such symmetry arises because of the imprecise nature of the reference predicates, whenever we allow such imprecision to be introduced in our models through a type-1 or fuzzy modelling of the paired references.

On the other hand, and more importantly, once we are faced to the estimation problem (membership functions should be somehow estimated), we should be aware of a different imprecision issue, now associated to the difficulty of estimating exact membership degrees $\mu_s \in L$. Our knowledge may be not so rich to allow estimating exact degrees, and thus imprecise degrees or estimations may be allowed to model lack of knowledge or ignorance. Again, decision makers can find difficulties in choosing among opposites under a significant level of ignorance. However, such ignorance is not semantically generated from a pair of opposing references, but it derives from an insufficient or imperfect knowledge of the objects into consideration in relation with type-1 imprecise references.

Therefore, this second kind of imprecision, let us refer to it as type-2 imprecision, is associated to a representational ignorance as it refers to a lack of knowledge about the exact degree to be chosen within a (gradual) secondary scale L. The representation of type-2 imprecision leads to consider more complex secondary scales, as for example $\mathrm{L} = \mathrm{L}^I = \{[\underline{\mu}, \bar{\mu}] \subset [0,1]\}$, i.e. the lattice of interval-valued fuzzy sets. The width of the intervals is naturally associated to a measure of the involved lack of knowledge, in such a way that in this context a valuation $\mu_s(x) = [0, 1] \in L^I$ is to represent a total lack of knowledge or ignorance about the exact degree of verification of the primary category $s \in S$ for object x. Such a type-2 imprecision might apply to each one of the fuzzy predicates composing a paired structure, leading to a more complex structure.

5 Paired Structures in Logics and Knowledge Representation

In this section, we use the framework of paired structures to analyze the relationships between different logical and knowledge representation models. To this aim, we propose an example in which we apply the previous paired approach to the notions of *truth* and *falsehood* of classical logic. We restrict ourselves to a crisp setting (i.e. we assume $L = \{0, 1\}$) for simplicity and clarity of exposition, since the main point of our argument is not affected by the consideration or not of a fuzzy framework. Indeed, as we shall see later, our main point is that opposition, and thus paired structures, can act at both the logical and the representational level.

Therefore, let us consider two crisp poles $T = \textit{true}$ and $F = \textit{false}$, related through an opposition operator O, and defined for the purpose of this example on a universe of discourse

$$U = \{P(x) = \textit{John is tall}, \neg P(x) = \textit{John is not tall}\}$$

formed by two propositions specified in terms of a single object x in X (*John*), and a single property P (*tall*) and its negation $\neg P$ (to be read as *not-tall*).

Within the paired approach we should consider also the negation of the poles $NT = \textit{not-true}$ and $NF = \textit{not-false}$, defined from both a negation $n{:}\{0, 1\} \to \{0, 1\}$, such that $n(0) = 1$ and $n(1) = 0$, and the membership functions $\mu_T, \mu_F{:}U \to \{0, 1\}$, in such a way that $\mu_{NT} = n \circ \mu_T$ and $\mu_{NF} = n \circ \mu_F$.

Let us remark that we use two (or three) different symbols for negation, i.e. the symbol $\neg$ to refer to the negation of properties at the level of the propositions on which the poles apply, and the symbol N to refer to negation at the level of poles (which is in turn dependent on n). We assume all these negations to be involutive.

Then, within the paired approach, evidence regarding objects u in U is evaluated through pairs

$$\mu_{PAIRED}(u) = (\mu_T(u), \mu_F(u)) \tag{1}$$

and thus the following valuations of $P(x)$ = *John is tall* are available (we denote them through the symbols before the last double arrow):

- $P(x)$ is true $\leftrightarrow T_{P(x)} \leftrightarrow \mu_T(P(x)) = 1$ and $\mu_F(P(x)) = 0$;
- $P(x)$ is false $\leftrightarrow F_{P(x)} \leftrightarrow \mu_F(P(x)) = 1$ and $\mu_T(P(x)) = 0$;
- $P(x)$ is ambivalent $\leftrightarrow A_{P(x)} \leftrightarrow \mu_T(P(x)) = 1$ and $\mu_F(P(x)) = 1$;
- $P(x)$ is indeterminate $\leftrightarrow I_{P(x)} \leftrightarrow \mu_T(P(x)) = 0$ and $\mu_F(P(x)) = 0$;

Similarly, regarding $\neg P(x)$ the following valuations are available:

- $\neg P(x)$ is true $\leftrightarrow T_{\neg P(x)} \leftrightarrow \mu_T(\neg P(x)) = 1$ and $\mu_F(\neg P(x)) = 0$;
- $\neg P(x)$ is false $\leftrightarrow F_{\neg P(x)} \leftrightarrow \mu_T(\neg P(x)) = 0$ and $\mu_F(\neg P(x)) = 1$;
- $\neg P(x)$ is ambivalent $\leftrightarrow A_{\neg P(x)} \leftrightarrow \mu_T(\neg P(x)) = 1$ and $\mu_F(\neg P(x)) = 1$;
- $\neg P(x)$ is indeterminate $\leftrightarrow I_{\neg P(x)} \leftrightarrow \mu_T(\neg P(x)) = 0$ and $\mu_F(\neg P(x)) = 0$.

Here we are neither concerned with interpreting this paired logical framework nor with establishing its soundness. Rather, we use the formal framework it provides to study how different logics and/or formal models for knowledge representation may be obtained by assuming different principles and properties. However, let us observe that different well-founded paraconsistent logics and semantics can be developed from this general approach (see e.g., [23, 29]) (Fig. 1).

5.1 *Paired Representation of Classical Logic*

From the general logical framework allowed by a paired representation of truth and falsehood, classical logic can be obtained by assuming just two conditions. First, let us assume

$$\mu_F(P(x)) = \mu_T(\neg P(x)) \tag{2}$$

that is, falsehood of a proposition is equal to the truth of the $\neg$-negated proposition. And second, assume also

$$\mu_T(\neg P(x)) = \mu_{NT}(P(x)) \tag{3}$$

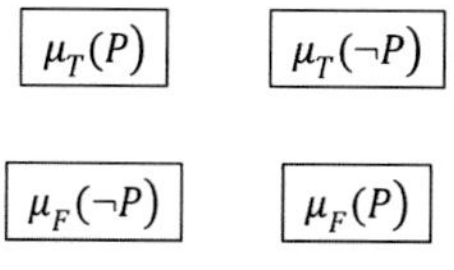

$\mu_T(P)$	$\mu_T(\neg P)$
$\mu_F(\neg P)$	$\mu_F(P)$

Fig. 1 Logical structure of a general paired approach. Truth and falsehood of a proposition P as well as truth of P and truth of its negation $\neg P$ are not related

that is, negation ¬ at the level of propositions is interchangeable with negation N at the level of the poles.

As a consequence of these two assumptions and the equality $\mu_{NT} = n \quad \mu_T$, the falsehood of P is equal to the non-truth of P, which in turn can be obtained from just the truth value of $P(x)$ through negation n, that is

$$\mu_F(P(x)) = n(\mu_T(P(x))). \tag{4}$$

Notice that (4) entails that the poles T and F are each other complement (i.e. $T = NF$ and $F = NT$). Moreover, the so obtained logic verifies both the *excluded middle principle* (*EMP*)

$$T_{P(x) \vee \neg P(x)} \leftrightarrow (T_{P(x)} \wedge F_{\neg P(x)}) \vee (F_{P(x)} \wedge T_{\neg P(x)}) \vee (T_{P(x)} \wedge T_{\neg P(x)})$$

and the no contradiction principle (NCP)

$$F_{P(x) \wedge \neg P(x)} \leftrightarrow (T_{P(x)} \wedge F_{\neg P(x)}) \vee (F_{P(x)} \wedge T_{\neg P(x)}) \vee (F_{P(x)} \wedge F_{\neg P(x)})$$

where ∨ and ∧ respectively represent the classical *OR* and *AND* connectives. Particularly, notice that the valuations $T_{P(x)} \wedge T_{\neg P(x)}$ and $F_{P(x)} \wedge F_{\neg P(x)}$ are not allowed in this framework.

In these conditions, a paired logical representation of the evidence available in the framework of classical logic for a proposition u in U is given by pairs

$$\mu_{CL}(u) = (\mu_T(u), \mu_F(u)) \tag{5}$$

such that $\mu_F(u) = n(\mu_T(u))$. Notice that as a consequence of the complementarity of T and F no neutral valuations are allowed, that is, classical logic does neither admit ambivalent nor indeterminate propositions. Obviously, the pair μ_{CL} is equivalent to the classical representation of the sets P and $\neg P$ through characteristic functions μ_P, $\mu_{\neg P}: X \rightarrow \{0, 1\}$ in such a way that $\mu_P(x) = \mu_T(P(x))$ and $\mu_{\neg P}(x) = \mu_F(P(x))$.

Let us remark that fuzzy representations of properties is basically grounded on the same principles (2) and (3), although the verification of *EMP* and *NCP* is dependent on the particular choice of fuzzy connectives ∨ and ∧ for representing *OR* and *AND*. Particularly, as shown for instance in [11], both principles hold simultaneously if only if n is a strong negation and ∨ and ∧ are Lukasiewicz-like operators (Fig. 2).

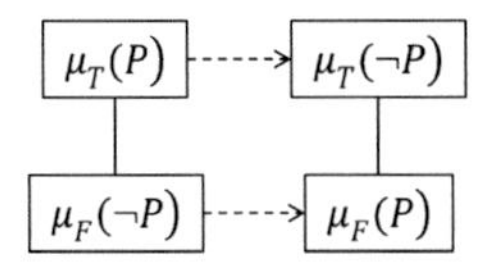

Fig. 2 In classical logic, truth of a proposition P is identified (*solid lines*) with falsehood of its negation $\neg P$, and the truth of P is related to the truth of $\neg P$ through a negation n (*dashed arrows*)

5.2 Paired Representation of Intuitionistic Logic

Notice that expression (2) above makes possible to interpret

$$\mu_{CL}(u) = (\mu_T(u), \mu_T(\neg u)) \tag{6}$$

for any proposition u in U. In a crisp setting, intuitionistic logic (in the sense of [3]) retains condition (2) and the interpretation in (6), but replaces condition (3) by the more general constraint

$$\mu_T(\neg P(x)) \leq \mu_{NT}(P(x)) \tag{7}$$

which asserts through (2) that the falsehood of a proposition is a more restrictive claim than the non-truth of the same proposition, that is, a proposition may be not-true while at the same time being not-false. Then, intuitionistic logic can be represented in a paired logical framework through pairs

$$\mu_{INT}(u) = (\mu_T(u), \mu_T(\neg u)) \tag{8}$$

such that $\mu_T(\neg u) \leq n(\mu_T(u))$ for any u in U.

It is important to notice that this setting allows a proposition $P(x)$ to be evaluated as indeterminate (and then so will be $\neg P(x)$), since now it is allowed that $\mu_T(P(x)) = 0$ and $\mu_F(P(x)) = \mu_T(\neg P(x)) = 0$ can simultaneously hold. This last entails that *EMP* does not hold in general in the framework of intuitionistic logic.

Notice also that, although the presence of indeterminacy violates the previous strong formulation of *NCP*, this principle still holds in the weaker sense of not allowing both $P(x)$ and $\neg P(x)$ to hold simultaneously, that is $\mu_{INT}(u) = (1, 1)$ is not an available valuation.

It is important to remark that in the case of intuitionistic logic, the logical poles T and F are no longer assumed to be each other complement, and thus they can respectively differ of *NF* and *NT*. Let us observe that the basic ideas holding in this crisp framework also hold in the fuzzy setting of [3] (Fig. 3).

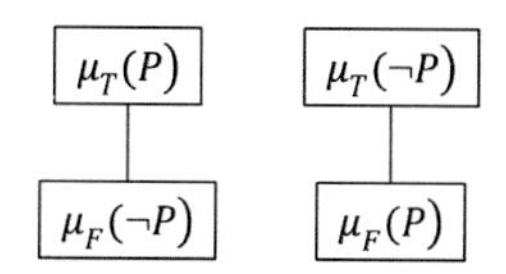

Fig. 3 In intuitionistic logic, truth of a proposition P is still identified (*solid lines*) with falsehood of its negation $\neg P$, but the truth of $\neg P$ can no more be obtained from that of P

5.3 Bipolar Knowledge Representation Models

Notice that in the previous examples we restricted ourselves to consider a universe of discourse formed by just two complementary propositions, $P(x)$ and $\neg P(x)$. Such a universe contains exclusively the needed propositions in order to study the logical dependence of a proposition and its negation, the first and main issue of any logical analysis as those of classical logic and intuitionistic logic.

However, bipolar models (in the sense of [8]) are not actually logical models referring to the poles T and F of classical logic, but rather bipolar models deal with knowledge representation in the presence of opposing arguments *under* the logical perspective of classical (or fuzzy) logic. That is, in this case we should consider the alternative universe

$$U = \{P(x) = \textit{John is tall},\ \neg P(x) = \textit{John is not tall},\\ Q(x) = \textit{John is short},\ \neg Q(x) = \textit{John is not short}\}$$

given by four propositions stated in terms of a single object x in X and a pair of properties (and the corresponding negations) sharing a certain kind of opposition. We may then assume that properties P and Q constitute a pair of poles related through an opposition operator ∂ in the sense of Definition 1, i.e. $Q = \partial(P)$.

Notice that we intentionally use two different symbols to differentiate the opposition operator acting at the level of logical poles (i.e. the operator O such that $F = O(T)$) from that acting at the level of the represented properties or predicates (i.e. the operator ∂). This also allows distinguishing between the neutral valuations arising at the logical level and those arising at the level of knowledge representation.

Particularly, we claim that most bipolar models actually admit neutral valuations at the level of the poles P and Q, but do not admit logical neutralities at the level of the poles T and F, similarly to classical logic. That is, bipolar models represent evidence through pairs

$$\mu_{BIP}(u) = (\mu_T(u), \mu_T(\partial u)) \tag{9}$$

but assume expressions (2) and (3) to hold regarding the logical relationships between a property and its negation in terms of logical truth and falsehood.

Thus, bipolar models enable to model opposite arguments at the level of knowledge representation, assuming two separate coordinates or dimensions (usually referred to as positive evidence and negative evidence, respectively) of knowledge representation, each of these dimensions in turn assuming a classical logic framework. That is, both *EMP* and *NCP* hold in each dimension regarding P and Q.

Moreover, these two dimensions may refer to logically independent properties (in the sense that P and Q may not be each other complement). Nevertheless, these two dimensions are not fully independent from a logical perspective since P and Q are related at the logical level through the opposition operator ∂, which relates the

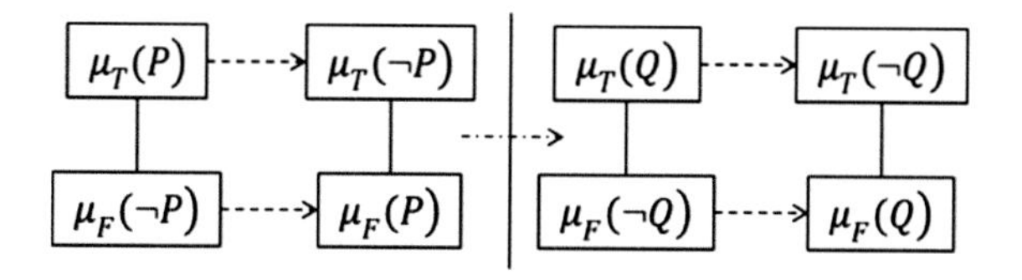

Fig. 4 In bipolar models, two separate dimensions of logical representation are employed to model opposite arguments, and are related through an opposition operator (*dot-dashed arrow*). Each of these dimensions assumes a classical logic structure

logical descriptions of P and Q (i.e. their membership functions μ_P and μ_Q) in order to guarantee that they meet the semantics of opposition (Fig. 4).

6 Conclusions

In our opinion, the proposed paired approach allows to state more precisely the relationships between intuitionistic and bipolar models. Particularly:

- Intuitionistic models rely on a different set of logical assumptions than bipolar models. The latter models assume a classical logical framework in each coordinate, while the former do not.
- Intuitionistic models are stated in terms of a proposition and its negation, while bipolar models work in terms of opposite properties. In other words, intuitionistic models introduce opposition at the level of logical poles but does not consider opposite (but complementary) properties, while bipolar models allow opposite properties but assumes the logical structure of classical logic at each opposite property.

Notice that these differences make possible to think that both approaches (intuitionistic and bipolar) could be in fact complementary, in the sense that we could allow introducing opposition at *both* levels of knowledge representation and logic, i.e. a paired structure at the level of the properties P to be represented and another paired structure at the level of the logical poles T and F that evaluates the verification of such properties. That is, nothing seems to forbid the formal specification of a model

$$\mu_{BIP-INT}(u) = ((\mu_T(u), \mu_T(\neg u)), (\mu_T(\partial u), \mu_T(\neg \partial u))) \tag{10}$$

in such a way that the evaluations $\mu_T(u)$ and $\mu_T(\neg u)$ as well as $\mu_T(\partial u)$ and $\mu_T(\neg \partial u)$ are related through a constraint similar to (7) (Fig. 5).

Here we do neither try to interpret this model nor to analyze its soundness, but we just claim it is formally possible in a general paired framework assuming paired structures on both the representational and logical levels. Moreover, we also claim that this model can be further extended from the crisp setting to allow representing

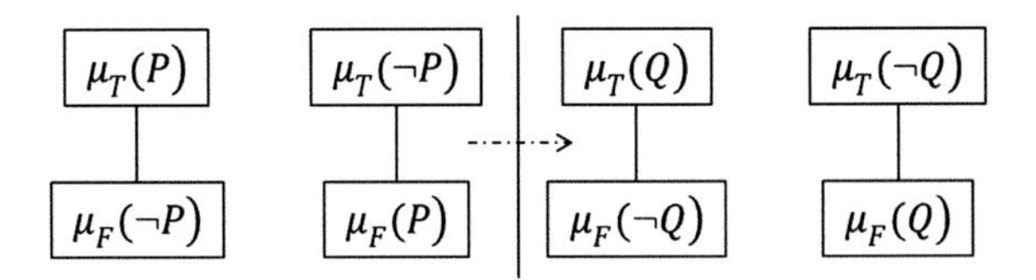

Fig. 5 The structure of a bipolar-intuitionistic model. Two dimensions of logical representation model opposite arguments related through opposition (*dot-dashed arrow*), but these dimensions assume an intuitionistic logical structure each

imprecise predicates and lack of knowledge, that is, type-1 and type-2 imprecision, at *both* the logical and the representational levels, although this is an issue that requires further study and research (for instance, both levels may have to share a similar lack of knowledge).

Anyway, we think that acknowledging that opposition may operate at two different levels (and that in fact many current opposition-based models operate either in one or the other), as well as that imprecision arises as a different issue from opposition, may be a useful approach to analyze the relationships between several models and acquire a better understanding of their representational power and the uncertainties they are able to express.

Acknowledgment This paper has been partially supported by grants TIN2012-32482 of the Government of Spain and S2013/ICCE-2845 of the Government of Madrid.

References

1. Amo, A., Gómez, D., Montero, J., Biging, G.: Relevance and redundancy in fuzzy classification systems. Mathw. Soft Comput. **8**, 203–216 (2001)
2. Amo, A., Montero, J., Biging, G., Cutello, V.: Fuzzy classification systems. Eur. J. Oper. Res. **156**, 495–507 (2004)
3. Atanassov, K.T.: Intuitionistic fuzzy sets. Fuzzy Sets Syst. **20**, 87–96 (1986)
4. Atanassov, K.T.: Answer to D. Dubois, S. Gottwald, P. Hajek, J. Kacprzyk and H. Prade's paper "Terminological difficulties in fuzzy set theory—the case of "Intuitionistic fuzzy sets". Fuzzy Sets Syst. **156**, 496–499 (2005)
5. Bustince, H., Fernandez, J., Mesiar, R., Montero, J., Orduna, R.: Overlap functions. Nonlinear Anal. **72**, 1488–1499 (2010)
6. Cacioppo, J.T., Berntson, G.G.: The affect system, architecture and operating characteristics—current directions. Psychol. Sci. **8**, 133–137 (1999)
7. Cacioppo, J.T., Gardner, W.L., Berntson, C.G.: Beyond bipolar conceptualizations and measures—the case of attitudes and evaluative space. Pers. Soc. Psychol. Rev. **1**, 3–25 (1997)
8. Dubois, D., Prade, H.: An introduction to bipolar representations of information and preference. Int. J. Intell. Syst. **23**, 866–877 (2008)
9. Dubois, D., Prade, H.: An overview of the asymmetric bipolar representation of positive and negative information in possibility theory. Fuzzy Sets Syst. **160**, 1355–1366 (2009)
10. Dubois, D., Prade, H.: Gradualness, uncertainty and bipolarity—making sense of fuzzy sets. Fuzzy Sets Syst. **192**, 3–24 (2012)

11. Fodor, J., Roubens, M.: Fuzzy Preference Modelling and Multicriteria Decision Support. Kluwer Academic, Dordrecht, Boston (1994)
12. Franco, C., Rodríguez, J.T., Montero, J.: Building the meaning of preference from logical paired structures. Knowl.-Based Syst. **83**, 32–41 (2015)
13. Kaplan, K.: On the ambivalence-indifference problem in attitude theory and measurement—a suggested modification of the semantic differential technique. Psychol. Bull. **77**, 361–372 (1972)
14. Likert, R.: A technique for the measurement of attitudes. Arch. Psychol. **140**, 1–55 (1932)
15. Montero, J.: Comprehensive fuzziness. Fuzzy Sets Syst. **20**, 79–86 (1986)
16. Montero, J.: Extensive fuzziness. Fuzzy Sets Syst. **21**, 201–209 (1987)
17. Montero, J., Gómez, D., Bustince, H.: On the relevance of some families of fuzzy sets. Fuzzy Sets Syst. **158**, 2429–2442 (2007)
18. Montero, J., Bustince, H., Franco, C., Rodríguez, J.T., Gómez, D., Pagola, M., Fernandez, J., Barrenechea, E.: Paired structures and bipolar knowledge representation, Working paper
19. Osgood, C.E., Suci, G.J., Tannenbaum, P.H.: The Measurement of Meaning. University of Illinois Press, Urbana (1957)
20. Rodríguez, J.T., Franco, C., Gómez, D., Montero, J.: Paired structures and other opposites-based models. Proc. of IFSA-EUSFLAT **1375–1381**, 2015 (2015)
21. Rodríguez, J.T., Franco, C., Montero, J.: On the semantics of bipolarity and fuzziness. Proc. Eurofuse **193–205**, 2011 (2011)
22. Rodríguez, J.T., Franco, C., Montero, J., Lu, J.: Paired structures in logical and semiotic models of natural language. Inf. Process. Manage. Uncertainty Knowl.-Based Syst. **Part II**, 566–575 (2014)
23. Rodríguez, J.T., Turunen, E., Ruan, D., Montero, J.: Another paraconsistent algebraic semantics for Lukasiewicz-Pavelka logic. Fuzzy Sets Syst. **242**, 132–147 (2014)
24. Ruspini, E.H.: A new approach to clus8tering. Inf. Control **15**, 22–32 (1969)
25. Slovic, P., Peters, E., Finucane, M.L., MacGregor, D.G.: Affect, risk, and decision making. Health Psychol. **24**, 35–40 (2005)
26. de Soto, A.R., Trillas, E.: On antonym and negate in fuzzy logic. Int. J. Intell. Syst. **14**, 295–303 (1999)
27. Trillas, E.: On the use of words and fuzzy sets. Inf. Sci. **176**, 1463–1487 (2006)
28. Trillas, E., Moraga, C., Guadarrama, S., Cubillo, S., Castineira, E.: Computing with antonyms. Forg. New Front. Fuzzy Pioneers I **217**, 133–153 (2007)
29. Turunen, E., Öztürk, M., Tsoukiàs, A.: Paraconsistent semantics for Pavelka style fuzzy sentential logic. Fuzzy Sets Syst. **161**, 1926–1940 (2010)

Suggestions to Make Dempster's Rule Convenient for Knowledge Combining

Ewa Straszecka

Abstract The paper deals with the Dempster's rule for the basic probability assignment combining from the point of view of a medical diagnosis support. The assignment determined on different sources of information is useful to establish symptoms weights, but often results of the combination are far from intuition. A modification of the Dempster's formula is proposed to make it possible to tune the resulting assignment according to the distance between the combined assignments. Properties of the proposed methods which are important for practical applications are shown on simulated data.

Keywords Dempster-Shafer theory of evidence · Knowledge combining

1 Introduction

The Dempster-Shafer theory of evidence [1] still remains one of the most important tools to represent and manage uncertainty in real-life problems of decision support. However, its flexibility is simultaneously an advantage and a drawback of applications. We benefit from neglecting dependence of focal elements as pieces of evidence, but the week point is a lack of indications in which way the basic probability assignment (bpa) for the elements should be determined. This is particularly crucial when the bpa expresses expert's knowledge, for instance in medical diagnosis. In this area a significance of a symptom is often roughly estimated and a combination of knowledge from different sources is necessary. Small changes of values of one bpa may be exaggerated after combination with another, also roughly determined bpa. Deficiencies of the classical Dempster's combination are pointed out in numerous papers, but no other general and satisfactory method is proposed so far, maybe except for the fuzzy rules aggregation, which however has its own disadvantages in

E. Straszecka (✉)
Institute of Electronics, Faculty of Automatic Control, Electronics and Computer Science, Silesian University of Technology, 16 Akademicka St., 44-100 Gliwice, Poland
e-mail: ewa.straszecka@polsl.pl

K.T. Atanassov et al. (eds.), *Novel Developments in Uncertainty Representation and Processing*, Advances in Intelligent Systems and Computing 401,
DOI 10.1007/978-3-319-26211-6_2

applications [10]. Various concepts of changing Dempster's rule of combination are valuable and important because of their solid theoretical background, e.g. [3, 4, 9]. Still, there is no suggestion for smooth tuning of the resulting bpa, if we discover that it does not entirely follow our intuition.

The aim of this paper is to consider introducing a factor to the combination that will play a role similar to the fuzzifying coefficient in clustering. This means that changing the factor we could approach soft or severe estimation of symptom weights. Common points of the Dempster-Shafer theory and pattern recognition are noticed in [2]. Thus, it could be profitable to use experience of the latter old and highly developed domain to solve some application problems in knowledge combining. In the presented method a coefficient based on the similarity of bpas is used in combining. It is worth to notice that this concept can be widen for a number of similarity definitions which can open field for further research.

The present paper tries to make several suggestions about tuning the bpa, which hopefully are general enough to be of use in practical solutions. The proposed method is verified through simulations of bpas designed for a model of a medical diagnosis.

2 An Interpretation of Basic Probability Assignments as Sources of Diagnostic Knowledge

2.1 *The Classical Dempster's Combination and a Medical Diagnosis*

In medical diagnosis the bpa may represent weights of symptoms. It is particularly convenient since dependence of symptoms can be neglected and subsets of symptoms which are not distinct can be freely considered. In this case the classical bpa definition [7] is:

$$m_d(f) = 0, \quad \sum_{S_i \in \mathbf{S}, i=1,\ldots,n} m_d(S_i) = 1, \tag{1}$$

where d denotes the diagnosis and S_i is its symptom—single or complex (i.e. a subset of symptoms). The false focal element f should be interpreted as the symptom that is never observed. Thus, the $m_d(S_i)$ is the weight of the rule: IF symptom is S_i THEN diagnosis is d.

Let us assume that the bpa is set by a physician using his experience. This is not rare and in many cases diagnostic indexes that are accepted as medical guidelines are not entirely based on statistic data, but also on subjective estimation [10]. It is also often that another physician has slightly different view on the diagnosis and make his own bpa. Then a combination of the bpas according to the classical Dempster's formula [1] is possible:

$$m_d(s_k) = \frac{\sum_{S_i \cap S_j = s_k} m_{d1}(S_i) m_{d2}(S_j)}{\sum_{S_i \cap S_j \neq f} m_{d1}(S_i) m_{d2}(S_j)}, \quad k = 1 \ldots n, \quad i = 1 \ldots n_1, j = 1 \ldots n_2. \tag{2}$$

It is obvious that the denominator in (2) stands for a normalization, such that (1) holds true. Let us denote it by:

$$m_d(s_k) = \left[\sum_{S_i \cap S_j = s_k} m_{d1}(S_i) m_{d2}(S_j) \right] \quad k = 1 \ldots n, i = 1 \ldots n_1, j = 1 \ldots n_2, \tag{3}$$

for simplicity. Usually, the symptoms in m_{d1} and m_{d2} are identically defined, so their number is stable: $n = n_1 = n_2$ and $s_i \equiv S_i$, $s_j \equiv S_j$ are identical in both sets of focal elements, but $s_i \cap s_j \neq \emptyset$, because complex symptoms usually include single symptoms which are individual focal elements in the same bpa. Let us notice, that if f is the symptom that is never observed, then the combination deficiency indicated by Zadeh [11], which occur when $m_1(a) = 0$; $m_1(b) = 0.1$; $m_1(c) = 0.9$ and $m_2(a) = 0.9$; $m_2(b) = 0.1$; $m_2(c) = 0$, is irrelevant for medical diagnosis. If $m_1(a) = 0$ then a should be a symptom that is never observed, so it does not occur in m_2, too. If a is a symptom that is not considered, for instance it is a result of a laboratory test which is not performed, its bpa value should not be equal to zero, though it could be low. Still, the classical Dempster's combination is too restrictive for applications for other reasons that illustrate the following example.

2.2 Example

Let us consider two bpas m_1 and m_2, defined for the following set of focal elements: $A = \{a, b, c, bc\}$. Assume at first that values of the bpas are set equal, i.e. 0.25, for the each element. Next let us change the bpa values, rising them by 0.05 for a and b as well as decreasing for c and ab, in m_1, while doing the opposite in m_2. Thus, at the end $m_1(a) = m_1(b) = 0.45$, $m_1(c) = m_1(bc) = 0.05$; $m_2(a) = m_2(b) = 0.05$, $m_2(c) = m_2(bc) = 0.45$. The combination is: $m_c(a) = \left[m_1(a)m_2(a)\right]$; $m_c(b) = \left[m_1(b)m_2(b) + m_1(b)m_2(bc) + m_2(b)m_1(bc)\right]$; $m_c(c) = \left[m_1(c)m_2(c) + m_1(c)m_2(bc) + m_2(c)m_1(bc)\right]$; $m_c(bc) = \left[m_1(bc)m_2(bc)\right]$; Results of calculations are in Fig. 1.

It can be noticed that though values of included bpas are at first equal, the resulting bpa shows big differences. We would prefer that $m_c(b)$ and $m_c(bc)$ would approach more the average value of the bpa than take extremely different values. Such phenomena usually contradict intuition. However, it is enough to add a modifying factor in the form of a power, i.e. $m_c^*(a) = \left[\left(m_1(a)m_2(a)\right)^\delta\right]$, for instance $\delta = 0.5$ in Fig. 1, to diminish changes of the final bpa.

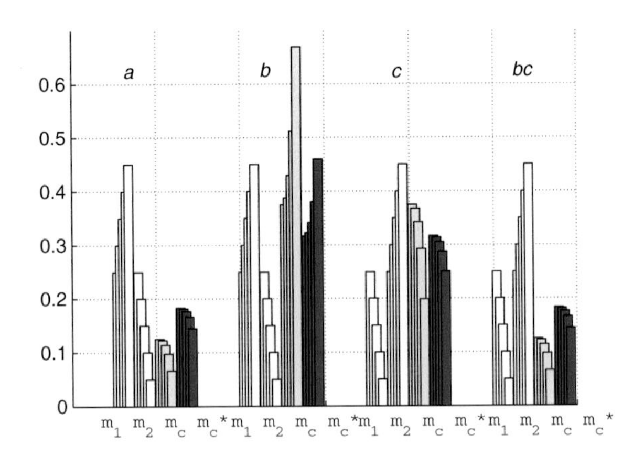

Fig. 1 Bpas resulting from the classical Dempster's formula (m_c) as well as from the modification (m_c^*), while gradual change of combined bpas (m_1 and m_2)

2.3 Tuning a Combination of Probability Assignments

If we choose a bpa parameter, we can allow for the classical Dempster's combination or its modification by the change of its value. It would be profitable to introduce a factor that can be decided according to the shape of combined bpas and in agreement with knowledge engineer's experience and intuition. Such a factor can be adjusted by investigation of similarity of the bpas. Let us consider tuning results of combination by means of a power coefficient, as it was suggested in Sect. 2.2. Thus:

$$m_c * (s_k) = \left[\left(\sum_{s_k = s_i \cap s_j} m_1(s_i) m_2(s_j) \right)^{\delta} \right] \tag{4}$$

It can be noticed, that for single focal elements indeed the combined bpas are both taken to the δ power, i.e. $\left[(m_1(s_i))^{\delta} (m_2(s_i))^{\delta} \right]$ which means that we simply moderate severe opinions of experts. The δ coefficient can be determined by means of similarity. Let us assume that the similarity is defined using a distance between the combined bpas. The distance can be formulated in several manners [6], for instance:

$$d(m_1, m_2) = \sqrt{(\mathbf{m_1} - \mathbf{m_2})' \mathbf{W} (\mathbf{m_1} - \mathbf{m_2})}, \tag{5}$$

where $\mathbf{m_1}$, $\mathbf{m_2}$ are vectors including bpa values. When $\mathbf{W}$ is equal to identity matrix (5) becomes the Euclidean distance d_e:

$$d_e(m_1, m_2) = \sqrt{(\mathbf{m_1} - \mathbf{m_2})' (\mathbf{m_1} - \mathbf{m_2})}. \tag{6}$$

If cardinality of focal elements is considered [5] then the Jaccard index [6] can be used, i.e. **W** elements are equal:

$$J(s_i, s_j) = \frac{\left|s_i \cap s_j\right|}{\left|s_i \cup s_j\right|} \tag{7}$$

In case of the example from Sect. 2.2, the Jaccard index matrix is:

$$\mathbf{B} = \begin{bmatrix} 1 & 0 & 0 & 0 \\ 0 & 1 & 0 & 0.5 \\ 0 & 0 & 1 & 0.5 \\ 0 & 0.5 & 0.5 & 1 \end{bmatrix},$$

and the distance is:

$$d_b(m_1, m_2) = \sqrt{(\mathbf{m_1} - \mathbf{m_2})'\mathbf{B}(\mathbf{m_1} - \mathbf{m_2})}. \tag{8}$$

Additionally, a measure of distance is introduced:

$$d_a(m_1, m_2) = \min(\left|\mathbf{m_1} - \mathbf{m_2}\right|) \tag{9}$$

to find out what is an influence of the distance definition on the similarity coefficient.

Thus, the combination (4) can be used with δ:

$$\delta = \frac{k}{2 - d_{\#}}, \; d < 2 \tag{10}$$

where $d_{\#}$ is d_e, d_b or d_a. The k parameter in (10) allows for precise tuning of the combination Fig. 4. The assumption $d < 2$ exclude the situation of extreme ("Zadeh's") conflict between the combined bpas. In the latter case, which the author thinks is irrelevant for practical problems of diagnosis support, the proposed modifications cannot be used.

There is a question why not to use the average [5] instead. The reason is that while we choose the type of the mean, for instance arithmetic, we cannot become "more strict" or "liberal" while combining distinct or similar bpas [8]. However, let us consider the average for a comparison of features expected from the combined bpa in applications. Then:

$$m_m(s_k) = \frac{1}{2}\left(m_1(s_k) + m_2(s_k)\right) \tag{11}$$

In this way four types of combinations are compared with the classical Dempster's rule: three based on distances (6)–(9) and δ (10), as well as the mean (11).

2.4 A Model of the Basic Probability Assignment in Medical Diagnosis

In medicine many symptoms are formulated by means of laboratory tests, for instance "test X result is low", "test Y result is normal". Let us use this formulation for the bpa determination. The point of norm (x_N) that indicates the symptom, is determined for ill (diagnosis 1) and healthy (diagnosis 2) populations (see Fig. 2). The norm point is set at the intersection of distributions of "ill" and "healthy". If patient's result is x, then $x < x_N$ means "test X result is low", while $x \geq x_N$ indicates "test X result is normal".

Let us assume that the diagnosis is based on values of three tests (variables): X, Y and Z. Tests Y and Z may be correlated, so they should be both considered as single focal elements as well as one complex focal element. The set of focal elements is described in Table 1 for the diagnosis 1. Results of the tests that are lower than the norm indicate diagnosis 1 (ill), which of course is a simplification. This assumption makes data simulation easier, but it does violate the generality of the diagnostic model. Symptoms for "healthy" (diagnosis 2) are formulated in analogy, and test results are "normal" when $x \geq x_N$, $y \geq y_N$, $z \geq z_N$.

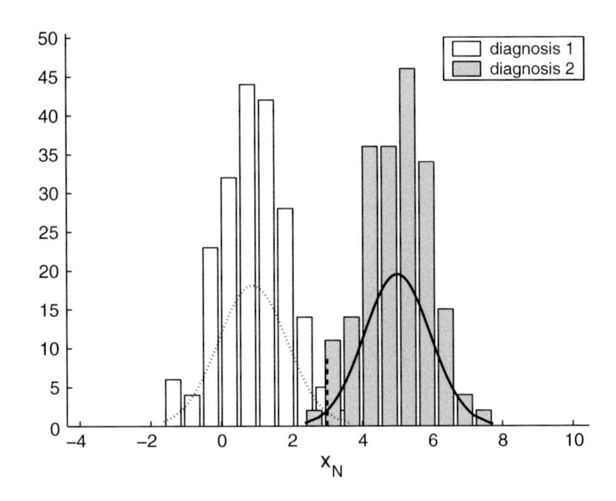

Fig. 2 An interpretation of laboratory test results: histograms and re-scaled normal distribution curves for patients with the diagnosis 1 (ill) and diagnosis 2 (healthy). The norm point is x_N

Table 1 Focal elements for diagnosis 1

Denotation	Heuristic meaning	Determination
s_1	"X is low"	$x < x_N$
s_2	"Y is low"	$y < y_N$
s_3	"Z is low"	$z < z_N$
s_4	"Y is low" and "Z is low"	$y < y_N$ and $z < z_N$

3 Simulations

Values of the X, Y, Z variables were generated by Matlab® normal distribution generator. Four kinds of normally distributed samples were simulated, with the mean ($\bar{\epsilon}$) and variance (σ), i.e. $N(\bar{\epsilon}_i, \sigma_i)$: $\epsilon_1 = 1$, $\sigma_1 = 1$, $\epsilon_2 = 1$, $\sigma_2 = 2$, $\epsilon_3 = 1$, $\sigma_3 = 3$, $\epsilon_4 = 5$, $\sigma_4 = 1$. Samples simulated for the X test were generated as not correlated with samples for Y and Z, while the latter were correlated. Each time the samples for diagnosis 2 (healthy) were characterized by the $N(\epsilon_4, \sigma_4)$ distribution (for all three tests). Samples for the diagnosis 1 (ill) had different distributions: $N(\epsilon_1, \sigma_1)$, $N(\epsilon_2, \sigma_2)$, $N(\epsilon_3, \sigma_3)$. Norm points were found from samples. From these samples basic probability assignments $m_{1d1}, m_{1d2}, m_{2d1}, m_{2d2}$ were calculated as normalized frequencies of occurrence of symptoms. Afterwards, m_{1d1}, and m_{2d1} were combined to obtain m_{d1}, while m_{1d2}, and m_{2d2} made m_{d2}. Calculations were performed for the same (200/200) and different (200/100) number of data in samples used for m_{1dj} and m_{2dj} $j = 1, 2$, determination. The cross-validation process was performed on 50 samples.

Combinations of m_{1dj} and m_{2dj} $j = 1, 2$ were performed to obtain m_c^* (4) with δ (10) for distances d_e (6), d_b (8), d_a (9) and the mean (11). It was necessary to define a criterion to estimate modifications of the classical combining. The modification results were compared to the bpa which was obtained if data from samples considered in one step of the cross-validation procedure were embedded in one dataset. This bpa was denoted as m_{all}. The criterion was the average of absolute differences between the result of a chosen modification and the m_{all} bpa, i.e.:

$$
\begin{aligned}
v_{ij\#} &= \sum_{k=1}^{4} \left| m_\#(s_k) - m_{all}(s_k) \right| \\
v_\# &= \frac{1}{2500} \sum_{i,j} v_{ij\#}
\end{aligned}
\tag{12}
$$

where # denoted e, b, a, m or c.

4 Results

During calculations it was stated that equal or different number of data in samples had negligible influence on the value of the difference criterion (12) (changes were smaller than 0.001). Thus, in the following only results for the different number of data in samples are discussed. Results of combinations are denoted in Figs. 3, 4 and 5 by the following indexes: e, b, a, m, while the index c stands for the classical Dempster rule.

It is observable that bpas obtained as outcomes of the proposed modifications bring values of v_e, v_b and v_a within the $[v_m, v_c]$ interval. Differences among used types of modifications are hardly influenced by the choice of the sample characteristics. It is illustrated in the Fig. 3. The diagrams (a) and (b) were obtained for X

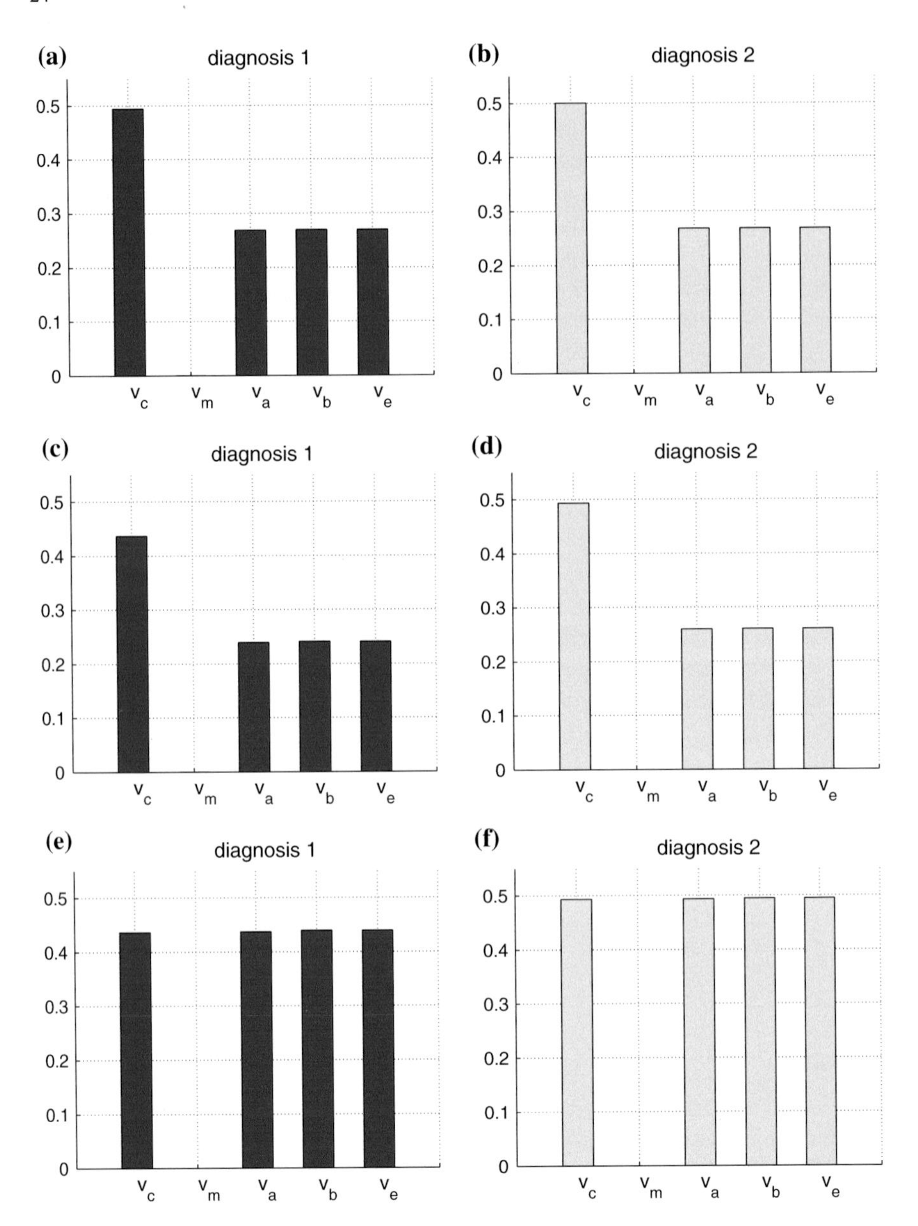

Fig. 3 The average difference (12) between the result of a chosen combination: $\mathtt{v}_{\mathtt{c}}$—classical, $\mathtt{v}_{\mathtt{m}}$—mean, $\mathtt{v}_{\mathtt{a}}$—minimum, $\mathtt{v}_{\mathtt{b}}$—Jaccard index, $\mathtt{v}_{\mathtt{e}}$—Euclidean; and the reference bpa—m_{all}. Diagrams **a–d** for $k = 1$, **e**, **f** for $k = 2$

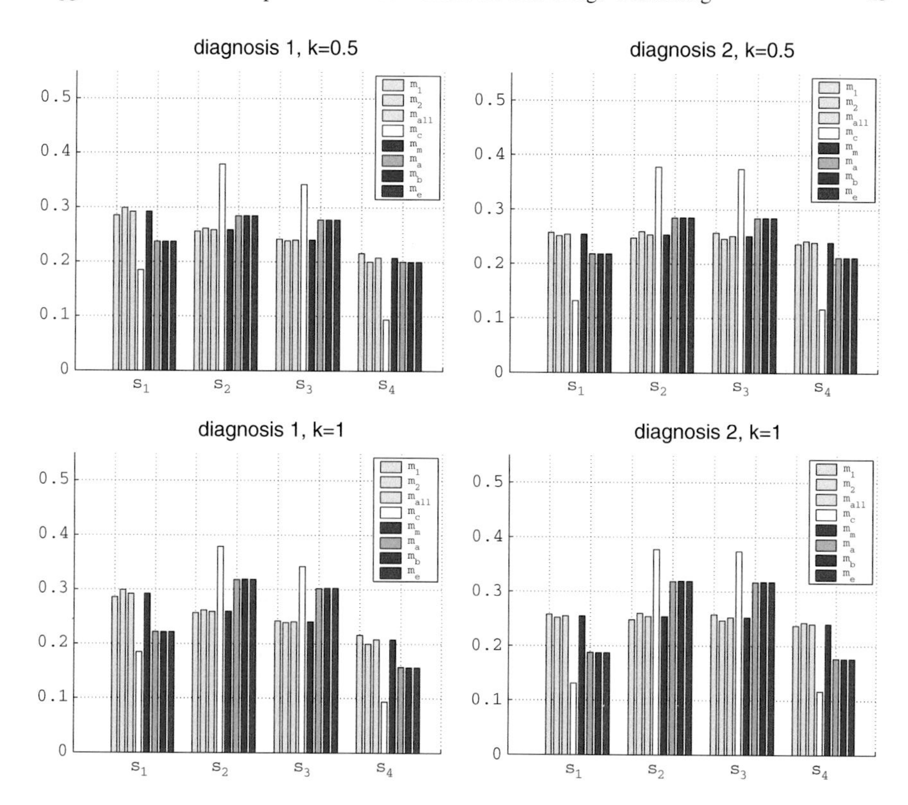

Fig. 4 Values of bpas obtained for different combination manners (part 1). Bars in the sequence denote: m_1, m_2, m_{all}, m_c, m_m, m_a, m_b, m_e

modeled by $N(\bar{\epsilon_3}, \sigma_3)$, Y by $N(\bar{\epsilon_2}, \sigma_2)$ and Z by $N(\bar{\epsilon_1}, \sigma_1)$, for the diagnosis 1 and X, Y, Z simulated by $N(\bar{\epsilon_4}, \sigma_4)$ for diagnosis 2. The diagrams (c) and (d) concern X modeled by $N(\bar{\epsilon_1}, \sigma_1)$, Y by $N(\bar{\epsilon_2}, \sigma_2)$ and Z by $N(\bar{\epsilon_3}, \sigma_3)$, for the diagnosis 1 and X, Y, Z simulated by $N(\bar{\epsilon_4}, \sigma_4)$ for diagnosis 2. Values of the $v_{\#}$ (12) are almost the same in the corresponding diagrams. This indicate that the method is universal enough to be used for a variety of diagnostic tests.

It is also effective, as the influence of the type of a modification as well as of its parameter k is visible. The v_m is close to zero (Fig. 3)—which is obvious for (11) definition. Yet, changing k in (10) we can tune the result of being closer to v_c or v_m. This is observable while comparing diagrams (a) and (e), as well as (b) and (f) in Fig. 3, determined for the same sample characteristics. Not only $v_{\#}$, but also values of bpas are regularly tuned along with the k (10) change. The Figs. 4 and 5 show values of bpas for different methods of combinations and various k. Sample characteristics for which bpas are calculated are the same as for diagrams (a), (b), (e), (f) in Fig. 3. For $k = 0.5$ the values of m_a, m_b and m_e are closer to m_m, while for $k = 2$ they are almost equal to m_c.

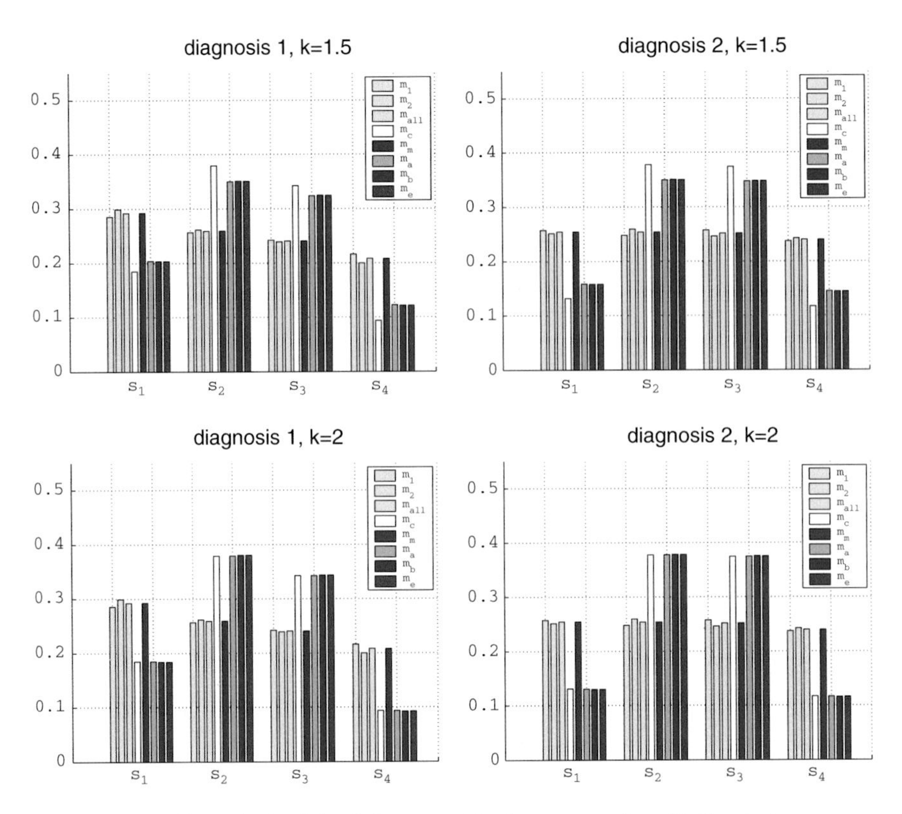

Fig. 5 Values of bpas obtained for different combination manners (part 2). Bars in the sequence denote: m_1, m_2, m_{all}, m_c, m_m, m_a, m_b, m_e

5 Discussion and Conclusions

In this paper modifications of the Dempster's rule for combining two bpas are proposed. Though the modifications do not eliminate "Zadeh's conflict", they can help to approach intuition while bpas combining. The idea comes from the author's experience from trials of diagnosis of combing information from different sources. In medical diagnosis support it is necessary to determine bpas on the basis of medical guidelines or expert's experience which are not very precise.

It should be noted that bpas combing is important not only from the point of view of calculation of the belief and plausibility of the diagnosis, but also as a tool for determination of symptoms weights. The right weights may help in planning diagnostic procedures, hence a moderate opinion is more valuable than two diverse judgments. Therefore, bpas combination is inevitable.

The presented modifications are based on the distance of combined bpas and generally aim at softening counter-intuitive effects while combining very similar or quite diverse bpas. A distance is chosen as a measure of similarity of the bpas. It is shown that the proposed concept works for several distance definitions. This may indicate that not only other manners of distance determination, but also other measures of similarity, not necessary based on distances, can be used to improve the Dempster's rule, by means of the introduced coefficient.

Simulations confirm that proposed methods make it possible to tune the resulting bpa towards mean or classical rule outcomes. Thus, it is possible to obtain results similar to the average, without explicit use of the mean. This allows for free change of final bpa to improve its conformity with intuition, whenever it is necessary.

The proposed modifications are numerically simple and do not show dependence on variables (symptoms) characteristics. Therefore, hopefully they can be used in many applications of medical diagnosis support.

References

1. Dempster, A.P.: A generalization of Bayesian inference. J. R. Stat. Soc. **30**(2), 205–247 (1968). Wiley, New Jersey
2. Denoeux, T.: A k-nearest neighbor classification rule based on the Dempster-Shafer theory. IEEE Trans. Syst. Man Cubernetics **25**(5), 804–813 (1995)
3. Denoeux, T.: Conjunctive and dijunctive combination of belief functions induced by nondistinct bodies of evidence. Artif. Intell. **172**, 234–264 (2008)
4. Dubois, D., Prade, H.: On the use of aggregation operations in information fusion process. Fuzzy Sets Syst. **142**, 143–161 (2004)
5. Han, D., Dezert, J., Yang, Y.: Evaluations of evidence combination rules in terms of statistical sensitivity and divergence. In: Proceedings of the 17th International Conference on Information Fusion (FUSION), pp. 1–7 (2014)
6. Jousselme, A.-L., Maupin, P.: Distances in evidence theory:comprehensive survaey and generalizatiuons. Int. J. Approx. Reason. **53**, 118–145 (2012)
7. Kacprzyk, J., Fedrizzi, M. (eds.): Advances in Dempster-Shafer Theory of Evidence. Wiley, New York (1994)
8. Murphy, C.K.: Combining belief functions when evidence conflicts. Decis. Support Syst. **29**(1), 1–9 (2000)
9. Smets, P.: Belief functions: the disjunctive rule of combination and the generalized Bayesian theorem. Int. J. Approx. Reason. **9**(1), 1–35 (1993)
10. Straszecka, E.: Measures of Uncertainty and Imprecision in Medical Diagnosis Support. Wydawnictwo Politechniki Slaskiej, Gliwice (2010)
11. Zadeh, L.A.: Review of mathematical theory of evidence by G. Shafer. AI Mag. **5**(3), 81–83 (1984)

On Partially Ordered Product Spaces

Považan Jaroslav and Riečan Beloslav

Abstract In the paper a very general system is presented including some known important structures, as continous effect algebras. As an illustration the generalization of the classical Poincaré theorem from ergodic theory is presented.

1 Introduction

In the paper we shall consider the notion of the binary product operation from the point of view probability theory. It is interesting that 15 years ago something similar was realized in multivalued algebras. In [12] from the point of view probability theory and in [11] from the point of view mathematical logic. And although the proposed definitions were different, they are equivalent. A review of applications of MV-algebras with product from the point of view probability theory has been presented in [16].

Probably inspired by the theory of MV-algebras with product the notion of a D-poset with product has been introduced by F. Kôpka in [9]. We have named the structure as Kôpka D-poset ([15]).

In the theory of Atanassov intuitionistic fuzzy sets the notion of the product of two IF-sets is natural as well as in fuzzy sets which are a special case of IF concept. Of course it is interesting that the product corresponds with the product of two real functions.

As an application a useful result will be presented in the general form: the Poincaré reccurence theorem. It says that if we have a good transformation $T : \Omega \to \Omega$ of a probability space (Ω, S, P) and any set $A \in S$, then almost every element of A

P. Jaroslav
Faculty of Natural Sciences, Matej Bel University, Tajovského 40,
Banská Bystrica, Slovakia
e-mail: jaroslav.povazan@umb.sk

R. Beloslav (✉)
Mathematical Institute, Slovak Academy of Sciences, Štefánikova 49,
Bratislava, Slovakia
e-mail: beloslav.riecan@umb.sk

K.T. Atanassov et al. (eds.), *Novel Developments in Uncertainty Representation and Processing*, Advances in Intelligent Systems and Computing 401,
DOI 10.1007/978-3-319-26211-6_3

will return to A infinitely many times. If $\omega \in A$ is such a good element, then to every $k \in \mathbb{N}$ there is $n > k$, such that $T^n(\omega) \in A$ i.e. $\omega \in T^{-n}(A)$. The set of such elements $\omega \in A$ which will not return has probability zero:

$$P\left(A \setminus \bigcup_{n=k}^{\infty} T^{-n}(A)\right) = 0$$

for any $k \in \mathbb{N}$. Of course, we assume that T is measure preserving, i.e. $P\big(T^{-1}(A)\big) = P(A)$ for any $A \in \mathcal{S}$.

2 Product Posets

We shall consider the following definition.

Definition 1 Let P be a partially ordered set $(P, \leq)$ with a binary operation $\star : P \times P \to P$. We shall say that $(P, \leq, \star)$ is a product space if the following conditions hold:

1. $\star$ is commutative and associative,
2. $a \star b \leq a$ for any $a, b \in P$,
3. if $a_n \in P, a_n \geq a_{n+1}$ $(n = 1, 2, \ldots)$, then there exists $\bigwedge_{n=1}^{\infty} a_n \in P$.

Example 1 Let P be a σ-algebra of subsets of a set X. For $A, B \in P$ put

$$A \leq B \Leftrightarrow A \subset B$$

and

$$A \star B = A \cap B.$$

Evidently if $A_n \in P, A_n \supset A_{n+1}$ $(n = 1, 2 \ldots)$ then

$$\bigcap_{n=1}^{\infty} A_n \in P.$$

Example 2 A natural generalization of the previous example is the system P of all measurable (with respect to a σ-algebra) fuzzy subsets of X i.e. ([17, 18])

$$P = \{f : X \to [0, 1]\,; f \text{ is measurable}\}.$$

Here $f \leq g$ is defined by a natural way and $f \star g$ is usual product of functions.

Example 3 Consider the Atanassov IF-theory ([1, 2]). Recall that an IF-subset of a set X is a pair $A = \big(\mu_A, \nu_A\big)$ of functions

$$\mu_A : X \to [0,1], \nu_A : X \to [0,1]$$

such that

$$\mu_A + \nu_A \leq 1.$$

We call μ_A the membership function, ν_A the non membership function and

$$A \leq B \Leftrightarrow \mu_A \leq \mu_B, \nu_A \geq \nu_B.$$

If $A = (\mu_A, \nu_A), B = (\mu_A, \nu_A)$ are two IF-sets, then we define

$$A \star B = (\mu_A \cdot \mu_B, 1 - (1 - \nu_A) \cdot (1 - \nu_B)) = (\mu_A \cdot \mu_B, \nu_A + \nu_B - \nu_A \cdot \nu_B).$$

Example 4 In [14] it was proved that any IF-algebra can be embedded to an MV-algebra. By the Mundici theorem MV-algebra can be defined as an interval $[0, u]$ in an l-group $(G, + \leq)$, i.e. such an additive group $(G, +)$ which is a distributive lattice satisfying the implication $a \leq b \Rightarrow a + c \leq b + c$. So the MV-algebra is the interval $P = \{x \in G, 0 \leq x \leq u\}$, where $u > 0$ is an element and the two following operations $\oplus, \odot$ are given

$$a \oplus b = (a + b) \wedge u, a \odot b = (a + b - u) \vee 0$$

Of course in [12] and independently in [11], there was introduced the notion of the MV-algebra with product $\star$, what is a binary operation which is commutative, associative and satisfies the following two identities:

$$u \star a = a$$
$$a \star (b \odot (u - c)) = (a \star b) \odot (u - a \star c).$$

We must to prove that $a \star d \leq a$ for any $a, d \in P$. Put in the second identity $b = u, c = u - d$. Then

$$a \star (b \odot (u - c)) = a \star (u \odot d) = a \star d$$

On the other hand

$$(a \star b) \odot (u - a \star c) = (a \star u) \odot (u - a \star (u - d)) = a \odot (u - a \star (u - d)) \leq a,$$

since $a \odot v \leq a$ for any v.

Example 5 The notion of a D-poset was introduced in [10] (see also [7, 8]). It is a partially ordered set $(P, \leq)$ with a partial binary operation $-$. Here $a - b$ is defined if and only if $b \leq a$. It is assumed that the following properties hold:

$$a \le b \Rightarrow b - a \le b, b - (b - a) = a$$
$$a \le b \le c \Rightarrow c - b \le c - a, c - a - (c - b) = b - a.$$

In [9] there was defined the product operation. It has been applied in [15].

3 States

In the classical case, if we have a probability $P : S \to [0, 1]$ on a σ-algebra, then for pairwise disjoint sets $A_1, \ldots, A_n \in S$ $\left(\text{i.e } A_i \cap A_j = \emptyset \text{ for } i \neq j\right)$ we have

$$\sum_{i=1}^{n} P(A_i) = P\left(\bigcup_{i=1}^{n} A_i\right) \le P(\Omega) = 1.$$

Therefore we shall introduce the following definition.

Definition 2 Let $\mathbb{P} = (P, \le, \star)$ be a product space
A mapping $m : P \to [0, 1]$ is a state if the following properties hold:

- $\forall a, b \in P, a \le b \Rightarrow m(a) \le m(b)$,
- $a_1, \ldots, a_n \in P, m(a_i \star a_j) = 0$ for $i \neq j \Rightarrow \sum_{i=1}^{n} m(a_i) \le 1$.

Example 6 Let P be a σ-algebra of subsets of X and $m : P \to [0, 1]$ be a probability measure, $A \star B = A \cap B$. Evidently $A \subset B$ implies $m(A) \le m(B)$. If $m(A_i \cap A_j) = 0$ for $i \neq j$, then

$$\sum_{i=1}^{n} m(A_i) = m\left(\bigcup_{i=1}^{n} A_i\right) \le 1,$$

hence m satisfies also the condition 2.

Example 7 Consider a family P of measurable IF-sets $A = (\mu_A, \nu_A)$. The probability has been defined constructively by P. Grzegorzewski and A. Mrowka ([6]), descriptively by B. Riečan ([14]) as a mapping $m : P \to [0, 1]$. In [3, 4] there was proved that to any state $m : P \to [0, 1]$ there exist probability measures $\lambda : S \to [0, 1]$, $\kappa : S \to [0, 1]$ and $\alpha \in \mathbb{R}$ such that

$$m((\mu_A, \nu_A)) = \int \mu_A \mathrm{d}\lambda + \alpha \left(1 - \int (\mu_A + \nu_A)\, \mathrm{d}\kappa\right).$$

The mapping $m : P \to [0, 1]$ is additive because it is σ-additive.

Example 8 Let P be an MV-algebra, $m : P \to [0, 1]$ be a state. It was intensively studied in [16]. And again m is additive, because it is σ-additive.

4 Dynamical Systems

The classical dynamical system is a quadrillup $(\Omega, \mathcal{S}, P, T)$, where $(\Omega, \mathcal{S}, P)$ is a probability space, and $T : \Omega \to \Omega$ is a measure preserving map, i.e.

$$P\big(T^{-1}(A)\big) = P(A),$$

whenever $A \in \mathcal{S}$. (Of course, T is measurable, i.e. $A \in \mathcal{S}$ implies $T^{-1}(A) \in \mathcal{S}$). Define $\tau : \mathcal{S} \to \mathcal{S}$ by the equality

$$\tau(A) = T^{-1}(A), A \in \mathcal{S}.$$

Then $\tau : \mathcal{S} \to \mathcal{S}$ is a mapping such that

$$P(\tau(A)) = P(A)$$

for any $A \in \mathcal{S}$. It leads to the following definition.

Definition 3 Let $\mathbb{P} = (P, \leq, \star)$ be a product poset. The family $(P, \leq, \star, \tau)$ will be called a p-dynamical system, if $m : P \to [0, 1]$ is a state and $\tau : P \to P$ is such mapping that

- $a \leq b \Rightarrow \tau(a) \leq \tau(b)$,
- $\tau(a \star b) = \tau(a) \star \tau(b)$
- $m(\tau(a)) = m(a)$

for any $a, b \in P$

Theorem 1 *Let $\mathbb{P} = (P, \leq, \star, \tau)$ be a p-dynamical system, $a \mapsto \neg a$ be a unary operation such that $m(a \star \neg a) = 0$ for any $a \in P$. Put*

$$a_n = \tau(\neg a) \star \cdots \star \tau^n(\neg a), b_k = \bigwedge_{n=k}^{\infty} a_n$$

Then

$$m\big(a \star b_k\big) = 0$$

for any $k \in \mathbb{N}$.

Proof Put

$$b = a \star b_1.$$

Then $b \leq a$ and also $b \leq a_n \leq \tau^n(\neg a)$ for any n, and since τ is monotone, we have $\tau^n(b) \leq \tau^n(a)$, so

$$b \star \tau^n(b) \leq \tau^n(a) \star \tau^n(\neg a) = \tau^n(a \star \neg a).$$

Therefore for any n

$$m(b \star \tau^n(b)) \leq m(\tau^n(a) \star \tau^n(\neg a)) = m(\tau^n(a \star \neg a)) = m(a \star \neg a) = 0. \quad (1)$$

Let $i, j \in \mathbb{N}, i \neq j$, e.g. $i < j$. Then

$$m\big(\tau^i(b) \star \tau^j(b)\big) = m\big(\tau^i\big(b \star \tau^{j-i}(b)\big)\big) = m\big(b \star \tau^{j-i}(b)\big) = 0 \quad (2)$$

by 1. Since 2 holds for any $i \neq j$, we obtain by the weak additivity

$$\sum_{i=1}^{n} m\big(\tau^i(b)\big) \leq 1.$$

But $m\big(\tau^i(b)()\big) = m(b)$. Therefore

$$nm(b) = \sum_{i=1}^{n} m\big(\tau^i(b)\big) \leq 1.$$

But the inequality $m(b) \leq \frac{1}{n}$ for any n implies $m(b) = 0$, hence

$$m\left(a \star \bigwedge_{n=1}^{\infty} \big(\tau(\neg a) \star \tau^2(\neg a) \star \cdots \star \tau^n(\neg a)\big) \right) = 0 \quad (3)$$

Consider now τ^k instead of τ. Then

$$m\left(a \star \bigwedge_{n=1}^{\infty} \big(\tau^k(\neg a) \star \tau^{2k}(\neg a) \star \cdots \star \tau^{nk}(\neg a)\big) \right) = 0.$$

But

$$m\big(a \star b_k\big) = m\left(a \star \bigwedge_{n=k}^{\infty} \big(\tau(\neg a) \star \tau^2(\neg a) \star \cdots \star \tau^n(\neg a)\big) \right) \leq$$
$$\leq m\left(a \star \bigwedge_{n=1}^{\infty} \big(\tau^k(\neg a) \star \tau^{2k}(\neg a) \star \cdots \star \tau^{nk}(\neg a)\big) \right) = 0$$

5 Conclusions

We presented many examples of partially ordered product spaces. Evidently our general Poincaré theorem can be applied in each of the examples. By this method one can obtain new versions of the Poincaré reccurrence theorem in different algebraic systems. It is also hopefull to study other problems of ergodic theory and probability theory in the partially ordered product spaces.

Acknowledgments The support of the grant VEGA 1/0621/1 is kindly announced.

References

1. Atanassov, K.T.: Intuitionistic Fuzzy Sets: Theory and Applications. Studies in Fuzzines and Soft Computing. Physica Verlag, Heidelberg (1999)
2. Atanassov, K.T.: On Intuitionistic Fuzzy Sets. Springer, Berlin (2012)
3. Ciungu, L., Riečan, B.: General form of probabilities on IF-sets. In: Fuzzy Logic and Applications. Proceedings of the WILF Palermo, pp. 101-107 (2009)
4. Ciungu, L., Riečan, B.: Representation theorem of probabilities on IFS-events. Inf. Sci. **180**, 793–798 (2010)
5. Foulis, D.J., Bennet, M.K.: Effect algebras and unsharp quantum logics. Found. Phys. **24**, 1331–1352 (1994)
6. Grzegorzewski, P., Mrowka, E.: Probability of intuitionistic fuzzy events. In: Grzegorzewski, P. et al. (eds) Soft Methods in Probability, Statistics and Data Analysis, pp. 105–115 (2002)
7. Chovanec, F.: Difference Posets and Their Graphical Representation (in Slovak). Liptovský Mikuláš (2014)
8. Chovanec, F., Kôpka, F.: D-posets. In: Riečan, B., Neubrunn, T. (eds.) Integral, Measure and Ordering, pp. 278–311. Kluwer, Dordrecht (1997)
9. Kôpka, F.: Quasiproduct on boolean D-posets. Int. J. Theor. Phys. **47**, 26–35 (2008)
10. Kôpka, F., Chovanec, F.: D-posets. Math. Slovaca **44**, 21–34 (1994)
11. Montagna, F.: An algebraic approach to propositional fuzzy logic. J. Logic Lang. Inform. **9**, 91–124 (2000)
12. Riečan, B.: On the product MV-algebras. Tatra Mt. Math. Publ. **16**, 143–149 (1999)
13. Riečan, B.: Variations on a Poincaré theorem. Fuzzy Sets Syst. **232**, 39–45 (2013)
14. Riečan, B.: Analysis of fuzzy logic models. In: Koleshko, V.M. (ed.) Intelligent Systems, INTECH 2012, pp. 219–244 (2012)
15. Riečan, B., Lašová, L.: On the probability on the Kôpka D-posets. In: Atanassov, K. et al. (eds) Developments in Fuzzy Sets, intuitionistic Fuzzy Sets, Generalized Nets and related Topics I, Warsaw 2010, pp. 167–176 (2010)
16. Riečan, B., Mundici, D.: Probability on MV-algebras. In: Pap, E. (ed.) Handbook of Measure Theory II, pp. 869–910. Elsevier, Heidelberg (2002)
17. Zadeh, L.A.: Fuzzy sets. Inf. Control **8**, 338–358 (1965)
18. Zadeh, L.A.: Probability measures on fuzzy sets. J. Math. Abal. Appl. **23**, 421–427 (1968)

Various Kinds of Ordinal Sums of Fuzzy Implications

Paweł Drygaś and Anna Król

Abstract In this contribution new ways of constructing of ordininal sum of fuzzy implications are presented. Moreover, some of their properties are examined, in particular neutral property, identity property, and ordering property.

Keywords Fuzzy implication · Triangular norm · Ordinal sum

1 Introduction

Fuzzy implications are one of the most important fuzzy connectives in many applications such as fuzzy reasoning and fuzzy control. For that reason new families these of connectives are the object of examination. In [5], Mesiar and Mesiarová obtained a special class of ordinal sum implications (called, R-ordinal sum in [2]) by residual implications of ordinal sum of t-norms. In [6] Su et al. introduced the concept of ordinal sum implications similar to the construction of ordinal sum of t-norms. First, in Sect. 2, we recall basic definitions and results concerning t-norms and fuzzy implications including the construction of ordinal sums of these fuzzy connectives. Next, in Sects. 3–6, we propose new constructions of ordinal sums of fuzzy implications and examine some of its properties. At the end we present further research directions for the ordinal sum of implications.

P. Drygaś (✉) · A. Król
Interdisciplinary Centre for Computational Modelling, Faculty of Mathematics and Natural Sciences, University of Rzeszów, Pigonia 1, 35-310 Rzeszów, Poland
e-mail: paweldrs@ur.edu.pl

A. Król
e-mail: annakrol@ur.edu.pl

K.T. Atanassov et al. (eds.), *Novel Developments in Uncertainty Representation and Processing*, Advances in Intelligent Systems and Computing 401,
DOI 10.1007/978-3-319-26211-6_4

2 Preliminaries

Here we present the notions of t-norms and fuzzy implications, as well we recall some of the constructions of ordinal sums of these fuzzy connectives.

2.1 Triangular Norms

First, we recall some information about t-norms.

Definition 1 ([4]) A triangular norm T is an increasing, commutative associative operation $T : [0,1]^2 \to [0,1]$ with neutral element 1.

Example 1 ([4]) Well-known t-norms are:

$$T_M(x,y) = \min(x,y), \qquad T_P(x,y) = xy,$$

$$T_L(x,y) = \max(x+y-1,0), \qquad T_D(x,y) = \begin{cases} x, & \text{if } y = 1 \\ y, & \text{if } x = 1 \\ 0, & \text{otherwise} \end{cases},$$

$$T_{nM}(x,y) = \begin{cases} 0, & \text{if } x+y \le 1 \\ \min(x,y), & \text{otherwise} \end{cases}.$$

Theorem 1 (cf. [4]) *For an operation $T : [0,1]^2 \to [0,1]$ the following items are equivalent* (Fig. 1)*:*

(i) T is a continuous t-norm.
(ii) T is uniquely representable as an ordinal sum of continuous Archimedean t-norms, i.e., there exists a uniquely determined (finite or countably infinite) index set I, a family of uniquely determined pairwise disjoint open subintervals (a_i, b_i) of $[0,1]$ and a family of uniquely determined continuous Archimedean t-norms $(T_i)_{i \in I}$ such that

$$T(x,y) = \begin{cases} T_i(x,y) & \textit{if } (x,y) \in (a_i, b_i]^2 \\ \min(x,y) & \textit{otherwise} \end{cases}.$$

2.2 Fuzzy Implications

Here we recall the notion of fuzzy implication, its possible properties, as well as the class of R-implications.

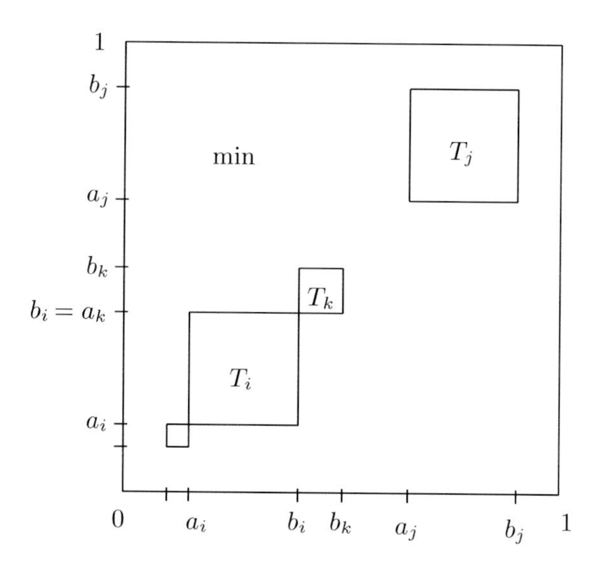

Fig. 1 The structure of ordinal sum of t-norms

Definition 2 ([1], *p. 2,* [3], *p. 21)* A function $I : [0,1]^2 \to [0,1]$ is called a fuzzy implication if it satisfies the following conditions:

(I1) decreasing in its first variable,
(I2) increasing in its second variable,
(I3) $I(0,0) = 1$,
(I4) $I(1,1) = 1$,
(I5) $I(1,0) = 0$.

Directly from the definition we obtain.

Corollary 1 *A fuzzy implication has a right zero element* 1 *and fulfils the condition*

$$I(0,y) = 1, \quad x, y \in [0,1].$$

We distinguish some other properties the fuzzy implication may have.

Definition 3 (*cf.* [1], *p. 9*) We say that a fuzzy implication I fulfils:

- the neutral property (NP) if

$$I(1,y) = y, \quad y \in [0,1], \tag{NP}$$

- the exchange principle (EP) if

$$I(x, I(y,z)) = I(y, I(x,z)), \quad x, y, z \in [0,1], \tag{EP}$$

- the identity principle (IP)

$$I(x,x) = 1, \quad x \in [0,1], \tag{IP}$$

- the ordering property (OP) if

$$I(x,y) = 1 \Leftrightarrow x \leqslant y, \quad x,y \in [0,1], \tag{OP}$$

- the property (CB) if

$$I(x,y) \geq y, \;\; x,y \in [0,1]. \tag{CB}$$

Example 2 ([1], *pp. 4, 5*) Operations I_0 and I_1 are the least and the greatest fuzzy implication, respectively, where

$$I_0(x,y) = \begin{cases} 1, & \text{if } x = 0 \text{ or } \; y = 1 \\ 0, & \text{else} \end{cases}, \; I_1(x,y) = \begin{cases} 0, & \text{if } x = 1, y = 0 \\ 1, & \text{else} \end{cases}.$$

The following are other examples of fuzzy implications.

$$I_{\text{ŁK}}(x,y) = \min(1-x+y,1), \; I_{\text{GG}}(x,y) = \begin{cases} 1, & \text{if } x \leq y \\ \frac{y}{x}, & \text{if} x > y \end{cases},$$

$$I_{\text{GD}}(x,y) = \begin{cases} 1, & \text{if } x \leq y \\ y, & \text{if } x > y \end{cases}, \quad I_{\text{RS}}(x,y) = \begin{cases} 1, & \text{if } x \leq y \\ 0, & \text{if } x > y \end{cases},$$

$$I_{\text{RC}}(x,y) = 1 - x + xy, \qquad I_{\text{YG}}(x,y) = \begin{cases} 1, & \text{if } x, y = 0 \\ y^x, & \text{if else} \end{cases},$$

$$I_{\text{DN}}(x,y) = \max(1-x,y), \qquad I_{\text{FD}}(x,y) = \begin{cases} 1, & \text{if } x \leq y \\ \max(1-x,y), & \text{if } x > y \end{cases},$$

$$I_{\text{WB}}(x,y) = \begin{cases} 1, & \text{if } x \leq 1 \\ y, & \text{if } x = 1 \end{cases}, \quad I_{\text{DP}}(x,y) = \begin{cases} y, & \text{if } x = 1 \\ 1-x, & \text{if } y = 0 \\ 1, & \text{if } x < 1, \; y > 0 \end{cases}.$$

Definition 4 A function $I \, : \, [0,1]^2 \to [0,1]$ is called a residual implication if there exists a t-norm T such that

$$I(x,y) = \sup\{t \in [0,1] \, : \, T(x,t) \leq y\}, \quad x,y \in [0,1].$$

Example 3 The following table shows R-implications obtained from basic t-norms presented in Example 1 by the above formula.

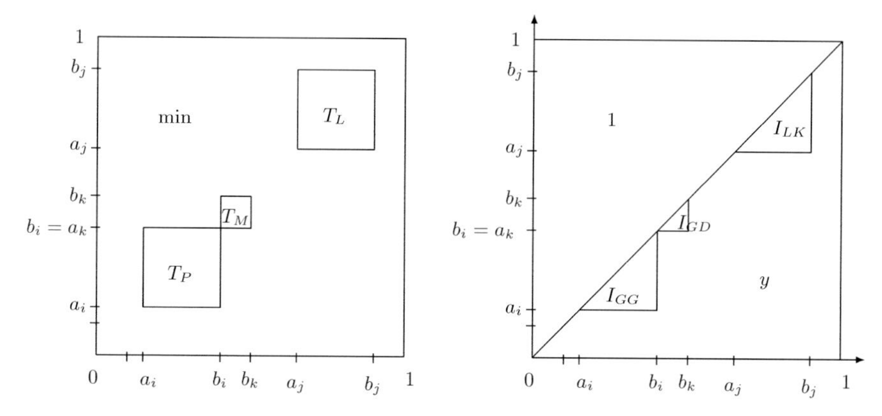

Fig. 2 The structure of R-implications of ordinal sum

t-norm T	R-implication I_T
T_M	I_{GD}
T_P	I_{GG}
T_L	I_{LK}
T_D	I_{WB}
T_nM	I_{FD}

Theorem 2 ([1, 5]) *If T is a continuous t-norm with the ordinal sum structure, then* (see Fig. 2)

$$I_T(x, y) = \begin{cases} 1, & \textit{if } x \leq y \\ a_k + (b_k - a_k) I_{T_k}\left(\frac{x-a_k}{b_k-a_k}, \frac{y-a_k}{b_k-a_k}\right), & \textit{if } x, y \in [a_k, b_k], x > y \\ y, & \textit{otherwise} \end{cases} .$$

3 Ordinal Sum I

The first approach to the construction of ordinal sum of implications is to obtain a structure similar to the structure of implications, which we will get from ordinal sum of t-norms (Fig. 3).

Definition 5 Let $\{I_k\}_{k \in A}$ be a family of implications and $\{(a_k, b_k)\}_{k \in A}$ be a family of pairwise disjoint subintervals of $[0, 1]$ with $a_k < b_k$ for all $k \in A$, where A is a finite or infinite index set. An ordinal sum of fuzzy implications $I : [0, 1]^2 \to [0, 1]$ is given by

$$I(x, y) = \begin{cases} 1, & \text{if } x \leq y \\ a_k + (b_k - a_k) I_k\left(\frac{x-a_k}{b_k-a_k}, \frac{y-a_k}{b_k-a_k}\right), & \text{if } x, y \in [a_k, b_k], x > y \\ y, & \text{otherwise} \end{cases} . \quad (1)$$

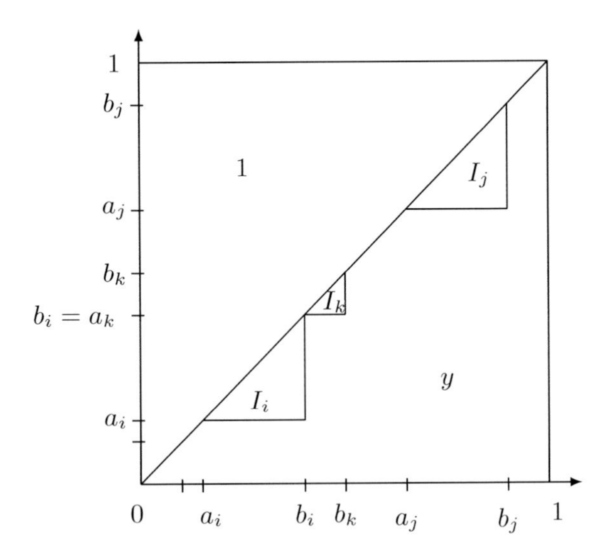

Fig. 3 The structure of ordinal sum I

Remark 1 Let us observe, that I given by (1) we can note as

$$I(x,y) = \begin{cases} a_k + (b_k - a_k)I_k\left(\frac{x-a_k}{b_k-a_k}, \frac{y-a_k}{b_k-a_k}\right), & \text{if } x, y \in [a_k, b_k], y < x \\ I_{GD}(x,y), & \text{otherwise} \end{cases}.$$

Lemma 1 *Let $\{I_k\}_{k\in A}$ be a family of fuzzy implications. Then I given by* (1) *satisfies (I2), (I3), (I4) and (I5).*

Proof First, let us consider the condition (I2). Let $y_1 < y_2$, $x, y_1, y_2 \in [0, 1]$.
If $x \in [a_k, b_k]$ for some $k \in A$, then we obtain the following cases

1. $y_2 < a_k$ or $x \le y_1$ or ($y_1 < a_k$ and $x \le y_2$). Then $I(x, y_1) = I_{GD}(x, y_1) \le I_{GD}(x, y_2) = I(x, y_2)$.
2. $y_1 < a_k \le y_2 \le x$. Then $I(x, y_1) = y_1 < a \le a_k + (b_k - a_k)I_k\left(\frac{x-a_k}{b_k-a_k}, \frac{y_2-a_k}{b_k-a_k}\right) = I(x, y_2)$.
3. $a_k \le y_1 \le y_2 \le x$. Then using monotonicity of I_k we have $I(x, y_1) = a_k + (b_k - a_k)I_k\left(\frac{x-a_k}{b_k-a_k}, \frac{y_1-a_k}{b_k-a_k}\right) \le a_k + (b_k - a_k)I_k\left(\frac{x-a_k}{b_k-a_k}, \frac{y_2-a_k}{b_k-a_k}\right) = I(x, y_2)$.
4. $a_k \le y_1 < x \le y_2$. Then $I(x, y_1) = a_k + (b_k - a_k)I_k\left(\frac{x-a_k}{b_k-a_k}, \frac{y_1-a_k}{b_k-a_k}\right) \le 1 = I(x, y_2)$.

In other cases we have similar situation as in 1.

Directly from (1) we have $I(0, 0) = I(1, 1) = 1$. So I fulfils (I3) and (I4). To prove (I5) let us consider two cases. If there exists $k \in A$ such that $[a_k, b_k] = [0, 1]$, then $I(1, 0) = I_k(1, 0) = 0$. Otherwise $I(1, 0) = y = 0$. □

Example 4 Let

$$I(x,y)=\begin{cases}1, & \text{if } x\le y\\ 0.5I_{RS}(2x,2y), & \text{if } x,y\in[0,0.5]\\ y, & \text{otherwise}\end{cases}$$

I does not fulfill (I1).

Theorem 3 *Let $\{I_k\}_{k\in A}$ be a family of fuzzy implications. Then ordinal sum of fuzzy implication satisfies (I1) if and only if I_k satisfies* (CB) *(see Definition 3) whenever $k\in A$ and $b_k<1$.*

Proof Let I_k satisfies (CB) for $k\in A$. Consider $x_1<x_2$, $x_1,x_2\in[0,1]$.
If $y\in[a_k,b_k]$ for some $k\in A$, then we obtain the following cases

1. $x_2\le y$ or $b_k<x_1$ or ($y_1\le y$ and $b_k<x_2$). Then $I(x_1,y)=I_{GD}(x_1,y)\ge I_{GD}(x_2,y)=I(x_2,y)$.
2. $x_1\le y<x_2\le b_k$. Then $I(x_1,y)=1\ge I(x_2,y)$.
3. $y<x_1<x_2\le b_k$. Then using monotonicity of I_k we have $I(x_1,y)=a_k+(b_k-a_k)I_k\left(\frac{x_1-a_k}{b_k-a_k},\frac{y-a_k}{b_k-a_k}\right)\ge a_k+(b_k-a_k)I_k\left(\frac{x_2-a_k}{b_k-a_k},\frac{y-a_k}{b_k-a_k}\right)=I(x_2,y)$.
4. $y<x_1\le b_k<x_2$. Then using (CB) we have

$$I(x_1,y)=a_k+(b_k-a_k)I_k\left(\frac{x_1-a_k}{b_k-a_k},\frac{y-a_k}{b_k-a_k}\right)\ge$$

$$a_k+(b_k-a_k)\frac{y-a_k}{b_k-a_k}=y=I(x_2,y).$$

In other cases we have similar situation as in 1. So, I satisfies (I1).

Now, let I satisfies (I1) and $k\in A$ such that $b_k<1$. Then taking $x,y\in[0,1]$ and $z>b_k$ we have $I_k(x,y)=\frac{I(a_k+(b_k-a_k)x,a_k+(b_k-a_k)y)-a_k}{b_k-a_k}\ge\frac{I(z,a_k+(b_k-a_k)y)-a_k}{b_k-a_k}=\frac{(a_k+(b_k-a_k)y)-a_k}{b_k-a_k}=y$. So, I_k fulfils condition (CB). □

Remark 2 If fuzzy implication satisfies (NP), then it satisfies (CB).

Theorem 4 *Let $\{I_k\}_{k\in A}$ be a family of implications satisfies* (CB) *and I be ordinal sum given by* (1).

(i) If $b_k<1$ for all $k\in A$ then I satisfies (OP).
(ii) If there exists $k_0\in A$ such that $b_k=1$ then I satisfies (OP) *if and only if $I_{k_0}(x,y)<1$ for all $y<x$.*
(iii) If $b_k<1$ for all $k\in A$ then I satisfies (NP).
(iv) If there exists $k_0\in A$ such that $b_{k_0}=1$ then I satisfies (NP) *if and only if I_{k_0} satisfies* (NP).
(v) I satisfies (IP).

Proof By (1) we have $I(x, y) = 1$ for $x \leq y$. Let us observe, that for $y \in [a_k, b_k], x > y$ we have $I(x, y) \in [a_k, b_k]$. So, $I(x, y) \leq b_k < 1$. In the rest of domain $I(x, y) = y < 1$. It means, that I satisfies (OP).

If there exists $k_0 \in A$ such that $b_{k_0} = 1$ then $I|_{\{(x,y)\in[a_{k_0},1]^2 : y<x\}}$ is isomorphic with $I_{k_0}|_{\{(x,y)\in[0,1]^2 : y<x\}}$. Therefore, I is different from the largest element in [0, 1] if and only if I_{k_0} is different from the largest element. So, *(ii)* is fulfilled.
The condition *(iii)* is a direct result from construction (1).

If there exists $k_0 \in A$ such that $b_{k_0} = 1$ and I_{k_0} satisfies (NP) then we have $I(1, y) = y$ for $y < a_{k_0}$. If $y \geq a_{k_0}$, then

$$I(1,y) = a_{k_0} + (1 - a_{k_0})I_{k_0}\left(\frac{1-a_{k_0}}{1-a_{k_0}}, \frac{y-a_{k_0}}{1-a_{k_0}}\right)$$
$$= a_{k_0} + (1 - a_{k_0})I_{k_0}\left(1, \frac{y-a_{k_0}}{1-a_{k_0}}\right) = a_{k_0} + (1 - a_{k_0})\frac{y-a_{k_0}}{1-a_{k_0}} = y.$$

It means, that I satisfies (NP).
Conversely, if I satisfies (NP) then for all $y \in [0, 1]$ we have

$$I_{k_0}(1, y) = \frac{I(1, a_{k_0} + (1 - a_{k_0})y) - a_{k_0}}{1 - a_{k_0}} = y,$$

i.e. I_{k_0} satisfies (NP). □

Remark 3 Let $\{I_k\}_{k\in A}$ be a family of implications satisfies (CB) and I be ordinal sum given by (1).

If there exists $k_0 \in A$ such that $b_k = 1$ and I_{k_0} satisfies (OP) then I satisfies (OP).

4 Ordinal Sum II

Let us note that in the ordinal sum given by (1) we use only the values $I_k(x, y)$ for $x > y$. Now we will present the construction similar to the ordinal sum of t-norms, which uses the value of I_k in the whole unit square.

Definition 6 ([6]) Let $\{I_k\}_{k\in A}$ be a family of implications and $\{[a_k, b_k]\}_{k\in A}$ be a family of pairwise disjoint close subintervals of [0, 1] with $0 < a_k < b_k$ for all $k \in A$, where A is a finite or infinite index set. The mapping $I : [0, 1]^2 \to [0, 1]$ given by

$$I(x,y) = \begin{cases} a_k + (b_k - a_k)I_k\left(\frac{x-a_k}{b_k-a_k}, \frac{y-a_k}{b_k-a_k}\right), & \text{if } x, y \in [a_k, b_k] \\ I_{GD}(x, y), & \text{otherwise} \end{cases} \quad (2)$$

we call ordinal sum of fuzzy implications $\{I_k\}_{k\in A}$ (Fig. 4).

It may be that I given by (2) is not an implication.

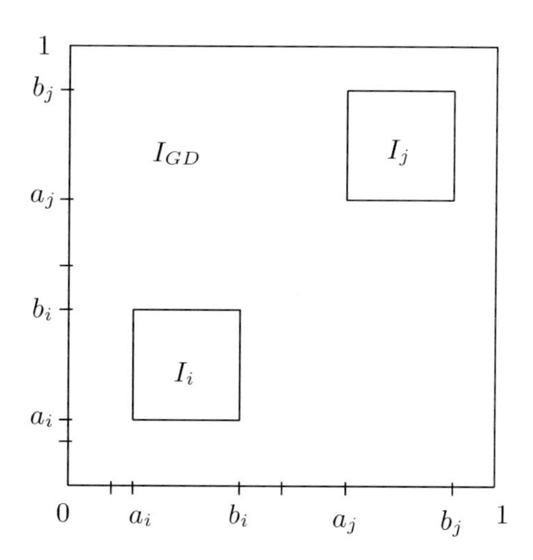

Fig. 4 The structure of ordinal sum II

Example 5 ([6]) Let

$$I(x,y) = \begin{cases} \frac{1}{4} + \left(\frac{1}{2} - \frac{1}{4}\right) I_{RS}\left(\frac{x-\frac{1}{4}}{\frac{1}{2}-\frac{1}{4}}, \frac{x-\frac{1}{4}}{\frac{1}{2}-\frac{1}{4}}\right) & \text{if } (x,y) \in [\frac{1}{4}, \frac{1}{2}]^2, \\ I_{GD}(x,y) & \text{otherwise.} \end{cases}$$

It is easy to see that $I\left(\frac{1}{2}, \frac{1}{3}\right) = \frac{1}{4} < \frac{1}{3} = I\left(\frac{3}{4}, \frac{1}{3}\right)$, i.e. I does not satisfy (I1).

The next theorem gives out the conditions that I given by (2) satisfies (I1).

Theorem 5 ([6]) *Let $\{I_k\}_{k\in A}$ be a family of implications. Then ordinal sum of implication given by* (2) *satisfies (I1) if and only if I_k satisfies* (CB) *whenever $k \in A$ and $b_k < 1$.*

Theorem 6 ([6]) *Let $\{I_k\}_{k\in A}$ be a family of implications satisfying* (CB) *and I be ordinal sum given by* (2)

(*i*) *If there exists $k \in A$ such that $b_k < 1$, then I satisfies neither* (IP) *nor* (OP).
(*ii*) *If card$A = 1$ with $a_1 > 0$, $b_1 = 1$ then I satisfies* (IP) *if and only if I_1 satisfies* (IP).
(*iii*) *If card$A = 1$ with $a_1 > 0$, $b_1 = 1$, then I satisfies* (OP) *if and only if I_1 satisfies* (OP).
(*iv*) *If $b_k < 1$ for all $k \in A$ then I satisfies* (NP).
(*v*) *If there exists $k_0 \in A$ such that $b_{k_0} = 1$, then I satisfies* (NP) *if and only if I_{k_0} satisfies* (NP).

5 Ordinal Sum III

As we can see, not every implication can be used in construction (1) and (2). Below we present a structure in which they can be used any implications.

Definition 7 Let $\{I_k\}_{k\in A}$ be a family of implications and $\{[a_k, b_k]\}_{k\in A}$ be a family of pairwise disjoint close subintervals of $[0, 1]$ with $0 < a_k < b_k$ for all $k \in A$, where A is a finite or infinite index set. The mapping $I : [0, 1]^2 \to [0, 1]$ given by Fig. 5

$$I(x, y) = \begin{cases} a_k + (b_k - a_k) I_k \left(\frac{x-a_k}{b_k-a_k}, \frac{y-a_k}{b_k-a_k} \right), & \text{if } x, y \in [a_k, b_k] \\ I_{RS}(x, y), & \text{otherwise} \end{cases} \tag{3}$$

we call ordinal sum of fuzzy implications $\{I_k\}_{k\in A}$, where

$$I_{\mathrm{RS}}(x, y) = \begin{cases} 1, & \text{if } x \le y \\ 0, & \text{if } x > y. \end{cases}$$

Theorem 7 ([6]) *Let $\{I_k\}_{k\in A}$ be a family of implications. Then ordinal sum of implication given by* (3) *is an implication.*

In the same way as in Theorems 4 and 6 we obtain the following.

Theorem 8 *Let $\{I_k\}_{k\in A}$ be a family of implications and I be ordinal sum given by* (3)

(i) If there exists $k \in A$ such that $b_k < 1$, then I satisfies neither (IP) *nor* (OP).
(ii) If card$A = 1$ with $a_1 > 0$, $b_1 = 1$ then I satisfies (IP) *if and only if I_1 satisfies* (IP).

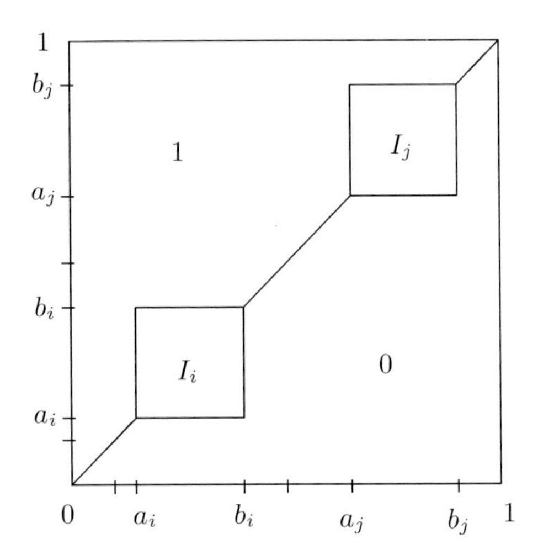

Fig. 5 The structure of ordinal sum III

(*iii*) *If* $cardA = 1$ *with* $a_1 > 0$, $b_1 = 1$, *then* I *satisfies* (OP) *if and only if* I_1 *satisfies* (OP).
(*iv*) I *satisfies* (NP) *if and only if* $cardA = 1$ *with* $a_1 = 0$, $b_1 = 1$ *and* I_1 *satisfies* (NP).

6 Ordinal Sum IV

In both construction (2) and (3) the intervals $[a_k, b_k]$ must be separable. This means that we are unable to construction implications in which the values $I(x, x)$ for $x \in]0, 1[$ depend on the components implications I_k. Below we present construction that solves this problem.

Definition 8 Let $\{I_k\}_{k\in A}$ be a family of implications and $\{(a_k, b_k)\}_{k\in A}$ be a family of pairwise disjoint subintervals of $[0, 1]$ with $a_k < b_k$ for all $k \in A$, where A is a finite or infinite index set. Ordinal sum $I : [0, 1]^2 \to [0, 1]$ is given by (Fig. 6)

$$I(x,y) = \begin{cases} a_k + (b_k - a_k)I_k\left(\frac{x-a_k}{b_k-a_k}, \frac{y-a_k}{b_k-a_k}\right), & \text{if } x, y \in (a_k, b_k] \\ 1, & \text{if } x \le y \\ 0, & \text{otherwise} \end{cases} \tag{4}$$

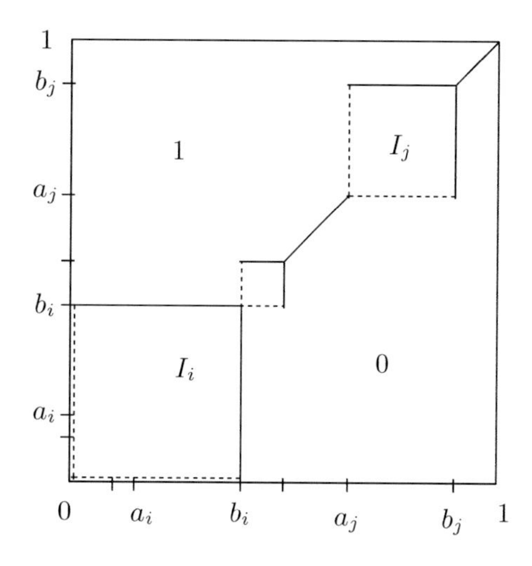

Fig. 6 The structure of ordinal sum IV

Example 6 Let

$$I(x,y) = \begin{cases} 1, & \text{if } x \leq y \\ 0.5I_{RC}(2x, 2y), & \text{if } x, y \in (0, 0.5] \\ 0.5 + 0.1I_{LK}(10x - 5, 10y - 5), & \text{if } x, y \in (0.5, 0.6] \\ 0, & \text{otherwise} \end{cases}$$

I is an implication.

Theorem 9 *Let $\{I_k\}_{k \in A}$ be a family of implications. Then ordinal sum of implication is an implication.*

Theorem 10 *Let $\{I_k\}_{k \in A}$ be a family of implications and I be ordinal sum given by* (4)

(i) *If there exists $k \in A$ such that $b_k < 1$, then I satisfies neither* (IP) *nor* (OP).
(ii) *If $cardA = 1$ with $a_1 \geq 0$, $b_1 = 1$ then I satisfies* (IP) *if and only if I_1 satisfies* (IP).
(iii) *If $cardA = 1$ with $a_1 \geq 0$, $b_1 = 1$, then I satisfies* (OP) *if and only if I_1 satisfies* (OP).
(iv) *I satisfies* (NP) *if and only if $cardA = 1$ with $a_1 = 0$, $b_1 = 1$ and I_1 satisfies* (NP).

7 Conclusion

In this paper we introduce the concept of three various constructions of ordinal sum of implications. Each construction has different properties. Basic properties thus obtained implications have been examined.

In future work, we will examine which properties of components implications are preserved by introduced ordinal sums. We will examine also, which construction give the implications not belonging to known classes.

Acknowledgments This work was partially supported by the Centre for Innovation and Transfer of Natural Sciences and Engineering Knowledge in Rzeszów, through Project Number RPPK.01.03.00-18-001/10. The first author has been partially supported by research fellowship within "UR—modernity and future of region" funded from European Social Fund, Human Capital, national Cohesion Strategy (contract no. UDA-POKL.04.01.01-00-068/10-00).

References

1. Baczyński, M., Jayaram, B.: Fuzzy Implications. Springer, Berlin (2008)
2. Durante, F., Klement, E.P., Mesiar, R., Sempi, C.: Conjunctors and their residual implicators: characterizations and construction methods. Mediterr. J. Math. **4**, 343–356 (2007)
3. Fodor, J., Roubens, M.: Fuzzy Preference Modelling and Multicriteria Decision Support. Kluwer, Dordrecht (1994)

4. Klement, E.P., Mesiar, R., Pap, E.: Triangular Norms. Kluwer, Dordrecht (2000)
5. Mesiar, R., Mesiarova, A.: Residual implications and left-continuous t-norms which are ordinal sums of semigroups. Fuzzy Sets Syst. **143**, 47–57 (2004)
6. Su, Y., Xie, A., Liu, H.: On ordinal sum implications. Inform. Sci. **293**, 251–262 (2015)

A-Poset with Multiplicative Operation

Daniela Kluvancová

Abstract In this paper we will prove that the new structure called A-poset, defined by Frič and Skřivánek (Generalized random events, 2015) is equivalent to D-posets and effect algebras. In next section we introduce a multiplicative operation on A-postes and prove that these two structures are isomorphic. In the last part of this paper we try to build probability theory on A-posets.

Keywords A-poset · Effect algebra · Partial ordering

1 Introduction

The paper was motivated by the D-poset theory (see [2, 4]). Of course it was based on Mundici's characterization of MV-algebras by the help of ℓ-groups [8].

It is interesting that independently there was introduced the notion of MV-algebras with product.

More generally, in [4] was introduced the notion of the D-poset with product and it was named as Kôpka's D-poset and applied in the probability theory. In [9], M. Paulínyová studied effect algebras and she showed that these two algebraic structures are isomorphic.

In the paper we study an algebraic structure called A-poset. It was introduced by Frič and Skřivánek [3] and they proved A-posets and D-posets are isomorphic. In this paper we give an original proof that A-posets are isomorphic to effect algebras. Moreover we introduced the notion of the product on any A-poset and show that A-posets with product are isomorphic to the effect algebras with product. Therefore Kôpka's D-posets, A-posets with product and effect algebras with product are isomorphic structures.

D. Kluvancová (✉)
Faculty of Natural Sciences, Matej Bel University,
Tajovského 40, Banská Bystrica, Slovakia
e-mail: daniela.kluvancova@umb.sk

K.T. Atanassov et al. (eds.), *Novel Developments in Uncertainty Representation and Processing*, Advances in Intelligent Systems and Computing 401,
DOI 10.1007/978-3-319-26211-6_5

We introduce some basic facts about the probability theory on A-posets with product including the central limit theorem.

Recall that the probability of IF-sets [1] can be embedded to MV-algebras [11] hence any result presented in this paper can be applied to the probability on IF-sets too.

2 A-Posets and Effect Algebras

Definition 2.1 Let $\leq$ be a relation on a set A. If $\leq$ is:

reflective: $a \leq a \quad \forall a \in A$;
antisymetric: if $a \leq b$ and $b \leq a$ then $a = b, \quad a, b \in A$;
transitive: if $a \leq b$ and $b \leq c$ then $a \leq c, \quad a, b, c \in A$

then $\leq$ is called partial ordering on A.

Definition 2.2 Let A be a nonempty set and relation $\leq$ be a partial ordering. Then a pair $(A, \leq)$ is called a partially ordered set or shortly poset.

Definition 2.3 An A-poset is a structure $A = (A, \leq, +, 0, 1)$, where A is a poset with the least element 0 and the greatest element 1 and $+ : A \times A \to A$ is a partial binary operation on D satisfying the conditions:

(A_1) if $a + b \in A$ then $b + a \in A$ and $a + b = b + a$, $a, b \in A$;
(A_2) if $(a + b) + c \in A$, then $a + (b + c) \in A$ and $(a + b) + c = a + (b + c)$, $a, b, c \in A$;
(A_3) $\forall a \in A\ \exists$ unique $a' \in A$ such that $a + a' = 1$;
(A_4) if $a + b \in A$, $a_1 \leq a$ and $b_1 \leq b$, then $a_1 + b_1 \in A$ and $a_1 + b_1 \leq a + b$, a_1, b_1, $a, b \in A$.

Definition 2.4 An effect algebra E is an algebraic structure $E = (E, \oplus, 0, 1)$, where 0, 1 are fixed elements, $\oplus : E \times E \to E$ is partial binary operation and following properties hold:

(E_1) if $a \oplus b \in E$ then $b \oplus a \in E$ and $a \oplus b = b \oplus a$, $a, b \in E$;
(E_2) if $(a \oplus b) \oplus c \in E$ then $a \oplus (b \oplus c) \in E$ and $(a \oplus b) \oplus c = a \oplus (b \oplus c)$, $a, b, c \in E$,
(E_3) $\forall a \in E\ \exists$ unique $a' \in E$ such that $a \oplus a' = 1$;
(E_4) if $a \oplus 1 \in E$ then $a = 0$.

Theorem 2.1 *An effect algebra and an A-poset are isomorphic structures.*

We will show, that for every A-poset there exists a corresponding effect algebra and for every effect algebra we can find a corresponding A-poset.

Let A be any given A-poset and we try to find equivalent effect algebra. We will need following lemmas.

Lemma 2.1 *For every* $a \in A$: $a + 1' = a$

Proof Let $(a + a') \in A$, then there exists exactly one element $(a + a')'$ such that

$$(a + a')' + (a + a') = 1,$$

and according to the definition of A-poset we can write

$$\begin{aligned} ((a + a')' + a) + a' &= 1 \\ (a + a') + a = (a')' &= a \\ 1' + a &= a \end{aligned}$$

□

Lemma 2.2 *Let A be an A-poset. Then* $0 + 0 = 0$.

Proof From Lemma 2.1 we have $a + 1' = a$ for every $a \in A$. Let $a = 0$, then $0 + 1' = 0$. Using this and condition A_4 we get:

$$(0 + 1') \in A, \quad 0 \le 0, \quad 0 \le 1' \quad \textit{then} \quad 0 + 0 \in A \quad \textit{and} \quad 0 + 0 \le 0 + 1' = 0 \le a$$

for every $a \in A$. Thus $0 + 0 = 0$. □

Lemma 2.3 *Let A be an A-poset. Then* $0 + 1 = 1$.

Proof We use the previous lemma. From axiom A_3 we have $0 + 0' = 1$ and $0 + 0 = 0$, so

$$\begin{aligned} (0 + 0) + 0' &= 1 \\ 0 + (0 + 0') &= 1 \\ 0 + 1 &= 1. \end{aligned}$$

□

Lemma 2.4 *Let A be an A-poset. Then* $1' = 0$ *and* $0' = 1$.

Proof We have $0 + 0' = 1$ and $0 + 1 = 1$, thus $0' = 1$.
An element $1'$ we can write as $(0')' = 0 = 1'$. □

Theorem 2.2 *Let A be an A-poset. If* $(a + 1) \in A$, *then* $a = 0$.

Proof We use axiom A_4. Let $(a + 1) \in A$. We have $a \le a$ and $a' \le 1 \Rightarrow (a + a') \in A$ and $1 = (a + a') \le a + 1$. Thus we have $a + 1 = 1$ and from axiom A_3 $a' = 1$, so $a = (a')' = 1' = 0$. □

Axioms A_1, A_2, A_3 from the definition of A-poset are equivalent to axioms E_1, E_2, E_3 from the definition of effect algebra and the axiom E_4 is now proved so we find an equivalent effect algebra to any given A-poset A.

Now, let E be an effect algebra and we want to derive a corresponding A-poset.

Lemma 2.5 *For every $a \in E$ there it holds $a \oplus 0 = a$.*

Proof Let $a \in A$, then exist exactly one a' such that $a \oplus a' = 1$. Consider $a = 1$. An element $(1 \oplus a') \in E$ and $(1 \oplus a') = 1$ thus from E_4 $a' = 0$, so $1 \oplus 0 = 1$.
Let $b \in E$, $b \oplus b' = 1$ and holds:

$$1 = 1 \oplus 0 = (b \oplus b') \oplus 0 = (b \oplus 0) \oplus b' \Rightarrow (b \oplus 0) = (b')' = b.$$ □

Now we can define a partial ordering $\leq$ on any effect algebra E as follows:

$$a \leq b \quad \Longleftrightarrow \exists c \in E \quad a \oplus c = b.$$

Theorem 2.3 *An ordered pair $(E, \leq)$ where E is an effect algebra, is a poset.*

Proof We need to show that the relation $\leq$ defined on E is a partial ordering.

1. Let $a \in E$, $a = a + 0 \leq a$, thus $a \leq a$.
2. Let $a, b \in A$ and $a \leq b$, $b \leq a$. There exist $c, d \in E$ such that $a \oplus c = b$, $b \oplus d = a$. Thus we have:

$$a = b \oplus d = (a \oplus c) \oplus d = a \oplus (c \oplus d) \Longrightarrow c \oplus d = 0,$$

and next

$$d' = d' \oplus 0 = d' \oplus (d \oplus c) = (d' \oplus d) \oplus c = 1 \oplus c.$$

The element $(1 \oplus c) \in E$, thus from the axiom E_4, $c = 0$. Now we have:

$$0 = c \oplus d = 0 \oplus d = d \Longrightarrow d = 0.$$

Hence $a = b \oplus 0 = b$.
3. Let $a, b, c \in E$ and let $a \leq b$ and $b \leq c$. There exist $d, e \in E$ such that $a \oplus d = b$, $b \oplus e = c$. Now we have:

$$c = b \oplus e = (a \oplus d) \oplus e = a \oplus (e \oplus d) \Longrightarrow a \leq c.$$

We proved that the relation $\leq$ satisfy all axioms of partial ordering, thus $(E, \leq)$ is a poset. □

Theorem 2.4 *An element $(a \oplus b) \in E \Longleftrightarrow b \leq a'$.*

Proof Let $(a \oplus b) \in E$. Then

$$(a \oplus b) \oplus (a \oplus b)' = 1 = a \oplus (b \oplus (a \oplus b)') \Longrightarrow$$
$$\Longrightarrow a' = b \oplus (a \oplus b)' \Longrightarrow b \leq a'.$$

On the other hand, let $a, b \in E$ and $b \leq a'$. Then exists $c \in E$, $b \oplus c = a'$. Now we can compute and substitute:

$$1 = a \oplus a' = a \oplus (b \oplus c) = (a \oplus b) \oplus c \Longrightarrow (a \oplus b) \in E. \qquad \square$$

Let $a \in E$. Equivalently as for A-posets, for elements from effect algebras there hold properties $a + 1' = 1$, $1' = 0$ and then $a + 0 = 0$.

Theorem 2.5 *The Theorem 2.4 is equivalent to the axiom A_4 of the definition of an A-poset.*

Proof 1. Let axiom A_4 holds and let $a, b \in E, a \leq b'$. We have $(b \oplus b') \in E, b \leq b$. Thus $a \oplus b \leq b \oplus b'$ and $a \oplus b \in E$, so $b \leq a'$.
2. Let assume that Theorem 2.4 holds. Let $a + b \in E \Longrightarrow a \leq b'$ and $b \leq a'$. Let $a_1, b_1 \in E, a_1 \leq a, b_1 \leq b$. Then $b' \leq b_1'$. Now we get:

$$a_1 \leq a \leq b' \leq b_1' \Longrightarrow a_1 \leq b_1'. \tag{1}$$

Thus $(a_1 \oplus b_1) \in E$ $\square$

So we introduced partial ordering on effect algebras and we showed, that it is possible to derive an A-poset from any effect algebra.

Thus Theorem 2.1 is proved.

3 Algebraic Structures with Multiplicative Operation

Definition 3.1 An effect algebra with the multiplicative operation $\mathbf{E} = (E, \oplus, \bullet, 0, 1)$ is an algebraic structure, where $\mathbf{E} = (E, \oplus, 0, 1)$ is an effect algebra and $\bullet : E \times E \to E$ is a commutative associative binary operation satisfying:

(e1) $\forall a \in E$: $a \bullet 1 = a$;
(e2) $\forall a, b \in E$ there exist b, c such that $b' = c \oplus d$ and $a = a \bullet b \oplus c$.

Definition 3.2 An A-poset with the multipicative operation $\mathbf{A} = (A, \leq, +, \star, 0, 1)$ is an algebraic structure, where $\mathbf{A} = (A, \leq +, 0, 1)$ is an A-poset and $\star : A \times A \to A$ is a commutative associative binary operation satisfying:

(a1) $\forall a \in E$: $a \star 1 = a$;
(a2) $a \star b \leq a$ and $a \leq a \star b + b'$, $\forall a, b \in E$.

Theorem 3.1 *An effect algebra with the multiplicative operation and an A-poset with the multiplicative operation are isomorphic algebraic structures.*

Proof First we will prove, that for each effect algebra with multiplicative operation there exists corresponding A-poset with multiplicative operation. Let E be an effect algebra with multiplicative operation. We want to derive both axioms of A-poset with multiplicative operation.

1. Let $a \in E$. Multiplication with fixed element 1 is defined by *e1* as $a \bullet 1 = a$ and we denote it $a \star 1 = a$.
2. Let $a, b \in E$. Then because of condition *a2* there exist $c, d \in E$ such that $a = a \bullet b + c$. Definition of partial ordering on effect algebras implies that $a \bullet b \leq a$, so $a \star b \leq a$.
   Now we know, that $a \bullet b \leq a$ and $\bullet$ is a commutative operation, so $a \bullet b = b \bullet a \leq b$. From reflexivity if follows $b' \leq b'$. Because of the axiom *E4*, the element $a \bullet b \oplus b' = f \in E$ and let $b' = c \oplus d$. Thus we have:

$$a \bullet b \oplus c \oplus d = f \Longrightarrow a = a \bullet b \oplus c \leq f = a \bullet b \oplus b' \Rightarrow a \leq a \bullet b \oplus b'.$$

   Now we just change the notation $a \leq a \star b + b'$.

Conversely, let A be an A-poset with multiplicative operation. We will show that we can find corresponding effect algebra with multiplicative operation.

1. Let $a \in A$. Multiplication with element 1 is defined as $a \star 1 = 1$ and we denote it $a \bullet 1 = 1$.
2. Let a, $b \in A$, then $a \star b \leq a$, so there exists $c \in A$ such that $a * b + c = a$, thus $a \bullet b \oplus c = a$.
   Next we know $a \star b = b \star a \leq b$. Then $a \star b \leq a$ and $a \star b \leq b$. Now we can use axiom A_4:

$$a \star b \leq b, \quad b' \leq b' \Longrightarrow a \star b + b' \in A \quad \textit{and} \quad a \star b + b' \leq b + b' = 1.$$

   There holds $a \leq a \star b + b'$, thus there exists $d \in A$ such that $a + d = a \star b + b'$ and we can substitute $(a \star b + c) + d = a \star b + b'$, so $c + d = b'$. We change notation
   $c \oplus d = b'$.

Hence we proved that for each A-poset with multiplicative operation there exist corresponding effect algebra with multiplicative operation. □

4 Probability on A-Posets

In the previous section we proved, that A-posets and effect algebras are equivalent systems. Similarly as on effect algebras we can build the probability theory on A-posets. We will need two important mappings equivalent to probability measure and random variable.

Definition 4.1 A state on an A-poset A is any mapping $m : A \to [0, 1]$

1. $m(1) = 1, \quad \mathrm{m}(0) = 0$,
2. $a_n \nearrow a \Longrightarrow m(a_n) \nearrow m(a), \forall a_n, a \in A$,
3. $a_n \searrow a \Longrightarrow m(a_n) \searrow m(a), \forall a_n, a \in A$.

Definition 4.2 Let $\mathcal{J} = \{(-\infty, t), t \in \mathbb{R}\}$. An observable on an A-poset A is any mapping $x : \mathcal{J} \to A$ satisfying the following properties:

1. $B_n \nearrow \mathbb{R} \Longrightarrow x(B_n) \nearrow 1$,
2. $B_n \searrow \emptyset \Longrightarrow x(B_n) \searrow 0$,
3. $B_n \nearrow B \Longrightarrow x(B_n) \nearrow x(B)$.

Theorem 4.1 *Let $x : \mathcal{J} \to A$ be an observable, $m : A \to [0, 1]$ be a state. Then the mapping $F : \mathcal{J} \to [0, 1]$ defined by formula*

$$F(t) = m(x((-\infty, t))) = m_x((-\infty, t)), \quad t \in \mathbb{R}$$

is a distribution function.

Proof Let $t_n \nearrow t$. Then $(-\infty, t_n) \nearrow (-\infty, t)$, and by 3. of Definition 4.2 we get $x((-\infty, t_n)) \nearrow x((-\infty, t))$. From 2. of Definition 4.1 we have

$$F(t_n) = m(x((-\infty, t_n))) \nearrow m(x((-\infty, t))) = F(t),$$

hence F is left continuous in any $t \in \mathbb{R}$.

Similarly, let $t_n \nearrow \infty$, then $(-\infty, t_n) \nearrow \mathbb{R}$, so $x((-\infty, t_n)) \nearrow 1$. Now we have

$$F(t_n) = m(x((-\infty, t_n))) \nearrow m(1) = 1 \Longrightarrow \lim_{x \to \infty} F(x) = 1.$$

Finally, let $t_n \searrow -\infty$, then $(-\infty, t_n) \searrow \emptyset$ and $x((-\infty, t_n)) \searrow 0$. Thus

$$F(t_n) = m(x((-\infty, t_n))) \searrow m(0) = 0 \Longrightarrow \lim_{x \to -\infty} F(x) = 0. \qquad \square$$

Definition 4.3 An observable $x : \mathcal{J} \to A$ is called integrable if there exists

$$E(x) = \int_{\mathbb{R}} t dm_x(t).$$

Observable x is square integrable if the dispersion

$$\sigma^2 = \int_{\mathbb{R}} (t - E(x))^2 dm_x(t) = \int_{\mathbb{R}} t^2 dm_x(t) - E(x)^2$$

exists.

Very important notion in probability theory is an independence. Start with a classical probability space $(\Omega, \mathcal{S}, P)$, where Ω is a nonempty set, $\mathcal{S}$ is a σ-algebra of subsets of Ω and $P : \mathcal{S} \to \langle 0, 1 \rangle$ is a probability measure. Consider two random variables ξ, η. These random variables are independent, if for all $A, B \in \mathcal{B}(\mathbb{R})$, $P(\xi^{-1}(A) \cap \eta^{-1}(B)) = P(\xi^{-1}(A)) \cdot P(\eta^{-1}(B))$.

Next consider random vector $T : \Omega \to \mathbb{R}^2$, $T(\omega) = (\xi(\omega), \eta(\omega))$ and probability distribution $P_T : \mathcal{B}(\mathbb{R}^2) \to \langle 0, 1\rangle$ determined by $P_T(A) = P(T^{-1}(A)) = P(h(A))$ where $f : \mathcal{B}(\mathbb{R}^2) \to \mathcal{S}$ is given by $f(A) = T^{-1}(A)$, for each $A \in \mathcal{B}(\mathbb{R}^2)$. Because $\xi^{-1}(A) \cap \eta^{-1}(B) = T^{-1}(A \times B)$, for independent random variables ξ, η there holds:

$$P_T(A \times B) = P(\xi^{-1}(A)) \cdot P(\eta^{-1}(B)) = P_\xi(A) \cdot P_\eta(B) = (P_\xi \times P_\eta)(A \times B).$$

We can write $P_T = P_\xi \times P_\eta$ and $P_T = P \circ T^{-1} = P \circ f$ thus ξ, η are independent if and only if there exists $f : \mathcal{B}(\mathbb{R}^2) \to \mathcal{S}$ such that $P \circ f = P_\xi \times P_\eta$.

There exists exactly one probability measure $\lambda_{F_\xi} \times \lambda_{F\eta} : \mathcal{B}(\mathbb{R}^2) \to \langle 0, 1\rangle$ such that for all $A, B \in \mathcal{B}(\mathbb{R})$ holds

$$\lambda_{F_\xi} \times \lambda_{F\eta}(A \times B)$$

Now we want to define the sum of two random variables ξ, η. We need to define $g : \mathbb{R}^2 \to \mathbb{R}$ such that $g(v, w) = v + w$. Hence

$$\xi + \eta = g(\xi, \eta) = g \circ T,$$

and

$$(\xi^{-1} + \eta^{-1})(A) = (g \circ T)^{-1}(A) = T^{-1}(g^{-1}(A)), \quad A \in \mathcal{B}(\mathbb{R}).$$

Now we can apply all above on A-posets.

Definition 4.4 Let $x_1, \dots, x_k : \mathcal{J} \to A$ be observables, $m : A \to \langle 0, 1\rangle$ be a state $\Delta_t^k = \{(u_1, \dots, u_k) \in \mathbb{R}^k; u_1 + \cdots u_k < t\}$, $\mathcal{M}_k = \{\Delta_t^k, t \in \mathbb{R}\}$. The observables are independent if there exists a mapping $h_k : \mathcal{M}_k \to A$ with the following properties:

1. $t_i \nearrow t \Longrightarrow h_k(\Delta_{t_i}^k) \nearrow h_k(\Delta_t^k)$,
2. $h_k(\bigcup_{t=1}^{\infty} \Delta_t^k) = 1$
3. $h_k(\bigcap_{t=-1}^{-\infty} \Delta_t^k) = 0$
4. $m(h_k(\Delta_t^k)) = \lambda_{x_1} \times \lambda_{x_2} \times \cdots \times \lambda_{x_k}(\Delta_t^k)$.

Let $h : \mathcal{B}(\mathbb{R}^2) \to A$ be a join observable. The sum of observables $x_1, x_2 : \mathcal{J} \to A$ is defined by:

$$(x + y)(A) = h(g^{-1}(A)), \quad A \in \mathcal{J}.$$

If $g(v, w) = v + w$ then $g^{-1}((-\infty, t)) = \{(u, v); u + v < t\} = \Delta_t$. Define $\mathcal{M} = \{\Delta_t, t \in \mathbb{R}\}$ thus $h : \mathcal{M} \to A$. The sum of observables $(x + y) : \mathcal{J} \to A$ is defined by the formula

$$(x + y)((-\infty, t)) = h(\Delta_t).$$

More generally, $x_1, \ldots, x_k : \mathcal{J} \to A$ be independent observables and the mapping $y_n : \mathcal{J} \to \mathcal{A}$ is defined by $y_n = \sum_{i=1}^{n} x_i$ i.e.

$$y_n(-\infty, t) = (\sum_{i=1}^{n} x_i)(-\infty, t) = h(\Delta_t^n), \quad t \in \mathbb{R}.$$

Then the mapping y_n is an observable and it is quite obvious from the definition of observable. We call it a sum of observables.

In [6] it has been proved, that in Kôpka D-posets exists the mapping h_k satisfying all properties of the previous Definition. Because of Kôpka D-poset is equivalent structure as A-poset with multiplicative operation, mapping h_k must to exist on this structure too.

Now we are able to formulate the central limit theorem.

Theorem 4.2 *Let $(x_i)_{i=1}^{\infty}$ be independent, square integrable observables, $E(x_1) = E(x_2) = \cdots = a$, $\sigma(x_1) = \sigma(x_2) = \cdots = \sigma$. Then*

$$\lim_{n\to\infty} m\left((\frac{\sqrt{n}}{\sigma} \sum_{i=1}^{n} x_i - a)((0, t)) \right) = \frac{1}{\sqrt{2\pi}} \int_{-\infty}^{t} e^{-\frac{u^2}{2}} du = \Phi(t),$$

for all $t \in \mathbb{R}$.

Proof Let $P_n : \mathcal{B}(\mathbb{R}^n) \to \langle 0, 1\rangle$ be defined by $P_n = \lambda_{x_1} \times \lambda_{x_2} \times \cdots \times \lambda_{x_n}$. Then P_n is a consistent system of probability measures, so

$$P_n(A \times R) = P_{n-1}(A), \quad A \in \mathcal{B}(\mathbb{R}), \quad n \in \mathbb{N}$$

and there exists $P : \sigma(C) \to \langle 0, 1\rangle$ such that

$$P(\pi_n^{-1}(B)) = P_n(B) = \lambda_{x_1} \times \lambda_{x_2} \times \cdots \times \lambda_{x_n}(B), \quad B \in \mathcal{B}(\mathbb{R}), \quad n \in \mathbb{N}$$

where

$$C = \{A \subset \mathbb{R}^N A = \pi_n^{-1}(B), B \in \mathcal{B}(\mathbb{R}), \quad n \in \mathbb{N}\}.$$

For each $n \in \mathbb{N}$ we can define a canonical projection function $\eta_n : \mathbb{R}^N \to \mathbb{R}$ given by

$$\eta_n((u_i)_{i=1}^{\infty}) = u_n.$$

Therefore

$$(\sum_{i=1}^{n} x_i)(-\infty, t) = h_n(\Delta_t^n) = \lambda_{x_1} \times \lambda_{x_2} \times \cdots \times \lambda_{x_n}(\Delta_t^n)$$
$$= P_n(\Delta_t^n) = \{\omega; \eta_1(\omega) + \cdots + \eta_n(\omega) < t\}.$$

Thus we get

$$m\Big(\frac{\sqrt{n}}{\sigma}(\sum_{i=1}^{n} x_i - a)((-\infty, t))\Big) = m\Big((\sum_{i=1}^{n} x_i)((-\infty, \frac{\sqrt{n}}{\sigma}(t+a))\Big)$$

$$= P\Big(\{\omega; \frac{\sqrt{n}}{\sigma}(\sum_{i=1}^{n} \eta_i - a) < t\}\Big).$$

Then by the classical limit theorem we obtain

$$\lim_{n\to\infty} m\Big((\frac{\sqrt{n}}{\sigma}\sum_{i=1}^{n} x_i - a)((-\infty, t))\Big)$$

$$= \lim_{n\to\infty} m\Big((\frac{\sqrt{n}}{\sigma}\sum_{i=1}^{n} \eta_i - a)((-\infty, t))\Big) = \Phi(t).$$

□

5 Conclusion

In this paper we introduced just basic notions of probability theory and central limit theorem. Of course it may be extended to the other concept, for example conditional probability.

References

1. Atanassov, K.T.: Intuitionistic Fuzzy Sets: Theory and Applications. Studies in Fuzziness and Soft Computing. PhysicaVerlag, Heidelberg (1999)
2. Foulis, D., Bennett, M.: Effect algebras and unsharp quantum logics. Found. Phys. **24**, 1325–1346 (1994)
3. Frič, R., Skřivánek, V.: Generalized Random Events (2015)
4. Kôpka, F., Chvanec, F.: D-posets. Math. Slovaca **44**, 21–34 (1994)
5. Kôpka, F.: Quasi product on Boolean D-posets. Int. J. Theor. Phys. **47**, 26–35 (2008)
6. Lašová, L., Rieèan, B.: On the probability theory on the Kôpka D-posets. Dev. Fuzzy Sets, Intuitionistic Fuzzy Sets, Gen. Nets Relat. Topics **1**, 167–176 (2010)
7. Montagna, F.: An algebraic approach to propositional fuzzy logic. J. Logic Lang. Inf. **9**, 91–124 (2000) (Mundici, D., et al. (eds.)) Special Issue on Logics of Uncertainty
8. Mundici, D.: Interpretation of AFC*-algebras in Łukasiewicz sentential calculus. J. Funct. Anal. **56**, 889–894 (1986)
9. Paulínyová, M.: D-posets and effect algebras. Notes Intuitionistic Fuzzy Sets **20**, 32–40 (2014)
10. Rieèan, B.: Analysis of fuzzy logic models. Intell. Syst. 219–244 (2012)
11. Rieèan, B.: On the product MV-algebras. Tatra Mt. Math. Publ. **16**, 143–149 (1999)

Method for Uncertainty Measurement and Its Application to the Formation of Interval Type-2 Fuzzy Sets

Mauricio A. Sanchez, Oscar Castillo and Juan R. Castro

Abstract This paper proposes a new method for directly discovering the uncertainty from a sample of discrete data, which is then used in the formation of an Interval Type-2 Fuzzy Inference System. A Coefficient of Variation is used to measure the uncertainty on a finite sample of discrete data. Based on the maximum possible coverage area of the Footprint of Uncertainty of Gaussian membership functions, with uncertainty on the standard deviation, which then are modified according to the found index values, obtaining all antecedents in the process. Afterwards, the Cuckoo Search algorithm is used to optimize the Interval Sugeno consequents of the Fuzzy Inference System. Some sample datasets are used to measure the output interval coverage.

1 Introduction

Uncertainty, as it is currently perceived, is still something of a mistified topic. Being defined as something that is doubtful or unknown, in which by nature cannot be directly measured, therefore showing a first problem in making use of it. Although by nature, uncertainty is an unknown, it has not stopped engineers, scientists, mathematicians, etc. from using it. That is, although directly not known, an approximate of it can be modeled and used, improving the models in which it is used. By using uncertainty in a model, that model will improve its resilience, thus obtaining a better model in the end.

M.A. Sanchez (✉) · J.R. Castro
Autonomous University of Baja California, Tijuana, Mexico
e-mail: mauricio.sanchez@uabc.edu.mx

J.R. Castro
e-mail: jrcastror@uabc.edu.mx

O. Castillo
Tijuana Institute of Technology, Tijuana, Mexico
e-mail: ocastillo@hafsamx.org

K.T. Atanassov et al. (eds.), *Novel Developments in Uncertainty Representation and Processing*, Advances in Intelligent Systems and Computing 401,
DOI 10.1007/978-3-319-26211-6_6

Most current literature on uncertainty [1–6] is mainly based on having previous knowledge of the confidence interval around certain measurements, which translates into what is the probable uncertainty which exists within certain measurements, usually expressed with the plus-minus symbol ± (e.g. 10.4 ± 0.02, with in interval representation of [10.38,10.42]).

As for models with uncertainty, there exists a logic which directly manages uncertainty, this being Interval Type-2 Fuzzy Logic (IT2 FL) [7], which infers Interval Type-2 Fuzzy Sets (IT2 FS) and ultimately obtains an interval or a crisp value [8]. IT2 FS manage uncertainty directly into its logic by means of confidence intervals [9], the best solution could be anywhere within such interval, and as such is an excellent tool for directly applying and inference when dealing with uncertainty. And as stated, the output interval can be used as the end result and a defuzzification process can be computed upon such interval in the case that a crisp value is required, and not an interval.

In this paper, a link is proposed between a measure of dispersion and uncertainty, which is ultimately used in the formation of IT2 FS. The platform for the model is created by a Fuzzy C-Means algorithm [10], afterwards using the Coefficient of Variation is used to calculate the Fingerprint Of Uncertainty (FOU) of each individual IT2 FS in the antecedents of the Interval Type-2 Fuzzy Inference System (IT2 FIS), and finally, a Cuckoo Search algorithm [11] is used to optimize Interval Type-2 Sugeno linear consequents [12]. The proposed method can be categorized as a hybrid algorithm because it requires multiple steps/algorithm to work in sequence for the final result to be obtained.

This paper is divided into three sections, the first is a brief introduction to the definition of Interval Type-2 Fuzzy Sets; the following section describes in detail both the premises and the proposed method; afterwards, some experimental results are shown and discussed which asses the viability of the proposed method; finally, concluding remarks are given as well as a couple of open questions as future work.

2 Interval Type-2 Fuzzy Sets

With the introduction of Fuzzy Sets in 1965 [13], it improved upon formal hard logic, where instead of only having two choices of truth values {0,1}, any value between [0,1] was now possible. This set an unprecedented involvement in research that up to today is still very strong, first came Type-1 Fuzzy Sets [14], which can only represent vagueness or imprecision, later came Interval Type-2 Fuzzy Sets, which could now, apart from vagueness, also represent a degree of uncertainty (which is the focus of the proposed method in this paper), although recently General Type-2 Fuzzy Sets [15] are starting to gain traction in research, is still far from maturity when compared to Type-1 or Interval Type-2 Fuzzy Sets.

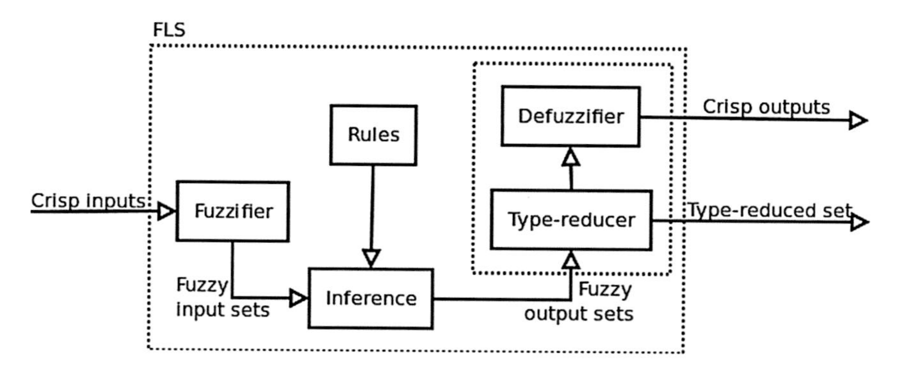

Fig. 1 Block diagram describing an IT2 FLS. With a cripst input, two outputs are possible, a confidence interval in which any possible point within such interval is a correct answer, or a crisp value, in the case a singel real number is required as output

By nature, IT2 FS directly integrate uncertainty into its reasoning. This behavior is best applied in the case of when it is expected to deal with uncertainty in the system that it is modeling, or when certain confidence intervals (uncertainty) are known a priori to designing the IT2 FIS.

The most general descriptive form of an IT2 FIS is through a block diagram, as shown in Fig. 1, which describes the basic inner functions of the complete inference. The Fuzzifier block may or may not transform the crisp input into a FS, this is chosen depending on the intended behavior of the system; the Inference block takes from the Rules block and reasons upon each input's compatibility; the Type-reducer block processes the outputs into an interval; finally, the Defuzzifier block reduces the interval from the previous block and obtains a single real number.

An IT2 FS $\widetilde{A}$ is represented by $\underline{\mu}_{\tilde{A}}(x)$ and $\overline{\mu}_{\tilde{A}}(x)$ which are the lower and upper membership functions respectively of $\mu_{\tilde{A}}(x)$, and is expressed as $\widetilde{A}=\int_{w^l \in \left[\underline{\mu}_{\tilde{F}_k^l}(x_k),\, \overline{\mu}_{\tilde{F}_k^l}(x_k)\right]} 1/w^l$. Where, $x \in X$, k is the kth antecedent, and l the lth rule. A sample IT2 FS is shown in Fig. 2, here a Gaussian membership function with uncertainty in the standard deviation.

The representation for rules in an IT2 FIS is formated as shown in Eq. 1, where, $l=1,\ldots,M$ rules, $p=1,\ldots,q$ inputs, $\widetilde{F}$ is an antecedent IT2 FS, and $\widetilde{G}$ a consequent IT2 FS.

$$R^l\text{: IF } x_1 \text{ is } \widetilde{F}_1^l \text{ and } \ldots \text{ and } x_p \text{ is } \widetilde{F}_p^l\text{, THEN } y \text{ is } \widetilde{G}^l \tag{1}$$

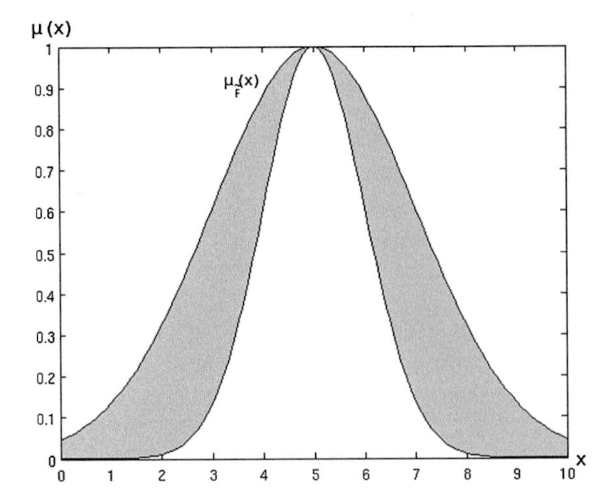

Fig. 2 Sample IT2 FS membership function. A Gaussian membership function is shown which has uncertainty through the standard deviation

3 Proposed Method for Measuring Uncertainty

Before giving a detailed description of the proposed method, the input data must first be defined. As a starting point, a dataset is required, these data pairs, defined as Eq. 2, where φ is a set of ordered input values, and γ is a set of ordered output values, such that Γ forms a tuple of ordered sets of inputs with their respective outputs.

$$\Gamma = \langle \varphi, \gamma \rangle \tag{2}$$

Having a dataset Γ, first some pre-processing must be done in order to obtain the required inputs to the proposed method, this process is executed in order to acquire a description of the IT2 FIS, that is, to obtain the rule description ω as well as each membership function's base description, and the set of data pairs which affected the formation of each membership function $\gamma \in \Gamma$. As this is are the required inputs $\{\omega, \gamma\}$ to the proposed method, a Fuzzy C-Means (FCM) algorithm was chosen to process the raw dataset Γ into the listed required inputs $\{\omega, \gamma\}$. The FCM provides a description of rules by means of a center for each membership function for each rule. Although the FCM can define consequents for the rules in a Fuzzy System, only the antecedents are used. As the other required input is a set of data pair sets which affected the definition of each center, this can be obtained from the partition matrix that is given by the FCM; for each data pair there exists a membership value [0,1] which defines how much a certain data pair belongs to a cluster, or rule of the found FIS, to simplify building the sets of data pairs, a simple competitive rule is used: *the cluster with the highest value decides that said data pair belongs to its formation set*. With both required inputs obtained, the proposed method can now begin.

3.1 Dispersion in Data

Data dispersion in a sample of data pairs can be interpreted as a case of uncertainty. An example of varying degrees of dispersion is shown in Fig. 3, where low, medium, and high data dispersion, in relation to its center can, be perceived.

When there is low dispersion of data samples near it representative center point, most data points are bound by only a small distances as the standard deviation is very small. This being interpreted directly into uncertainty in data, low dispersion is low uncertainty because its numerical evidence concludes that there is near zero possibility that further singular samples will fall far from the central point, unto which all previous numerical evidence is very close to. In the case of medium data dispersion, although there is a concentration of numerical evidence near its central point, there are still data points farther from its center, this leads to knowing that although future reading might obtain evidence which is far from the center, the probabilities of this occurring is low when compared to having future readings fall near the center, although not as near in the case of lower dispersion, this behavior points to having a medium amount of uncertainty. On the extreme case of high

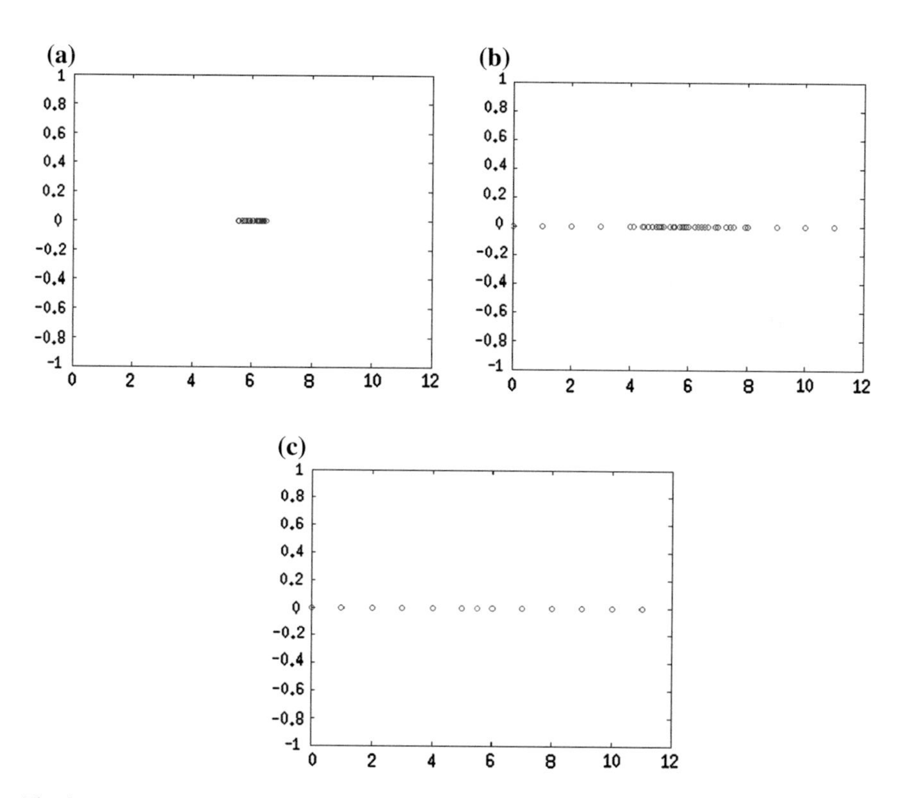

Fig. 3 Example of data dispersion. **a** Low data dispersion, **b** medium data dispersion, and **c** high data dispersion

dispersion, where every sampled data point is evenly distributed throughout the range, the available numerical evidence gives way to conclude that any future sample may equally land on any section, therefore a high amount of uncertainty exists.

3.2 Relation Coefficient of Variation with Uncertainty

For the purpose of converting dispersion into uncertainty, a measure is first required which can identify a degree of dispersion in a given set, preferably a normalized value, and as such requirement, the Coefficient of Variation c_v, shown in Eq. 3, where σ is the standard deviation, and μ is the mean of the set.

$$c_v = \frac{\sigma}{\mu} \tag{3}$$

This coefficient has some limitations which can be avoided by applying some modifications. First, c_v should only be computed on non-negative values, for the case of existing negative values, the solution is to remap all values unto the positive side of the axis. Second, if μ has a value of 0 (zero), this would case an error in computation, the solution is to add ε, which is a very small value, assuring a non-division by zero. Another note on the behavior of c_v, is that in normal distributions, values of [0,1] are most likeley to be obtained, but non-normal distributions can obtain values above 1. Fortunately, with the FCM, all calculated sets are normal distributions, so this is a non-issue with the current implementation.

With the known limitations of c_v, an equation which modifies a set D is proposed which addresses the issue negative values, shown in Eq. (4), where if a value exists that is negative then the absolute value of the minimum is added to the set, thus remapping all values into the domain of positive values.

$$\text{IF}(\exists x \in D), \{x | x < 0\} \text{ THEN } D = D + |\min(D)| \tag{4}$$

In addition, a modification of Eq. (3) to surpress a possible division by zero, as shown in Eq. (5), where ε is a very small value.

$$c_v = \frac{\sigma}{\mu + \varepsilon} \tag{5}$$

To express a relation dispersion-uncertainty, when dealing with IT2 FS, the Footprint of Uncertainty (FOU) is used. This relation is a direct proportion $FOU \propto c_v$. When there is low dispersion, there is a small FOU, when there is a medium amount of dispersion, there is a medium amount of FOU, and when there exists a high amount of dispersion, there is a high amount of FOU. This is better expressed in Fig. 4, where varying degrees of a measure of dispersion has been

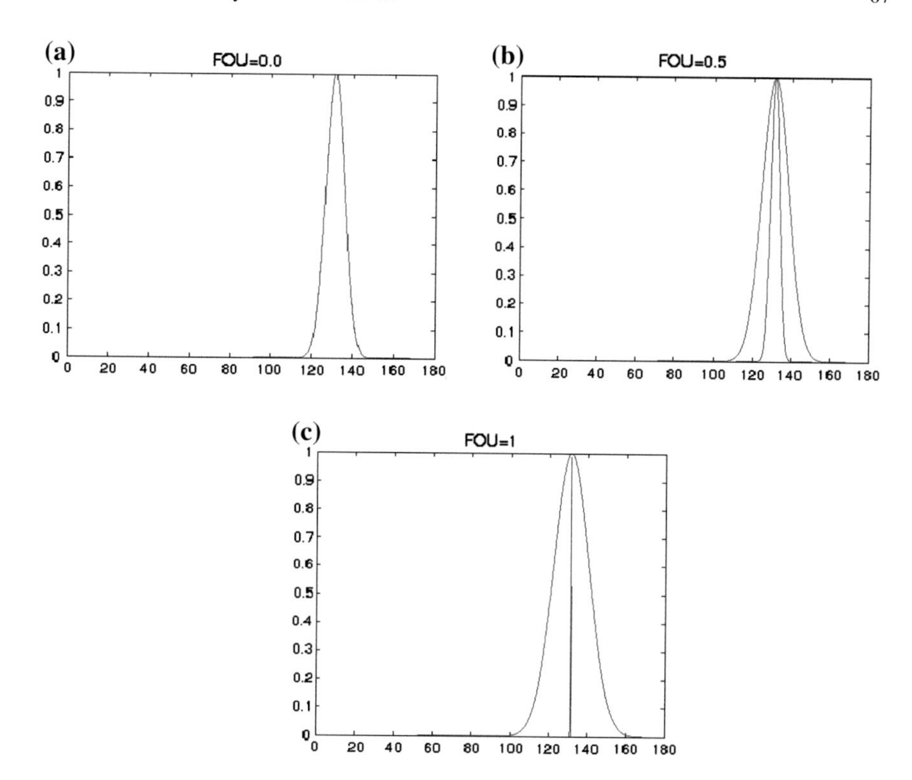

Fig. 4 Examples of varying degrees of FOU. Where **a** FOU = 0, **b** FOU = 0.5, and **c** FOU = 1

converted into a FOU which directly forms an IT2 FS, explained in the following sub-section.

3.3 Proposed Method

To form IT2 FS for the antecedents of a FIS, the first step is to obtain rule configuration, and data pair sets for each inputs on each rule, via a FCM algorithm. Afterwards each set of data pairs is worked on independently of each other. First, a standard deviation σ is found for the set in relation to its μ, which was found by the FCM, then the c_v is calculated. This value is now used to search for the optimal FOU area in an IT2 FS. Considering Fig. 4c, this would be the highest possible area. The initial search is done by first considering the highest possible area and the σ whichs was already calculated, with discrete small steps a search is performed for the FOU value which equals c_v. The smallest value is set as $\sigma_1 = \sigma_2$, shown in Fig. 4a. Each increment step λ affects σ as shown in Eq. (6), this is done iteratively while $\sigma_i \leq \|\mu, \sigma_0\|$.

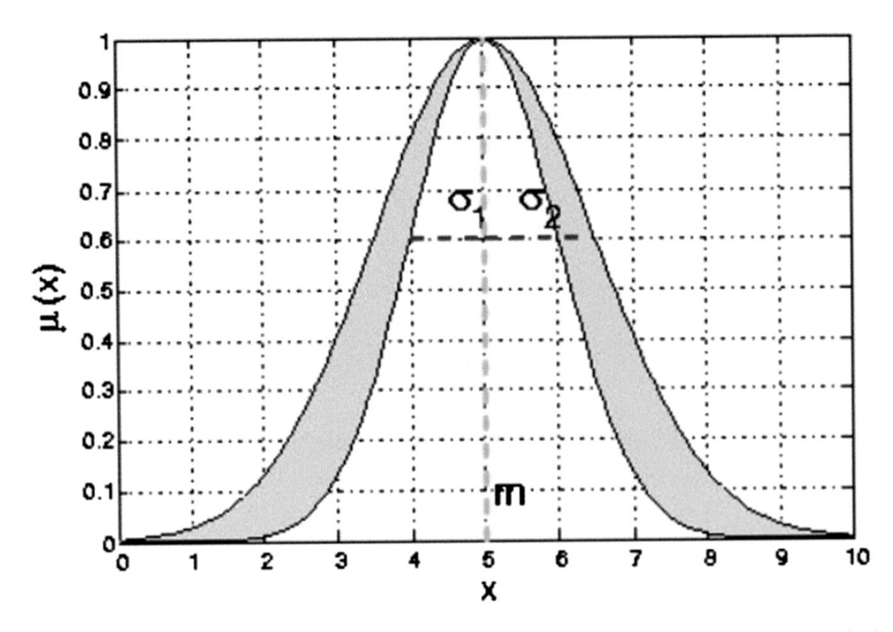

Fig. 5 IT2 FS represented by a Gaussian membership function with uncertainty in the standard deviation

$$\sigma \pm \lambda \tag{6}$$

Once the search has found the values of σ_1 and σ_2 which represent the desired FOU, the IT2 FS can be formed. Which has the form of Fig. 5, this can be formed with the values which have been calculated, by the FCM, μ, and by the proposed method, σ_1 and σ_2. This concludes the proposed method for building the antecedents of an IT2 FIS.

3.4 IT2 Sugeno Fuzzy Consequents

The proposed method only obtains the IT2 FS for the antecedents of a FIS, the next required step is to obtain the consequents of the FIS. This is done by optimizing the IT2 Sugeno linear parameters via a Cuckoo Search algorithm. Although any other optimization algorithm can be used.

4 Experimental Results and Discussion

To test the proposed method, various datasets were used. The validation method was to verify that the interval output of the IT2 FIS had good coverage of the reference targets and at the same time not overreaching too far with the output interval.

Among the used datasets, three were used. A synthetic dataset of a 5th Order curve [16], with 1 input (x) and 1 output (y), and 94 total samples. And two real datasets; engine behavior [16], with 2 inputs (fuel rate, speed) and 2 outputs (torque, nitrous oxide emissions), and 1199 total samples; and Hahn1 [16], with 1 input (temperature) and 1 output (thermex), with 236 total samples.

4.1 Experimental Results

The obtained IT2 FIS for each dataset is shown in Figs. 6, 7, 8, making emphasis on the FOU of the individual membership functions in the antecedents, where varying dregrees of uncertainty can be seen.

As for the output coverage for each dataset, using 40 % training and 60 % training, Table 1 show the summary of the obtained coverage results.

The last set of results show graphical representations of the respective outputs for each dataset, these are shown in Figs. 9, 10, 11, 12. Where the blue points represent the output targets, and the lower and green lines represent the coverage of the FOU.

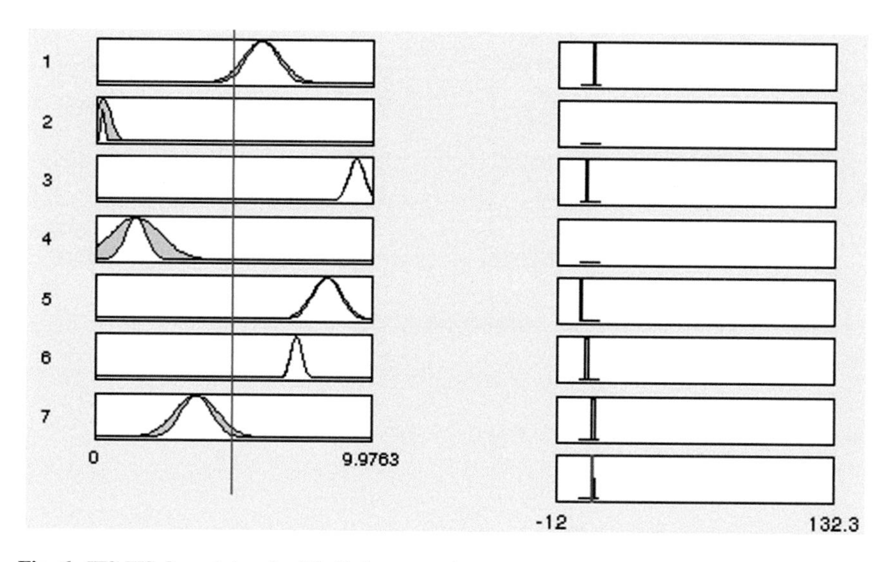

Fig. 6 IT2 FIS for solving the 5th Order curve dataset

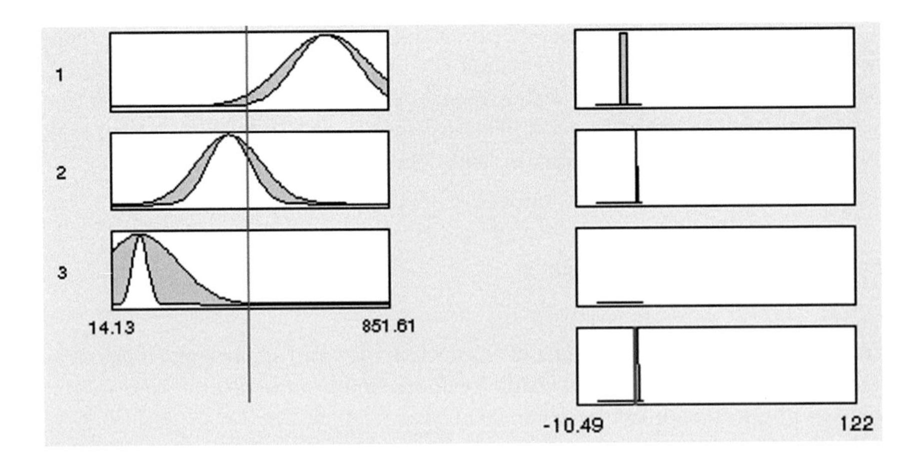

Fig. 7 IT2 FIS for solving the Hahn1 dataset

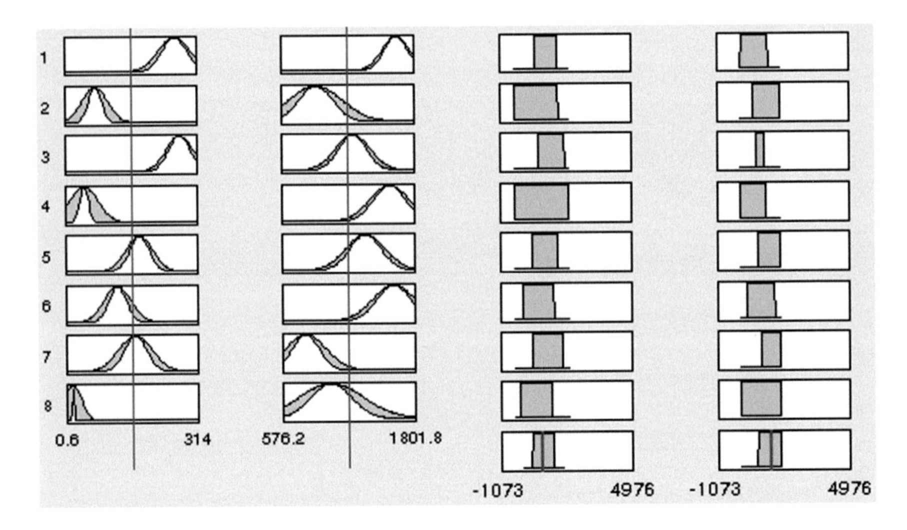

Fig. 8 IT2 FIS for solving the Engine behavior dataset

Table 1 Obtained output coverage results for the chosen datasets

Dataset name	Coverage %
5th Order curve	100
Hahn1	100
Engine behavior	99.88/99.66

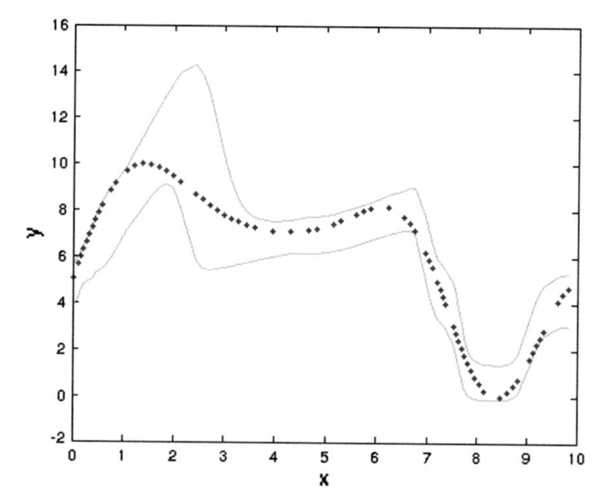

Fig. 9 Output coverage for the 5th Order dataset

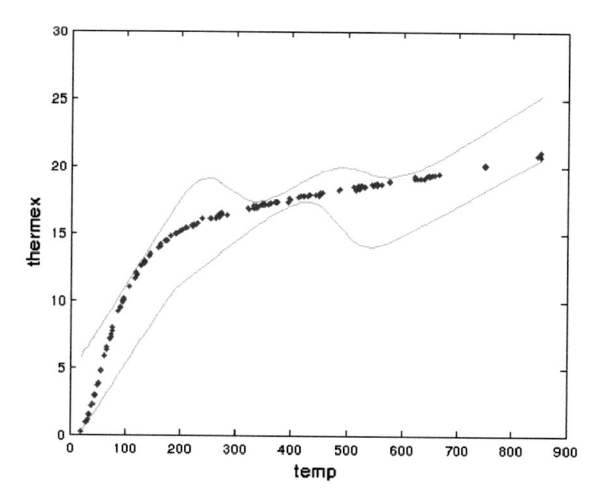

Fig. 10 Output coverage for the Hahn1 dataset

4.2 Results Discussion

With the obtained results, two facets of discussion arise, on the individual level and on the general level. On the individual level; for the 5th Order curve, a full coverage of the target is achieved although there are spikes where the curve changes slope, this is caused by the linear consequents which cannot follow abrupt changes in the curve grade due to the small amount of rules used for this FIS. A solution would be to use more rules to compensate, but this would also cause an additional, and unnecessary, complexity in the system. Yet the overall behavior is acceptable as there is sufficient coverage as well as a controlled width of the output uncertainty. For the Hahn1 solution, it has the same curve behavior of the 5th Order curve, where with only three rules there is a pronounced visible behavior in the linear

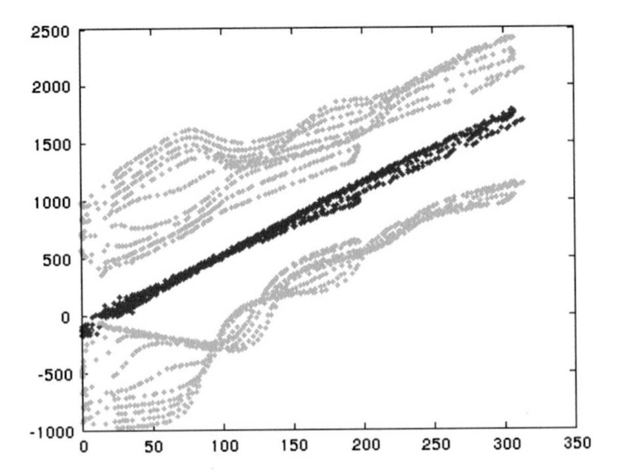

Fig. 11 Output coverage for the Hahn1 dataset. For the first output of the FIS

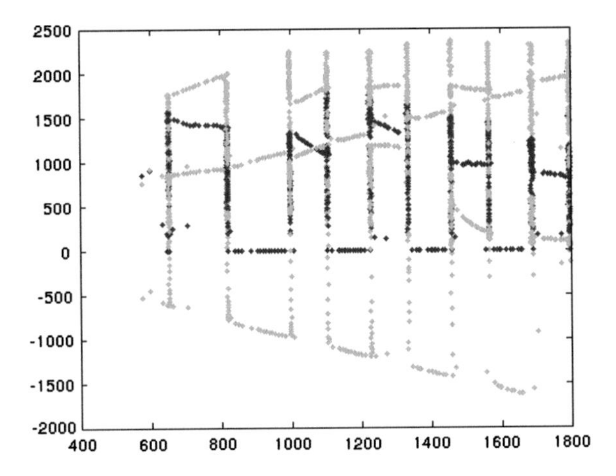

Fig. 12 Output coverage for the Hahn1 dataset. For the second output of the FIS

output of the consequents. Yet there is good coverage, of 100 %, of the target outputs via the controlled output uncertainty. Finally, for the engine, having two outputs, each ones behavior was slightly different. The first output has a more predictable behavior by better following the output targets with its coverage of 99.88 % of reference targets, whereas the second output's behavior is not as linear, such that it holds a tendency to expand as the x axis increases, although it has a coverage of 99.66 % of its reference targets. It must be noted that this specific behavior is more in line with how the Cuckoo Search algorithm optimized the consequents, because the spreads on each individual consequent control the output interval behavior. The solution would be to adjust the Cuckoo Search for better performance or use another optimization algorithm that obtains a better solution.

On the general level of the obtained results, the formed antecedents give a good representation of uncertainty based on the dispersion of the individual sets which affected the creation of the rule configurations found by the FCM. It also depicts a behavior that IT2 FS are not always necessary, with low to no dispersion, and a T1 FS would be more than enough.

Being dependent on other algorithms can limit the general performance of the proposed method. Yet it also adds more possibilities, such as interchanging clustering algorithms to one that can obtain better rule configurations and belonging sets to be used by the proposed method. As for the optimization of the IT2 Sugeno linear consequents, there is a vast amount of optimization algorithms which could also be used for acquiring better results and thus improving the output interval behavior.

5 Conclusion and Future Work

5.1 Conclusions

With the suggested relation dispersion-uncertainty, direct uncertainty extraction is possible from existing data. This relation is found through the Coefficient of Variation, an existing equation used to measure the amount of dispersion in a set, this measure is a normalized value between 0 and 1, that altough higher values than 1 are possible, this is only for non-normal distributions, which, for the purposed application, are non existent considering that the sets are created by a clustering algorithm which only groups in normal distributions of data.

The application shown in this paper, of forming IT2 FS through the suggested equation, which relates dispersion-uncertainty, finds this relation based on the maximum possible achievable FOU, valued at 1, and relates to the maximum possible Coefficient of Variation, in a normal distribution, valued also at 1. This relation of dispersion-uncertainty-FOU is the main contribution of this paper.

With a deeper examination of the experimental results, there is much dependence on the FCM algorithm, where if such algorithm fails to provide a good model, the proposed method would fail also, since the proposed method depends on the performance of the clustering algorithm. Fortunately, if the FCM fails, other clustering algorithms could be used.

5.2 Future Work

Considering the limitation, as well as dependence, of the clustering algorithm, which other clustering or non-clustering techniques could be used to create a better pairing with the proposed method?

With the other high dependence on optimization algorithms for the consequent section of the IT2 FIS, what other optimization algorithm could be used to best pair with the proposed method?

In this paper an IT2 FS was formed, represented by a Gaussian membership function with uncertainty in the standard deviation. How would other IT2 FS membership function be adapted to use the proposed method?

How the area was directly correlated to the FOU by means of its maximum possible area was proposed. Is this the best approach?

Acknowledgement We thank the MyDCI program of the Division of Graduate Studies and Research, UABC, and Tijuana Institute of Technology the financial support provided by our sponsor CONACYT contract grant number: 314258.

References

1. Yu, X., Mehrotra, S.: Capturing Uncertainty in Spatial Queries over Imprecise Data, pp. 192–201. Springer, Heidelberg (2003)
2. Klir, G.J.: Uncertainty and Information: Foundations of Generalized Information Theory. Wiley-IEEE Press, New York (2005)
3. Weise, K., Woger, W.: A Bayesian theory of measurement uncertainty. Measure. Sci. Technol. **3**, 1–11 (1992)
4. Jurado, K., Ludvigson, S.C., Ng, S.: Measuring Uncertainty. Am. Econ. Rev. (AEA) **105**(3), 1177–1216 (2013)
5. Chen, G., Ying, M., Liu, Y.: Dealing with uncertainty and fuzziness in intelligent systems. Int. J. Intell. Syst. **24**, 223–225 (2009)
6. Klir, G.J., Wierman, M.J.: Uncertainty-Based Information. Physica-Verlag HD, Heidelberg (1999)
7. Mendel, J.M., John, R.I., Liu, F.: Interval Type-2 fuzzy logic systems made simple. IEEE Trans. Fuzzy Syst. **14**, 808–821 (2006)
8. Castillo, O., Melin, P.: Recent Advances in Interval Type-2 Fuzzy Systems. Springer, Berlin (2012)
9. Mo, H., Wang, F.-Y., Zhou, M., Li, R., Xiao, Z.: Footprint of uncertainty for type-2 fuzzy sets. Inf. Sci. (Ny) **272**, 96–110 (2014)
10. Bezdek, J.C., Ehrlich, R., Full, W.: FCM: the fuzzy c-means clustering algorithm. Comput. Geosci. **10**, 191–203 (1984)
11. Yang, X.-S.: Cuckoo Search via Lévy flights. 2009 World Congress on Nature & Biologically Inspired Computing (NaBIC). pp. 210–214. IEEE (2009)
12. Ying, H.: Interval Type-2 Takagi-Sugeno fuzzy systems with linear rule consequent are universal approximators. In: NAFIPS 2009–2009 Annual Meeting of the North American Fuzzy Information Processing Society, pp. 1–5. IEEE (2009)
13. Zadeh, L.A.: Fuzzy Sets. Inf. Control **8**, 338–353 (1965)
14. Melin, P., Castillo, O.: Fuzzy Modeling Fundamentals. Wiley Encyclopedia of Computer Science and Engineering. Wiley, New York (2007)
15. Mendel, J.: General type-2 fuzzy logic systems made simple: a tutorial. IEEE Trans. Fuzzy Syst. pp. 1–1 (2013)
16. The MathWorks, Inc., Natick, Massachusetts, U.S.: MATLAB Release 2013b, (2013)

A Proposal for a Method of Defuzzification Based on the Golden Ratio—GR

Wojciech T. Dobrosielski, Janusz Szczepański and Hubert Zarzycki

Abstract This article presents a proposal for a new method of defuzzification a fuzzy controller, which is based on the concept of the golden ratio, derived from the Fibonacci series [1]. The origin of the method was the observation of numerous instances of the golden ratio in such diverse fields as biology, architecture, medicine, and painting. A particular area of its occurrence is genetics, where we find the golden ratio in the very structure of the DNA molecule [2] (deoxyribonucleic acid molecules are 21 angstroms wide and 34 angstroms long for each full length of one double helix cycle). This fact makes the ratio in the Fibonacci series in some sense a universal design principle used by man and nature alike. In keeping with the requirements, the authors of the present study first explain the essential concepts of fuzzy logic, including in particular the notions of a fuzzy controller and a method of defuzzification. Then, they postulate the use of the golden ratio in the process of defuzzification and call the idea the Golden Ratio (GR) Method. In the subsequent part of the article, the proposed GR-based instrument is compared with the classical methods of defuzzification, including COG, FOM, and LOM. In the final part, the authors carry out numerous calculations and formulate conclusions which serve to classify the proposed method. At the end they present directions of further research.

Keywords Fuzzy logic · Fuzy sets · Fuzzy control system · Deffuzification · Fibonacci series

W.T. Dobrosielski (✉)
Casimir the Great University in Bydgoszcz, Institute of Technology,
ul. Chodkiewicza 30, 85-064 Bydgoszcz, Poland
e-mail: wdobrosielski@ukw.edu.pl

J. Szczepański
Institute of Fundamental Technological Research, Polish Academy of Sciences Warsaw,
ul. Pawinskiego 5B, 02-106 Warsaw, Poland
e-mail: jszczepa@ippt.gov.pl

H. Zarzycki
Wroclaw School of Information Technology,
ul. Wejherowska 28, 54-239 Wroclaw, Poland
e-mail: hzarzycki@horyzont.eu

K.T. Atanassov et al. (eds.), *Novel Developments in Uncertainty Representation and Processing*, Advances in Intelligent Systems and Computing 401,
DOI 10.1007/978-3-319-26211-6_7

1 Introduction

Curiosity, a desire to learn, and observations of traditional areas of science -all of these become the platforms on which new solutions are developed, which in turn result in new products being made. Another source of inspiration is construction of synergistic systems based mainly on concrete solutions in mechatronics. The authors of this paper make use of the well-known Fibonacci series, and in particular the golden ratio, as a method of defuzzification a fuzzy controller. The golden ratio method is also the subject of the works of Euclid, an author of theoretical works in mathematics where, in the authors' opinion, the use of the ratio affords a new look at the area of fuzzy logic. The use of this approach for defuzzification expands the range of existing instruments based on such methods as center of gravity, center of area, or methods involving maxima.

2 Fundamental Concepts

The introduction of fuzzy sets [3], along with its accompanying studies, have made their application widespread in the world around us. The impact of the concept of fuzzy sets is mainly seen in the use of solutions in control systems, robotics and decision-making systems. Among other applications, such solutions are also used in medicine, image processing or speech analysis. The essence of using such a solution is the ability to express imprecise phenomena, where the existing, classic mathematical models that operate on binary *yes/no* dichotomies have been redefined to include intermediate elements. In any field of application of fuzzy control models [4], the system is a classic block containing input and output zones. The input zones are provided with signals from sensors, and the outputs control their assigned actuators. The internal structure of a fuzzy controller is presented in Fig. 1, which shows such operations as fuzzification, inference, and defuzzification.

The first phase contains the operation of fuzzification. It is connected with calculating the degree of membership in the relevant fuzzy sets. Calculation of the resulting membership function concludes the inferencing stage based on input degrees of membership. Both operations (fuzzification and inferencing) contain a number of specific elements. The information referred to here is discussed more extensively in

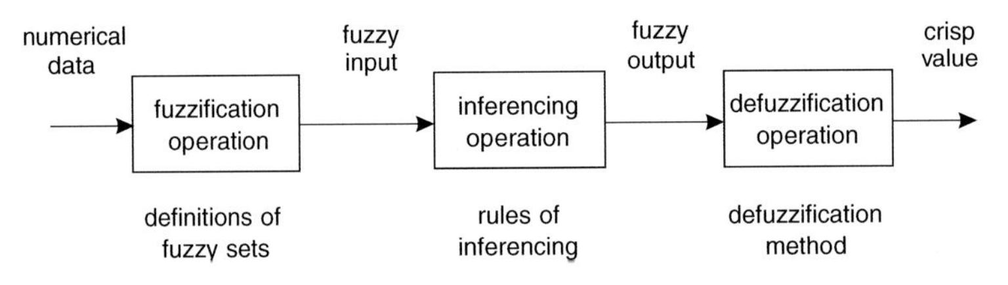

Fig. 1 The internal structure of a fuzzy controller

the available literature on the subject [5–10]. The final operation of the system is the defuzzification block. It is characterized by providing a real value in the output. This value will be the product of applying the method to the resultant membership function, enabling the activation of an actuator in the desired manner. In the remainder of this article, this transformation will be referred to as defuzzification, and it will also set the background for further work.

In specific terms, the defuzzification process comes down to the use of appropriate methods which allow reducing a fuzzy set to a single real value. In previous research, the main methods of defuzzification involved either operations on the maxima (FOM, LOM, MOM, RCOM), or on the field (COG, COA). For some problems, and for some systems, there is a need to develop new methods that address new trends in fuzzy logic.

The concept of a fuzzy set, formulated by Zadeh in the mid-sixties as the theory of fuzzy sets [3], was preceded the work of Łukaszewicz [11], in which the foundation of the then-dominant predicate calculus based on bivalent logic was replaced by multivalent logic. This contributed to the development of various derivative theories which broadened the area under discussion. One of such theories, proposed by Dubois and Prade [12] is a version of Zadeh's theory introducing a limited class of membership functions, along with two shape functions, L and R.

According to Zadeh's definition, a fuzzy set is:

Definition 1 Fuzzy set A, in a certain area X, is the set of pairs:

$$A = \{(x, \mu(x))\} \ \forall x \in X \tag{1}$$

where: $\mu_(x)$ membership function assigning to each element $\forall x \in X$ (of the assumed area of consideration of X) its degree of membership to set A, whereas:

$$\mu_A : X \to [0, 1] \ thus \ \mu_A(x) \in [0, 1] \tag{2}$$

Definition 2 The support for fuzzy set A in X is the non-fuzzy set supp(A), which we define as:

$$supp(A) = \{x : \mu_A(x) > 0, x \in X\} \tag{3}$$

Definition 3 The height of a fuzzy set A in X determines the maximum value accepted by membership function $\mu_A(x)$ in the whole X set, as follows:

$$h(A) = height(A) = hgt(A) = sup_{x \in X}(\mu(x)) \tag{4}$$

Definition 4 We define the core of fuzzy set A in X as:

$$core(A) = \{x : \mu_A(x) = 1, x \in X\} \tag{5}$$

A graphic interpretation of a fuzzy set, along with the respective Definitions 2, 3, 4 are shown in Fig. 2.

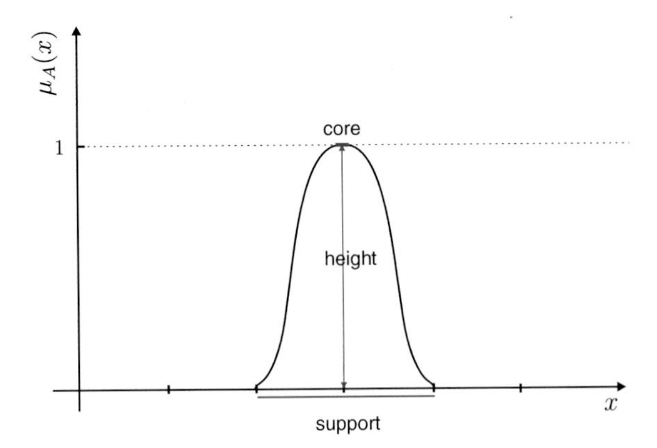

Fig. 2 A fuzzy set with its support, height, and core

The defuzzification process, as the final step in the three-stage model of fuzzy control, transforms the fuzzy set to a single real value that specifies the membership function. The expression below formally describes defuzzification:

Definition 5

$$W = \{f : X \to [0, 1]\} \to X \tag{6}$$

where: W is the defuzzification operator, f is the membership function, and X is the set of the universe specifying the membership functions.

A process may be characterized based on characteristics which are more desirable for a particular system [7, 8]. With respect to the nature of the system, we can talk about fuzzy a inference system or a fuzzy control system. For a fuzzy inference system a feature such as computational efficiency is of less importance than, say, for a fuzzy control system, where performance is a key parameter. Sources [13] and [14] introduced criteria of defuzzification for fuzzy numbers, against which individual methods of defuzzification were evaluated. The main conclusion drawn from the evaluations was that there is no universal defuzzification method. A method of defuzzification should be selected with its use in mind. For example maxima-based methods, which include LOM (Last Of Maxima), and FOM (First Of Maxima), are better suited for inference systems. The authors' previous studies demonstrated that distribution- and field-based methods are more appropriate for those applications where control systems are used. Among these methods we find COG (Center Of Gravity) and COA (Center Of Area).

3 Methods of Defuzzification

It is known that the process of defuzzification [15] reduces a fuzzy set to a single real value. The mechanism of this operation consists mainly in the use of the appropriate defuzzification method. Among the available methods we find the classical solutions:

FOM—first of maxima

FOM is a method based on selecting the smallest element from the core of set A, where the real value is presented in relation (7).

$$FOM(A) = min\ core(A) \tag{7}$$

LOM—last of maxima

In contrast, by selecting the largest value element from the core of set A, we are using LOM, whose formula is shown below:

$$FOM(A) = max\ core(A) \tag{8}$$

MOM—mean of maxima

Equation (9) presents the use of **FOM** and **LOM** as methods whose defuzzification values take into account the minimum and maximum elements of the core of fuzzy set A. The resultant value is the mean of the two methods.

$$MOM(A) = \frac{min\ core(A) + max\ core(A)}{2} \tag{9}$$

RCOM—random choice of maxima

This method is also referred to as defuzzification of the core, as the defuzzification value is always included in the core of the fuzzy set. The defuzzification value of this method is a random element $x \in core(A)$ calculated as the probability as in (15):

$$RCOM(A) = P(x) = \frac{\lambda(x)}{\lambda(core(A))} \tag{10}$$

where λ jest the Lebesgue measure in universe X.

MOS—mean of support

In MOS, the defuzzification value is the mean from the support of number A.

$$RCOM(A) = P(x) = \frac{supp(A)}{2} \tag{11}$$

COG—center of gravity

This method is most widespread—it is bases on determining the center of gravity of the system under consideration. In the process of defuzzification fuzzy number A, the method is expressed as Eq. (12)

$$COG(A) = \frac{\int_a^b x\mu_A(x)\mathrm{d}x}{\int_a^b \mu_A(x)\mathrm{d}x} \tag{12}$$

BADD—basic defuzzification distribution
A method of defuzzification proposed in [16], as an expansion of COG and MOM. The defuzzification value of fuzzy set A is obtained as:

$$BADD(A) = \frac{\int_a^b x\mu_A^\gamma(x)\mathrm{d}x}{\int_a^b \mu_A^\gamma(x)\mathrm{d}x} \tag{13}$$

Depending on parameter $\gamma \in [0, \infty]$, BADD may assume the following instances: when $\gamma = 0$, $BADD(A) = MOS(A)$; when $= 1$, $BADD(A) = COG(A)$; when $\gamma \to \infty$, $BADD(A) = MOM(A)$.

4 The Golden Ratio Method

In most cases, a natural way of representing information is to determine a particular order in a given set. Considerations on this issue are broadly presented in mathematics as number series, or numerical sequences. Focusing on the numerical sequence we think about a transformation, a function mapping the set of natural numbers in the set of real numbers. A special representative of this operation is the Fibonacci series. The series is based on the assumption that it starts with two ones, and each consecutive number is the sum of the previous two. The proposal for the Golden Ratio method of defuzzification is based on the proportion of the golden ratio. As a result of dividing each of the numbers by its predecessor, we always obtain quotients oscillating around the value of 1,618033998875 ...—the golden ratio number. The exact value of the limit is the golden number itself:

$$\lim_{n\to 0} \frac{k_{n+1}}{k_n} = 1,618033998875 \cdots = \Phi \tag{14}$$

The possibility of using this formula in the process of defuzzification will be another example of the universality of the method, as it is applied in the new domain of fuzzy logic theory. Calculation of the classical formula of the golden mean assumes that two values of line segments a, and b, are in golden ratio Φ to one another if:

$$\frac{a+b}{a} = \frac{a}{b} = \Phi \tag{15}$$

In this case, one method of finding the value of Φ is to transform the left-hand fraction of Eq. (15) into:

$$\frac{a+b}{a} = 1 + \frac{b}{a} = 1 + \frac{1}{\Phi}, \ where\ \frac{b}{a} = \frac{1}{\Phi} \tag{16}$$

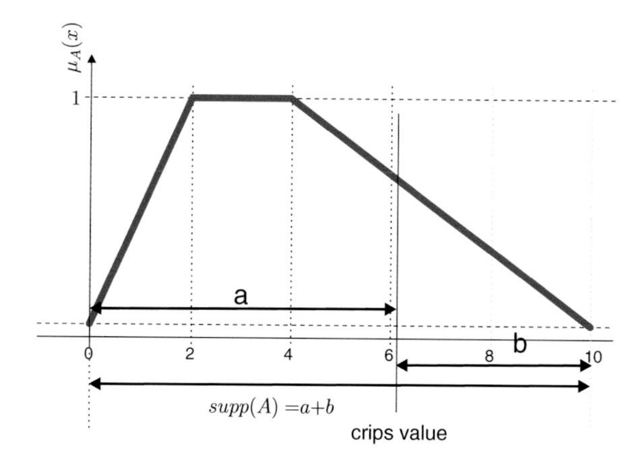

Fig. 3 Golden ration defuzzificaton value

Following subsequent transformations of Eq. (16) we obtain quadratic Eq. (17), for which we will calculate the roots.

$$\Phi^2 - \Phi - 1 = 0 \tag{17}$$

As appropriate, using transformations of the formula in (17) we obtain two square roots (18).

$$\Phi_1 = \frac{1 + \sqrt{5}}{2} \; or \; \Phi_2 = \frac{1 - \sqrt{5}}{2} \tag{18}$$

In view of the fact that the value of Φ must be positive, in our example we select the positive root, as in Eq. (19) (Fig. 3).

$$\Phi = \Phi_1 = \frac{1 + \sqrt{5}}{2} = 1,618033998875 \ldots \tag{19}$$

In sum, the ratio between two objects a, and b is called the golden ratio when the value of $\Phi = 1.61803398875 \ldots$

The method of the Golden Ratio (GR) for fuzzy number is the Eq. (20):

Definition 6

$$GA = \frac{supp(A)}{\Phi} \quad where \; \Phi = 1,618033998875 \ldots \tag{20}$$

where: *GR* is the defuzzification operator, *supp*(*A*) is support for fuzzy set *A* in universe *X*.

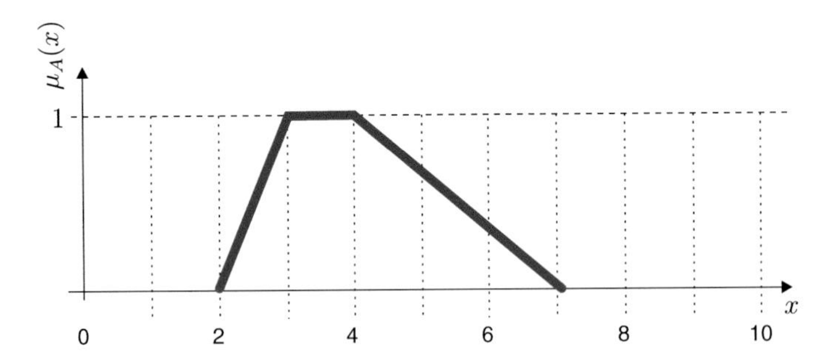

Fig. 4 Convex fuzzy number A

Table 1 Defuzzification measurement for a convex fuzzy number

Fuzzy number	Defuzzification method			
	GR	FOM	LOM	COG
A	5.09	3	4	4.11

5 Observation

The proposed Golden-Ratio-based method of defuzzification presented in section four was designed as a simple method for convex fuzzy numbers. In this section we will present a comparison of the proposed GR method with FOM GR, LOM and COG. The first two methods are from the group of maxima-based methods. In First Of Maxima, the defuzzification value is set for the first element of the core, and we understand the core as the segment of the domain in which the membership function reaches the maximum value of one. If the defuzzification value is related to the last element of the core, we are dealing with the method called LOM (Last Of maxima). The final method selected for the comparison is Center of Gravity (COG). For convex fuzzy number A, shown in Fig. 4, the calculation results are listed in Table 1.

6 Conclusions

Balance between extremes can be well implemented in architecture, or painting, as well as in fuzzy logic. An interpretation of this balance, according to the authors, is using Golden Ratio method (GR) in the process of defuzzification. The instrument of the golden mean, as proposed in this paper for fuzzy numbers may serve as another defuzzification method. As a mathematical apparatus that affords wide-ranging possibilities in description and processing of information, it becomes a new solution in constructing models of fuzzy controllers used as tools for inferencing or control. In further research on this issue, it is planned to carry out tests of the method in real control models. Future activities will concern implementation of the method in the measurement environment. A concurrent idea for further studies is the use of the golden ratio in relation to methods based on field membership function of a fuzzy number. It is planned to replace the base of the support, as shown in this study, with the ratio of the fields of a given fuzzy number. The future solution will in a certain way relate to the COG method, except that the balance of forces of the fields of a given fuzzy number will be related to the golden ratio. Another challenge will be to conduct studies of the proposed Golden Ratio method for OFN—Ordered Fuzzy Numbers [17, 18, 24].

References

1. Dunlap, R.A., Dunlap, R.: The Golden Ratio and Fibonacci Numbers. World Scientific, Singapore (1997)
2. Buldyrev, S.V., Goldberger, A.L., Havlin, S., Peng, C.K., Stanley, H.E.: Fractals in biology and medicine: from DNA to the heartbeat. Fractals in Science, pp. 49–88. Springer, Berlin (1994)
3. Zadeh, L.: Fuzzy sets. Inf. Control **8**(3), 338–353 (1965)
4. Yager, R.R., Filev, D.P.: Essentials of Fuzzy Modeling and Control. Wiley, New York (1994)
5. Zadeh, L.A.: Is there a need for fuzzy logic? Inf. Sci. **178**(13), 2751–2779 (2008)
6. Dubois, D., Prade, H.: Fuzzy elements in a fuzzy set. In; Proceedings of the IFSA, vol. 5, pp. 55–60 (2005)
7. Mahdiani, H., Banaiyan, A., Javadi, M.H.S., Fakhraie, S., Lucas, C.: Defuzzification block: new algorithms, and efficient hardware and software implementation issues. Eng. Appl. Artif. Intell. **26**(1), 162–172 (2013)
8. Sugeno, M.: An introductory survey of fuzzy control. Inf. Sci. **36**(1–2), 59–83 (1985)
9. Kacprzyk, J.: Fuzzy Sets in System Analysis. PWN, Warsaw (1986) (in Polish)
10. Kacprzyk, J., Yager, R.R.: Emergency-oriented expert systems: a fuzzy approach. Inf. Sci. **37**(1), 143–155 (1985)
11. Łukasiewicz, J.: Elements of Mathematical Logic, vol. 31. Macmillan, New York (1963)
12. Dubois, D., Prade, H.: Operations on fuzzy numbers. Int. J. Syst. Sci. **9**(6), 613–626 (1978)
13. Van Leekwijck, W., Kerre, E.E.: Defuzzification: criteria and classification. Fuzzy Sets Syst. **108**(2), 159–178 (1999)
14. Runkler, T., Glesner, M.: A set of axioms for defuzzification strategies towards a theory of rational defuzzification operators. In: Fuzzy Systems, 1993, Second IEEE International Conference on, IEEE, pp. 1161–1166 (1993)
15. Roychowdhury, S., Pedrycz, W.: A survey of defuzzification strategies. Int. J. Intell. Syst. **16**(6), 679–695 (2001)
16. Filev, D.P., Yager, R.R.: A generalized defuzzification method via bad distributions. Int. J. Intell. Syst. **6**(7), 687–697 (1991)
17. Kosiński, Witold: On defuzzyfication of ordered fuzzy numbers. In: Rutkowski, L., Siekmann, J., Tadeusiewicz, R., Zadeh, L.A. (eds.) Artificial Intelligence and Soft Computing—ICAISC 2004. Lecture Notes in Computer Science, vol. 3070, pp. 326–331. Springer, Heidelberg (2004)
18. Bednarek, T., Kosiński, W., Węgrzyn-Wolska, K.: On orientation sensitive defuzzification functionals. In: Artificial Intelligence and Soft Computing, pp. 653–664. Springer, Berlin (2014)
19. Chen, G., Pham, T.T.: Introduction to Fuzzy Sets, Fuzzy Logic, and Fuzzy Control Systems. CRC Press, Boca Raton (2000)
20. Goetschel, R., Voxman, W.: Elementary fuzzy calculus. Fuzzy Sets Syst. **18**(1), 31–43 (1986)
21. Gottwald, S.: Mathematical aspects of fuzzy sets and fuzzy logic: some reflections after 40 years. Fuzzy Sets Syst. **156**(3), 357–364 (2005)

22. Czerniak, J.: Evolutionary approach to data discretization for rough sets theory. Fundamenta Informaticae **92**(1–2), 43–61 (2009)
23. Flasiński, M.: Introduction to Artificial Intelligence. PWN, Warsaw (2011) (in Polish)
24. Kosiński, W., Prokopowicz, P., Ślęzak, D.: Ordered fuzzy numbers. Bull. Pol. Acad. Sci., Ser. Sci. Math. **51**(3), 327–338 (2003)
25. Angryk, R.A., Czerniak, J.: Heuristic algorithm for interpretation of multi-valued attributes in similarity-based fuzzy relational databases. Int. J. Approx. Reason. **51**(8), 895–911 (2010)
26. Xu, Z., Shang, S., Qian, W., Shu, W.: A method for fuzzy risk analysis based on the new similarity of trapezoidal fuzzy numbers. Expert Syst. Appl. **37**(3), 1920–1927 (2010)

A New Approach to the Equivalence of Relational and Object-Oriented Databases

Swietłana Lebiediewa, Hubert Zarzycki and Wojciech T. Dobrosielski

Abstract In the paper the condition for equivalence of problem-oriented databases (DB) models has been formulated. A data segment and problem-focused data manipulation language of database for multi-stage recognition of objects has been characterized. The relational and a corresponding object-oriented models of DB has been described. A few assertions regarding the equivalence of the relational and object DB models for recognition have been proved.

Keywords Equivalence of databases · Relational database · Object-oriented database · Problem-oriented databases · Data manipulations language

1 Formulation of the Problem

This paper regards the possibility of cooperation of databases having different models, that is, heterogeneous DB systems. This issue is related closely to the problem of equivalence of DB models. This problems has two aspects: theoretical—equivalence of the data structures (DS) and operations—and practical—possibility of cooperation between heterogeneous DB systems and with numerous computer solutions. Additionally creating some of computer systems is only possible thanks to proper DB interfaces. There are solutions in Artificial Intelligence [1], Business

S. Lebiediewa
Wroclaw School of Information Technology, ul. Wejherowska 28,
54-239 Wroclaw, Poland
e-mail: swietlana@lebiediewa.com

H. Zarzycki
Wroclaw School of Applied Informatics "Horyzont", ul. Wejherowska 28,
54-239 Wroclaw, Poland
e-mail: hzarzycki@horyzont.eu

W.T. Dobrosielski (✉)
Institute of Technology, Casimir the Great University in Bydgoszcz,
ul. Chodkiewicza 30, 85-064 Bydgoszcz, Poland
e-mail: wdobrosielski@ukw.edu.pl

K.T. Atanassov et al. (eds.), *Novel Developments in Uncertainty Representation and Processing*, Advances in Intelligent Systems and Computing 401,
DOI 10.1007/978-3-319-26211-6_8

Intelligence [2], Decision Support System [3], Expert System, ERP [4, 5], Rough Sets [6], GIS, Fuzzy Logic [7], etc. which must be based on such technical interfaces to properly cooperate with different DB. There are practical standard such as JDBS designed by Sun and ODBC designed by Microsoft, which allows to align databases of different models. The existing systems of distributed databases (DDB) are usually based on the relational DB model where the basis for division is relation [8]. As a result of distribution there takes place a horizontal or vertical division (fragmentation) of relations. We are going to be interested in DDB based on decomposition of a complex multistage decision making (MDM) system and not in the fragmentation of relations. In the MDM systems the basic unit of a DB is not a relation but a complex DS—segment [9]. There is not assumed any uniformity of equipment, HOP, DBMS or DB models. In the papers [9–11] there has been demonstrated the equivalence of the hierarchical, network-based and relational models of problem-oriented DB when it comes to multistage recognition due to basic DS and DML instructions. The aim of this paper is to demonstrate the equivalence of relational and object models of DB for multistage recognition [12] due to basic DS and DML instructions and therefore the possibility of cooperation of these DB systems.

2 Database for Multistage Recognition

Multistage recognition consists in decomposition of the decisional problem, that is, replacement of a one-time recognition with sequence of the so-called local recognitions carried out in particular nodes according to the given structure of the decision tree (DT), Fig. 1.

One of the basic tasks when designing DB for MDM is the creation of an appropriate user interface, namely a model of external and problem-oriented language of

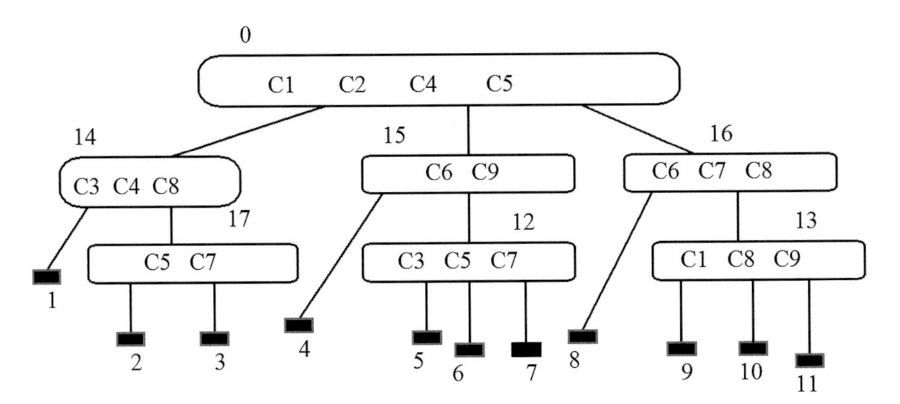

Fig. 1 Decision tree

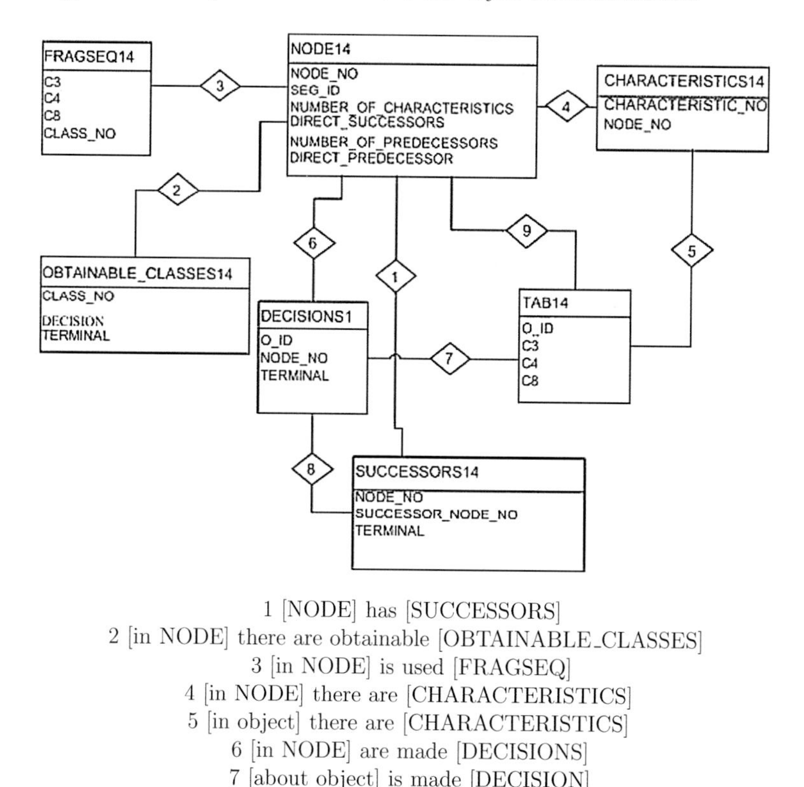

Fig. 2 E-R entity of the SEG* segment

access to the DB. The external model is a data segment, that is, a set of data to be recognized in a node. The segment includes information on the Node, Node Successors, Classes Obtainable from the Node, Characteristics measured in the Node and fragment of the Learning Sequence (LS) used in the Node (FRAGSEQ). The segment includes also information on objects recognized (OBJECTS) and decisions made. The E-R model of a segment of data has been presented on Fig. 2.

The problem-oriented language of access to the DB (DML) contains instructions used by the recognition algorithms. The following instructions are examples of DML instructions [9]:

READ CHARACTERISTICS *—provide the vector of characteristics used in the node *;
READ CLASSES*—which classes are obtainable in the node *;
READ DECISIONS *—provide all decisions made in the node *;
READ DECISIONS, O_ID *— what decisions were made in the node * regarding the object O_ID;

WRITE DECISIONS,O_ID *—write decision made in the node * regarding the object O_ID;
READ CONTROL, i *—To which node should the control be handed over if the result of the decision is an i class;
READ O_ID *, CHARACTERISTICS, NODE_NO—provide all the characteristics of the object O_ID and the decision made (result of recognition) in the node*.

The cooperation of the recognition algorithm (1) with the database segment is as follows:

Algorithm 1 ALGORITHM OF RECOGNIZING IN NODE*

Algorithm start
STEP 1. Read from the DB the identifier and the vector of features of the object being recognized.
STEP 2. Read the fragment of TS FRAGSEQ*, being utilized in the given node.
STEP 3. (calculation of the values of the function of recognizing the vector of features of the object g being recognized): for all the elements of relation FRAGSEQ*: take an element of TS, calculate the euclidean norm with the object being recognized, store the euclidean norm and the corresponding class number of the object.
STEP 4. Write the object identifier and the number k, being the result of recognizing in NODE*, to the relation DECISIONS* in the DB.
Algorithm stop

---

3 Relational Model

The relation diagram of DB reduced to a single segment (SEG*) has the following form:

Relation NODE*(NODE_NO, NUMBER_OF_CHARACTERISTICS, NUMBER_OF_SUCCESSORS, NUMBER_OF_PREDECESSORS, PREDECESSOR)
Relation OBTAINABLE_CLASSES* (CLASS_NO, DECISION, TERMINAL)
Relation SUCCESSORS* (NODE_NO, SUCCESSOR_NODE_NO, TERMINAL)
Relation CHARACTERISTICS* (C3, C4, C8)
Relation FRAQSEQ* (C3, C4, C8, CLASS_NO)
Relation OBJECTS* (O_ID, C3, C4, C8)
Relation DECISIONS* (O_ID, NODE_NO, TERMINAL)

Relation NASTÊPNIKI14 the numbers of direct successors node 14. The relationship CECHY14 includes a feature vector for the node 14 and the information about where each of the features has been (or will be) measured for the first time. Relation TAB14 (matrix measured features of objects) it includes the IDs of objects and features of the measured values of objects. Relation DECYZJE14 includes the recognition results in node 14. The relation OBTAINABLE_CLASSES_14 contains information about the numbers of classes reachable from the node 14 with information

CHARACTERISTICS14

CHARACTE RISTIC_NO	NODE_NO
C3	14
C4	0
C8	14

SUCCESSORS14

NODE_NO	SUCCESOR NODE_NO	TERMINAL
14	1	YES
14	17	NO

OBTAINABLE_CLASSES_14

CLASS NO	DECISION	TERMINAL
1	1	YES
2	17	**NO**
3	17	NO

DECISIONS14

O-ID	NODE_NO	TERMIN AL
001	1	YES
005	17	NO
007	17	NO

TAB14 (fragment)

O-ID	C3	C4	C8
001	2	6.24	125
005	1	6.86	131
007	1	6.35	145

SEQUENCES14

C3	C4	C8	CLASS_NO
3	6.58	128	1
4	6.31	145	3
1	6.12	137	3
2	6.75	186	2
5	6.43	129	1
3	6.13	135	2
2	6.02	112	1
4	6.21	122	3

Fig. 3 Segment_14 of database

on the number of a decision taken in the node. The decision is either the result of which is the class number recognition, or the number of the node to which it will be transferred to the control. Relation FRAGSEG14 includes a fragment of the CU used in the node CU 14.

On Fig. 3 there has been presented a fragment of segment 14.

In the relational model there are included executable instruction of the problem-oriented DML listed in item 2. As an example we will demonstrate the executability of instruction 1. READ STEROWANIE, i * (to which node should the control be handed over if the result of the decision is an i class) and 2. READ (O_ID) *, CHARACTERISTICS*, NODE_NO* (provide all the characteristics of the object O_ID and the decision made (result of recognition) in the node*).

1. RESTRICT ObtainableClasses WHERE ClassNo = i IF Terminal = "Yes" GO TO EndOfRecognition ELSE PRINT "Hand control over to DecisionNo" EndOfRecognition PRINT i END
2. JOIN ((RESTRICT Objects* WHERE Oid = object-id), (PROJECT Decisions *(Oid, NodeNo)) OVER Oid

The algorithms of performance of other instructions have been presented in [9].

4 Object-Oriented Model

Below C++ notation is used for the description of object-oriented DB scheme. The scheme is reduced to a single segment (SEG *):

class NODE14int NODE_NO; int NUMBER_OF_CHARACTERISTICS;
int NUMBER_OF_SUCCESSORS; int NUMBER_OF_PREDECESSORS;
NODE14* DIRECT_PREDECESSOR;;
class OBTAINABLE_CLASSES14 int CLASS_NO, int NODE_NO;
bool TERMINAL; READCLASSES(); READCONTROL();;
class SUCCESSORS14int NODE_NO; int SUCCESSOR_NODE_NO; bool TERMINAL;;
class OCCURRING_CHARACTERISTICS14int CHARACTERSTIC_NO;
int NODE_NO;;
class SEQUENCES14 int C3; int C4; int C8; int CLASS_NO;;
class OBJECTS14 int O_ID; int C3; int C4; intC8; READIDOB();;
class DECISIONS14 int O_ID; int NODE_NO; bool TERMINAL;
WRITEDECISIONS(); READDECISIONS(); READIDOBDECISIONS();;
where NODE14* DIRECT_PREDECESSOR—is a pointer to object of type NODE14 (predecessor to current object).

The following diagram presents the structure of the object-oriented database segment covering classes, attributes, methods, and the relationship between the classes (Fig. 4).

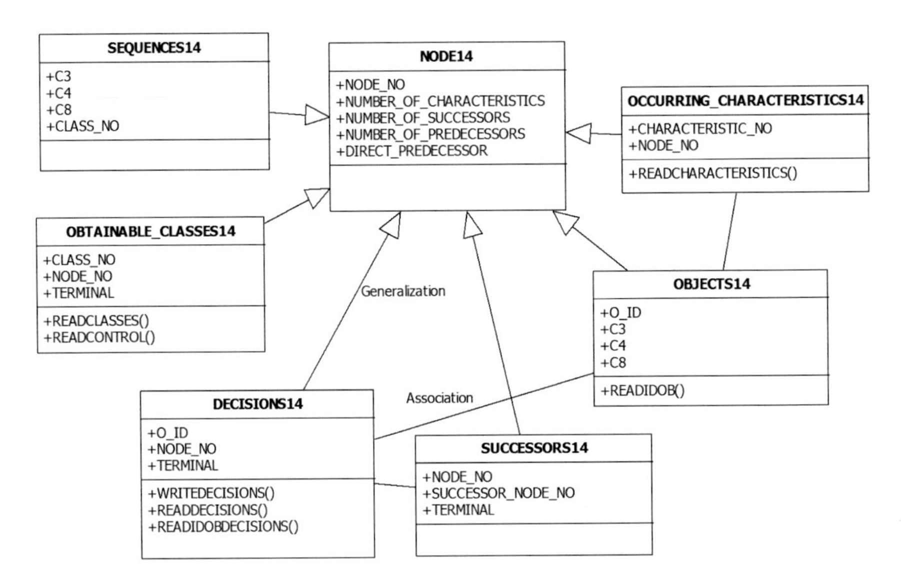

Fig. 4 Class diagram of the segment SEG*

Methods available in the classes:

READCHARACTERISTICS()—provide vector of characteristics used in a node specified by a given number;
READCLASSES()—which classes are obtainable in a node specified by a given number;
READDECISIONS()—provide all decisions made in a node specified by a given number;
READIDOBDECISIONS()—what decisions were made in the node specified by a given number regarding the object O_ID;
WRITEDECISIONS()—write decision made in the node specified by a given number regarding the object O_ID;
READCONTROL()—to which node should the control be handed over if the result of the decision is an i class;
READIDOB()—provide all the characteristics of the object O_ID and the decision made (result of recognition) in a node specified by a given number.

5 Equivalence of the Relational Model and Object-Oriented Model

We do state that the databases A and B are equivalent with regard to basic data structures if each basic structure of DB A can be mapped into a basic structure of DB B, and if each basic structure of DB B can be mapped into a basic structure of DB [13]. Similarly, the databases A and B are equivalent with regard to operations if each operation of DB A is executable in DB B, and if each operation of DB B is executable in DB A. What concerns the equivalence with reference do data structures, the hierarchical, network, and relational models of general-purpose databases (GPDBs) are equivalent [14]. In order to demonstrate the equivalence of the relational and object models of DB we will demonstrate the equivalence of basic DS and operation of the problem-oriented DML listed in subsection 2.

Assertion 1 The relational and object models of DB for multistage recognition are equivalent with regard to data structures.

Proof The main DS of relational and object models is a segment. We are going to demonstrate that there exists mapping that assigns to each element of segment of a relational model in a clear way an element of segment of an object-oriented model. Let us define on the element of segment of a relational model the mapping taking values in the object-oriented model in the following way:

NODE*, OBTAINABLE_CLASSES*, SUCCESSORS*, CHARACTERISTICS*, FRAGSEQ*, OBJECTS* and DECISIONS* relations of the relational model are mapped accordingly in the NODE, OBTAINABLE_CLASSES, SUCCESSORS, OCCURRING_CHARACTERISTICS, SEQUENCES, OBJECTS and

DECISIONS classes of the object model. This mapping is mutually unambiguous since the counterimages of the NODES, OBTAINABLE_CLASSES, SUCCESSORS, OCCURING_CHARACTERISTICS, SEQUENCES, OBJECTS and Decisions classes of the object model are, accordingly, NODES*, OBTAINABLE_CLASSES*, SUCCESSORS*, CHARACTERISTICS*, FRAGSEQ*, OBJECTS* AND DECISIONS* relations of the relational model.

Assertion 2 The relational and object-oriented models of DB for multistage recognition are equivalent with regard to operations of the problem-oriented DML.

Proof We are going to demonstrate that each operation of the problem-oriented DML carried out in a relational model is carried out also in the object model. READ CHARACTERISTICS* operation corresponds to the method READCHARACTERISTICS() of the class OCCURRING_CHARACTERISTICS, READ CLASSES * operation corresponds to the method READCLASSES of the class OBTAINABLE _CLASSES, WRITE DECISIONS, O_ID* operation operation corresponds to the method WRITE DECISIONS() of the class DECISIONS, READ DECISIONS* operation corresponds to the method READDECISIONS() of the class DECISIONS, READ DECISIONS, O_ID* operation corresponds to the method READIDOBDECISIONS() of the class DECISIONS, READ CONTROL, i* operation corresponds to the method READCONTROL() of the class OBTAINABLE_CLASSES, READ O_ID *, CHARACTERISTICS, NODE_NO operation corresponds to the method READIDOB() of the class TAB14.

Assertion 3 The relational and object models of DB for multistage recognition are equivalent with regard to data structures and operations of the problem-oriented DML.

Assertion 3 results from Assertions 1 and 2.

6 Concluding Remarks

The aim of the paper was to examine the condition for cooperation of problem-oriented databases (PODB) irrespective of the LDB model. The basis for the proposed approach is the user interface from PODB—a data segment (external model of the user) and problem-oriented DML. The equivalence of various models with regard to basic DS and DML instructions is of large practical significance. Firstly, the proposes approach ensures independence from hardware, software and DB model. There has been considered the equivalence of object-oriented and relational models of DB, but the IDEN PODB may be executed as an hierarchical or network-based BD. Secondly, the definition of POBD equivalence presented in subsection 4 is of general nature: the equivalence with regard to operations does not assume the identity of every instruction of a problem-oriented JMD in various DB models but within a certain subset of these instructions. For example, in each DB model for RW there is assumed execution of instruction of a problem-oriented DML used by the recognition algorithms in the node. On the other hand, certain queries for DB (e.g. queries regarding the DD structure or list of all the object belonging to a specific class) that are carried out in relational and network-based DB models cannot be carried out in a hierarchical model [9]. Thirdly, similar DS and problems related to them may occur in other similar IT systems, e.g. used for purposes of multistage identification or multilevel controlling [15]. In the paper [9–11] there was considered the equivalence of hierarchical, network based and relational models of REC DB. However, REC DB can be created as an ODB—(complex DB structure, connections between segments, heredity) or, after adding decision rules (recognition algorithms), as a deductive DB.

References

1. Flasiński, M.: Introduction to artificial intelligence [in polish]. PWN (2011)
2. Drelichowski, L., Zarzycki, H., Oszuścik, G., Lewandowski, R.: The supportive role of business intelligence tools for the analysis of budget balance in EU countries in turbulent environments. In: Studies and Proceedings of Polish Association for Knowledge Management, pp. 33–47 (2012)
3. Czerniak, J.M., Dobrosielski, W., Zarzycki, H., Apiecionek, Ł.: A proposal of the new owlant method for determining the distance between terms in ontology. In: Intelligent Systems' 2014, pp. 235–246. Springer (2015)
4. Zarzycki, H.: Enterprise resource planning systems selection, application, and implementation on the example of simple. ERP software package. Prace Naukowe Uniwersytetu Ekonomicznego we Wrocławiu (205), 281–291 (2011)
5. Zarzycki, H., Fronczak, E.: A practical approach to computer system architecture for an agrofood industry information center [in Polish]. In: Studies and Proceedings of Polish Association for Knowledge Management, vol. 33, pp. 248–255 (2010)
6. Czerniak, J.: Evolutionary approach to data discretization for rough sets theory. Fundam. Inform. **92**(1–2), 43–61 (2009)
7. Bednarek, T., Kosiński, W., Węgrzyn-Wolska, K.: On orientation sensitive defuzzification functionals. In: Artificial Intelligence and Soft Computing, pp. 653–664. Springer (2014)
8. Beynon-Davies, P.: Database systems [in Polish]. WNT (2005)
9. Lebiediewa, S.: Methodology of designing problem-oriented databases for systems of multistage decision making [in Polish]. PWr (1998)
10. Lebiediewa, S.: On the equivalence of problem-oriented databases. Int. J. Appl. Math. Comput. Sci. **9**(4), 965–977 (1999)
11. Lebiediewa, S., Węglorz, B.: Equivalence of problem-oriented models of heterogeneous databases [in Polish]. Contemporary Problems of Science. Legnica (2006)
12. Lebiediewa, S.: Computer system for multi-stage object recognition. Wroclaw Sch. Inf. Technol. Sci. Bull. **4** (2014)
13. Tsitchizris, D., Lochovsky, F.: Data Models. Prentice-Hall, Englewood Cliffs (1982)
14. Lien, Y.E.: On the equivalence of database models. J. ACM (JACM) **29**(2), 333–362 (1982)
15. Świątek, J.: On the two-level identification and their technical and biomedical applications. [in Polish]. In: Scientific Papers of the Institute of Control and System Engineering of Wroclaw Technical University, vol. 4(2) (1987)
16. Apiecionek, Ł., Czerniak, J.M., Zarzycki, H.: Protection tool for distributed denial of services attack. In: Beyond Databases, Architectures, and Structures. pp. 405–414, Springer (2014)
17. Angryk, R.A., Czerniak, J.: Heuristic algorithm for interpretation of multi-valued attributes in similarity-based fuzzy relational databases. Int. J. Approx. Reason. **51**(8), 895–911 (2010)

Part II
Intuitionistic Fuzzy Sets: Foundations, Tools and Techniques

Intuitionistic Fuzzy Implications and Klir-Yuan's Axioms

Nora Angelova and Krassimir Atanassov

Abstract During years of research, there have been defined 149 intuitionistic fuzzy implications. In the paper, it is checked which of these implications satisfy Klir-Yuan's axioms, whether as (classical) tautologies or as intuitionistic fuzzy tautologies.

Keywords Implication · Intuitionistic fuzzy logic · Klir-Yuan's axioms

1 Introduction

In a series of publications, (see, e.g. [1, 2]), 153 different operations "implication" in intuitionistic fuzzy propositional calculus, have been intorduced. 149 of them are non-parametricones. They have been found to generate 45 operations "negation" over intuitionistic fuzzy sets.

In the present paper, we study which of these implications and their respective netations satisfy which axioms of Klir-Yuan's axioms, and whether, doing so, they behave tautologies or as intuitionistic fuzzy tautologies.

In intuitionistic fuzzy propositional calculus, if x is a variable, then its truth-value is represented by the ordered couple

$$V(x) = \langle a, b \rangle,$$

so that $a, b, a + b \in [0, 1]$, where a and b are degrees of validity and of non-validity of x.

N. Angelova (✉) · K. Atanassov
Department of Bioinformatics and Mathematical Modelling, Institute of Biophysics and Biomedical Engineering Bulgarian Academy of Sciences, 105 Acad. G. Bonchev Str., Sofia 1113, Bulgaria
e-mail: nora.angelova@biomed.bas.bg

K. Atanassov
e-mail: krat@bas.bg

K.T. Atanassov et al. (eds.), *Novel Developments in Uncertainty Representation and Processing*, Advances in Intelligent Systems and Computing 401,
DOI 10.1007/978-3-319-26211-6_9

For the needs of the discussion below, we shall define the notion of tautology and of Intuitionistic Fuzzy Tautology (IFT, see, e.g. [1, 2]) by:

$$x \text{ is an IFT if and only if for } V(x) = \langle a, b\rangle \text{ it holds that } a \geq b,$$

and

$$x \text{ will be a (classical) tautology if and only if } a = 1 \text{ and } b = 0.$$

Obiously, each tautology is intuitionistic fuzzy tautology, but the opposite is not true.

2 On the Operations "Implication" and "Negation" in Intuitionistic Fuzzy Propositional Calculus

Below, we shall assume that for the two variables x and y the equalities: $V(x) = \langle a, b\rangle$ and $V(y) = \langle c, d\rangle$ $(a, b, c, d, a + b, c + d \in [0, 1])$ hold.

In some definitions, we need to use the auxiliary functions sg and $\overline{\text{sg}}$ defined by,

$$\text{sg}(x) = \begin{cases} 1 & \text{if } x > 0 \\ 0 & \text{if } x \leq 0 \end{cases}, \overline{\text{sg}}(x) = \begin{cases} 0 & \text{if } x > 0 \\ 1 & \text{if } x \leq 0 \end{cases}$$

The list of all currently existing intuitionistic fuzzy implications and negations are given in Tables 1 and 2, respectively.

The relations between the implications and negations are the following: negation $\neg_1$ is generated by implications $\rightarrow_1, \rightarrow_4, \rightarrow_5, \rightarrow_6, \rightarrow_7, \rightarrow_{10}, \rightarrow_{13}, \rightarrow_{61}, \rightarrow_{63}, \rightarrow_{64}, \rightarrow_{66}, \rightarrow_{67}, \rightarrow_{68}, \rightarrow_{69}, \rightarrow_{70}, \rightarrow_{71}, \rightarrow_{72}, \rightarrow_{73}, \rightarrow_{78}, \rightarrow_{80}, \rightarrow_{124}, \rightarrow_{125}, \rightarrow_{127}$; negation $\neg_2$ – by $\rightarrow_2, \rightarrow_3, \rightarrow_8, \rightarrow_{11}, \rightarrow_{16}, \rightarrow_{20}, \rightarrow_{31}, \rightarrow_{32}, \rightarrow_{37}, \rightarrow_{40}, \rightarrow_{41}, \rightarrow_{42}$; negation $\neg_3$ – by $\rightarrow_9, \rightarrow_{17}, \rightarrow_{21}$; $\neg_4$ – by $\rightarrow_{12}, \rightarrow_{18}, \rightarrow_{22}, \rightarrow_{46}, \rightarrow_{49}, \rightarrow_{50}, \rightarrow_{51}, \rightarrow_{53}, \rightarrow_{54}, \rightarrow_{91}, \rightarrow_{93}, \rightarrow_{94}, \rightarrow_{95}, \rightarrow_{96}, \rightarrow_{98}, \rightarrow_{134}, \rightarrow_{135}, \rightarrow_{137}$; $\neg_5$ – by $\rightarrow_{14}, \rightarrow_{15}, \rightarrow_{19}, \rightarrow_{23}, \rightarrow_{47}, \rightarrow_{48}, \rightarrow_{52}, \rightarrow_{55}, \rightarrow_{56}, \rightarrow_{57}$; $\neg_6$ – by $\rightarrow_{24}, \rightarrow_{26}, \rightarrow_{27}, \rightarrow_{65}$; $\neg_7$ – by $\rightarrow_{25}, \rightarrow_{28}, \rightarrow_{29}, \rightarrow_{62}$; $\neg_8$ – by $\rightarrow_{30}, \rightarrow_{33}, \rightarrow_{34}, \rightarrow_{35}, \rightarrow_{36}, \rightarrow_{38}, \rightarrow_{39}, \rightarrow_{76}, \rightarrow_{82}, \rightarrow_{84}, \rightarrow_{85}, \rightarrow_{86}, \rightarrow_{87}, \rightarrow_{89}, \rightarrow_{129}, \rightarrow_{130}, \rightarrow_{132}$; $\neg_9$ – by $\rightarrow_{43}, \rightarrow_{44}, \rightarrow_{45}, \rightarrow_{83}$; $\neg_{10}$ – by $\rightarrow_{58}, \rightarrow_{59}, \rightarrow_{60}, \rightarrow_{92}$; $\neg_{11}$ – by $\rightarrow_{74}, \rightarrow_{97}$; $\neg_{12}$ – by $\rightarrow_{75}$; $\neg_{13}$ – by $\rightarrow_{77}, \rightarrow_{88}$; $\neg_{14}$ – by $\rightarrow_{79}$; $\neg_{15}$ – by $\rightarrow_{81}$; $\neg_{16}$ – by $\rightarrow_{90}$; $\neg_{17}$ – by $\rightarrow_{99}$; $\neg_{18}$ – by $\rightarrow_{100}$; $\neg_{19}$ – by $\rightarrow_{101}$; $\neg_{20}$ – by $\rightarrow_{102}, \rightarrow_{108}$; $\neg_{21}$ – by $\rightarrow_{103}$; $\neg_{22}$ – by $\rightarrow_{104}$; $\neg_{23}$ – by $\rightarrow_{105}$; $\neg_{24}$ – by $\rightarrow_{106}$; $\neg_{25}$ – by $\rightarrow_{107}$; $\neg_{26}$ – by $\rightarrow_{109}, \rightarrow_{110}, \rightarrow_{111}, \rightarrow_{112}, \rightarrow_{113}$; $\neg_{27}$ – by $\rightarrow_{114}, \rightarrow_{115}, \rightarrow_{116}, \rightarrow_{117}, \rightarrow_{118}$; $\neg_{28}$ – by $\rightarrow_{119}, \rightarrow_{120}, \rightarrow_{121}, \rightarrow_{122}, \rightarrow_{123}$; $\neg_{29}$ – by $\rightarrow_{126}$; $\neg_{30}$ – by $\rightarrow_{128}$; $\neg_{31}$ – by $\rightarrow_{131}$; $\neg_{32}$ – by $\rightarrow_{133}$; $\neg_{33}$ – by $\rightarrow_{136}$; $\neg_{34}$ – by $\rightarrow_{138}$; $\neg_{35}$ – by $\rightarrow_{139}$; $\neg_{36}$ – by $\rightarrow_{140}$; $\neg_{37}$ – by $\rightarrow_{141}$; $\neg_{38}$ – by $\rightarrow_{142}, \rightarrow_{143}$; $\neg_{39}$ – by $\rightarrow_{144}, \rightarrow_{145}$; $\neg_{40}$ – by $\rightarrow_{146}, \rightarrow_{147}$; $\neg_{41}$ – by $\rightarrow_{148}, \rightarrow_{149}$; $\neg_{42}$– by $\rightarrow_{149}$; $\neg_{43}$ – by $\rightarrow_{150}$; $\neg_{44}$ – by $\rightarrow_{151}$, $\neg_{45}$– by $\rightarrow_{152}$.

Table 1 IF Implications

$\to_1$	$\langle \max(b, \min(a, c)), \min(a, d)\rangle$
$\to_2$	$\langle \overline{\text{sg}}(a-c), d.\text{sg}(a-c)\rangle$
$\to_3$	$\langle 1-(1-c).\text{sg}(a-c)), d.\text{sg}(a-c)\rangle$
$\to_4$	$\langle \max(b, c), \min(a, d)\rangle$
$\to_5$	$\langle \min(1, b+c), \max(0, a+d-1)\rangle$
$\to_6$	$\langle b+ac, ad\rangle$
$\to_7$	$\langle \min(\max(b, c), \max(a, b), \max(c, d)), \max(\min(a, d), \min(a, b), \min(c, d))\rangle$
$\to_8$	$\langle 1-(1-\min(b, c)).\text{sg}(a-c), \max(a, d).\text{sg}(a-c), \text{sg}(d-b))\rangle$
$\to_9$	$\langle b+a^2c, ab+a^2d\rangle$
$\to_{10}$	$\langle c.\overline{\text{sg}}(1-a)+\text{sg}(1-a).(\overline{\text{sg}}(1-c)+b.\text{sg}(1-c)), d.\overline{\text{sg}}(1-a)$ $+a.\text{sg}(1-a).\text{sg}(1-c)\rangle$
$\to_{11}$	$\langle 1-(1-c).\text{sg}(a-c), d.\text{sg}(a-c).\text{sg}(d-b)\rangle$
$\to_{12}$	$\langle \max(b, c), 1-\max(b, c)\rangle$
$\to_{13}$	$\langle b+c-b.c, a.d\rangle$
$\to_{14}$	$\langle 1-(1-c).\text{sg}(a-c)-d.\overline{\text{sg}}(a-c).\text{sg}(d-b), d.\text{sg}(d-b)\rangle$
$\to_{15}$	$\langle 1-(1-\min(b, c)).\text{sg}(a-c).\text{sg}(d-b)-\min(b, c).\text{sg}(a-c).\text{sg}(d-b),$ $1-(1-\max(a, d)).\text{sg}(\overline{\text{sg}}(a-c)+\overline{\text{sg}}(d-b))-\max(a, d).\overline{\text{sg}}(a-c).\overline{\text{sg}}(d-b)\rangle$
$\to_{16}$	$\langle \max(\overline{\text{sg}}(a), c), \min(\text{sg}(a), d)\rangle$
$\to_{17}$	$\langle \max(b, c), \min(a.b+a^2, d)\rangle$
$\to_{18}$	$\langle \max(b, c), \min(1-b, d)\rangle$
$\to_{19}$	$\langle \max(1-\text{sg}(\text{sg}(a)+\text{sg}(1-b)), c), \min(\text{sg}(1-b), d)\rangle$
$\to_{20}$	$\langle \max(\overline{\text{sg}}(a), \text{sg}(c)), \min(\text{sg}(a), \overline{\text{sg}}(c))\rangle$
$\to_{21}$	$\langle \max(b, c.(c+d)), \min(a.(a+b), d.(c^2+d+c.d))\rangle$
$\to_{22}$	$\langle \max(b, 1-d), 1-\max(b, 1-d)\rangle$
$\to_{23}$	$\langle 1-\min(\text{sg}(1-b), \overline{\text{sg}}(1-d)), \min(\text{sg}(1-b), \overline{\text{sg}}(1-d))\rangle$
$\to_{24}$	$\langle \overline{\text{sg}}(a-c).\overline{\text{sg}}(d-b), \text{sg}(a-c).\text{sg}(d-b)\rangle$
$\to_{25}$	$\langle \max(b, \overline{\text{sg}}(a).\overline{\text{sg}}(1-b), c.\overline{\text{sg}}(d).\overline{\text{sg}}(1-c)), \min(a, d)\rangle$
$\to_{26}$	$\langle \max(\overline{\text{sg}}(1-b), c), \min(\text{sg}(a), d)\rangle$
$\to_{27}$	$\langle \max(\overline{\text{sg}}(1-b), \text{sg}(c)), \min(\text{sg}(a), \overline{\text{sg}}(1-d))\rangle$
$\to_{28}$	$\langle \max(\overline{\text{sg}}(1-b), c), \min(a, d)\rangle$
$\to_{29}$	$\langle \max(\overline{\text{sg}}(1-b), \overline{\text{sg}}(1-c)), \min(a, \overline{\text{sg}}(1-d))\rangle$
$\to_{30}$	$\langle \max(1-a, \min(a, 1-d)), \min(a, d)\rangle$
$\to_{31}$	$\langle \overline{\text{sg}}(a+d-1), d.\text{sg}(a+d-1)\rangle$
$\to_{32}$	$\langle 1-d.\text{sg}(a+d-1), d.\text{sg}(a+d-1)\rangle$
$\to_{33}$	$\langle 1-\min(a, d), \min(a, d)\rangle$
$\to_{34}$	$\langle \min(1, 2-a-d), \max(0, a+d-1)\rangle$
$\to_{35}$	$\langle 1-a.d, a.d\rangle$
$\to_{36}$	$\langle \min(1-\min(a, d), \max(a, 1-a), \max(1-d, d)), \max(\min(a, d),$ $\min(a, 1-a), \min(1-d, d))\rangle$
$\to_{37}$	$\langle 1-\max(a, d).\text{sg}(a+d-1), \max(a, d).\text{sg}(a+d-1)\rangle$

(continued)

Table 1 (continued)

$\rightarrow_{38}$	$\langle 1-a+(a^2.(1-d)), a.(1-a)+a^2.d\rangle$
$\rightarrow_{39}$	$\langle (1-d).\overline{\text{sg}}(1-a)+\text{sg}(1-a).(\overline{\text{sg}}(d)+(1-a).\text{sg}(d)),$ $d.\overline{\text{sg}}(1-a)+a.\text{sg}(1-a).\text{sg}(d)\rangle$
$\rightarrow_{40}$	$\langle 1-\text{sg}(a+d-1), 1-\overline{\text{sg}}(a+d-1)\rangle$
$\rightarrow_{41}$	$\langle \max(\overline{\text{sg}}(a), 1-d), \min(\text{sg}(a), d)\rangle$
$\rightarrow_{42}$	$\langle \max(\overline{\text{sg}}(a), \text{sg}(1-d)), \min(\text{sg}(a), \overline{\text{sg}}(1-d))\rangle$
$\rightarrow_{43}$	$\langle \max(\overline{\text{sg}}(a), 1-d), \min(\text{sg}(a), d)\rangle$
$\rightarrow_{44}$	$\langle \max(\overline{\text{sg}}(a), 1-d), \min(a, d)\rangle$
$\rightarrow_{45}$	$\langle \max(\overline{\text{sg}}(a), \overline{\text{sg}}(d)), \min(a, \overline{\text{sg}}(1-d))\rangle$
$\rightarrow_{46}$	$\langle \max(b, \min(1-b, c)), 1-\max(b, c)\rangle$
$\rightarrow_{47}$	$\langle \overline{\text{sg}}(1-b-c), (1-c).\text{sg}(1-b-c)\rangle$
$\rightarrow_{48}$	$\langle 1-(1-c).\text{sg}(1-b-c), (1-c).\text{sg}(1-b-c)\rangle$
$\rightarrow_{49}$	$\langle \min(1, b+c), \max(0, 1-b-c)\rangle$
$\rightarrow_{50}$	$\langle b+c-b.c, 1-b-c+b.c\rangle$
$\rightarrow_{51}$	$\langle \min(\max(b, c), \max(1-b, b), \max(c, 1-c)),\ \max(1-\max(b, c),$ $\min(1-b, b), \min(c, 1-c))\rangle$
$\rightarrow_{52}$	$\langle 1-(1-\min(b, c)).\text{sg}(1-b-c), 1-\min(b, c).\text{sg}(1-b-c)\rangle$
$\rightarrow_{53}$	$\langle b+(1-b)^2.c, (1-b).b+(1-b)^2.(1-c)\rangle$
$\rightarrow_{54}$	$\langle c.\overline{\text{sg}}(b)+\text{sg}(b).(\overline{\text{sg}}(1-c)+b.\text{sg}(1-c)),$ $(1-c).\overline{\text{sg}}(b)+(1-b).\text{sg}(b).\text{sg}(1-c)\rangle$
$\rightarrow_{55}$	$\langle 1-\text{sg}(1-b-c), 1-\overline{\text{sg}}(1-b-c)\rangle$
$\rightarrow_{56}$	$\langle \max(\overline{\text{sg}}(1-b), c), \min(\text{sg}(1-b), (1-c))\rangle$
$\rightarrow_{57}$	$\langle \max(\overline{\text{sg}}(1-b), \text{sg}(c)), \min(\text{sg}(1-b), \overline{\text{sg}}(c))\rangle$
$\rightarrow_{58}$	$\langle \max(\overline{\text{sg}}(1-b), \overline{\text{sg}}(1-c)), 1-\max(b, c)\rangle$
$\rightarrow_{59}$	$\langle \max(\overline{\text{sg}}(1-b), c), (1-\max(b, c))\rangle$
$\rightarrow_{60}$	$\langle \max(\overline{\text{sg}}(1-b), \overline{\text{sg}}(1-c)), \min((1-b), \overline{\text{sg}}(c))\rangle$
$\rightarrow_{61}$	$\langle \max(c, \min(b, d)), \min(a, d)\rangle$
$\rightarrow_{62}$	$\langle \overline{\text{sg}}(d-b), a.\text{sg}(d-b)\rangle$
$\rightarrow_{63}$	$\langle 1-(1-b).\text{sg}(d-b), a.\text{sg}(d-b)\rangle$
$\rightarrow_{64}$	$\langle c+b.d, a.d\rangle$
$\rightarrow_{65}$	$\langle 1-(1-\min(b, c)).\text{sg}(d-b), \max(a, d).\text{sg}(d-b).\text{sg}(a-c)\rangle$
$\rightarrow_{66}$	$\langle c+d^2.b, b.d+d^2.a\rangle$
$\rightarrow_{67}$	$\langle b.\overline{\text{sg}}(1-d)+\text{sg}(1-d).(\overline{\text{sg}}(1-b)+c.\text{sg}(1-b)),$ $a.\overline{\text{sg}}(1-d)+d.\text{sg}(1-d).\text{sg}(1-b)\rangle$
$\rightarrow_{68}$	$\langle 1-(1-b).\text{sg}(d-b), a.\text{sg}(d-b).\text{sg}(a-c)\rangle$
$\rightarrow_{69}$	$\langle 1-(1-b).\text{sg}(d-b)-a.\overline{\text{sg}}(d-b).\text{sg}(a-c), a.\text{sg}(a-c)\rangle$
$\rightarrow_{70}$	$\langle \max(\overline{\text{sg}}(d), b), \min(\text{sg}(d), a)\rangle$
$\rightarrow_{71}$	$\langle \max(b, c), \min(c.d+d^2, a)\rangle$
$\rightarrow_{72}$	$\langle \max(b, c), \min(1-c, a)\rangle$
$\rightarrow_{73}$	$\langle \max(1-\max(\text{sg}(d), \text{sg}(1-c)), b), \min(\text{sg}(1-c), a)\rangle$

(continued)

Table 1 (continued)

$\rightarrow_{74}$	$\langle \max(\text{sg}(b), \overline{\text{sg}}(d)), \min(\overline{\text{sg}}(b), \text{sg}(d))\rangle$
$\rightarrow_{75}$	$\langle \max(c, b.(a+b)), \min(d.(c+d), a.(b^2+a)+a.b)\rangle$
$\rightarrow_{76}$	$\langle \max(c, 1-a), \min(1-c, a)\rangle$
$\rightarrow_{77}$	$\langle (1-\min(\overline{\text{sg}}(1-a), \text{sg}(1-c))), \min(\overline{\text{sg}}(1-a), \text{sg}(1-c))\rangle$
$\rightarrow_{78}$	$\langle \max(\overline{\text{sg}}(1-c), b), \min(\text{sg}(d), a)\rangle$
$\rightarrow_{79}$	$\langle \max(\overline{\text{sg}}(1-c), \text{sg}(b)), \min(\text{sg}(d), \overline{\text{sg}}(1-a))\rangle$
$\rightarrow_{80}$	$\langle \max(\overline{\text{sg}}(1-c), b), \min(d, a)\rangle$
$\rightarrow_{81}$	$\langle \max(\overline{\text{sg}}(1-b), \overline{\text{sg}}(1-c)), \min(d, \overline{\text{sg}}(1-a))\rangle$
$\rightarrow_{82}$	$\langle \max(1-d, \min(d, 1-a)), \min(d, a)\rangle$
$\rightarrow_{83}$	$\langle \overline{\text{sg}}(a+d-1), a.\text{sg}(a+d-1)\rangle$
$\rightarrow_{84}$	$\langle 1-a.\text{sg}(a+d+1), a.\text{sg}(a+d+1)\rangle$
$\rightarrow_{85}$	$\langle 1-d+d^2.(1-a), d.(1-d)+d^2.\rangle$
$\rightarrow_{86}$	$\langle (1-a).\overline{\text{sg}}(1-d)+\text{sg}(1-d).\overline{\text{sg}}(a+\min(1-d, \text{sg}(a))),$ $a.\overline{\text{sg}}(1-d)+d.\text{sg}(1-d).\text{sg}(a)\rangle$
$\rightarrow_{87}$	$\langle \max(\overline{\text{sg}}(d), 1-a), \min(\text{sg}(d), a)\rangle$
$\rightarrow_{88}$	$\langle \max(\overline{\text{sg}}(d), \text{sg}(1-a)), \min(\text{sg}(d), \overline{\text{sg}}(1-a))\rangle$
$\rightarrow_{89}$	$\langle \max(\overline{\text{sg}}(d), 1-a), \min(d, a)\rangle$
$\rightarrow_{90}$	$\langle \max(\overline{\text{sg}}(a), \overline{\text{sg}}(d)), \min(d, \overline{\text{sg}}(1-a))\rangle$
$\rightarrow_{91}$	$\langle \max(c, \min(1-c, b)), 1-\max(b, c)\rangle$
$\rightarrow_{92}$	$\langle \overline{\text{sg}}(1-b-c), \min(1-b, \text{sg}(1-b-c))\rangle$
$\rightarrow_{93}$	$\langle (1-\min(1-b, \text{sg}(1-b-c)), \min(1-b, \text{sg}(1-b-c))\rangle$
$\rightarrow_{94}$	$\langle c+(1-c)^2.b, (1-c).c+(1-c)^2.(1-b)\rangle$
$\rightarrow_{95}$	$\langle \min(b, \overline{\text{sg}}(c))+\text{sg}(c).(\overline{\text{sg}}(1-b)+\min(c, \text{sg}(1-b))),\ (\min(1-b, \overline{\text{sg}}(c))$ $+\min(1-c, \text{sg}(c), \text{sg}(1-b))\rangle$
$\rightarrow_{96}$	$\langle \max(\overline{\text{sg}}(1-c), b), \min(\text{sg}(1-b), 1-c)\rangle$
$\rightarrow_{97}$	$\langle \max(\overline{\text{sg}}(1-c), \text{sg}(b)), \min(\text{sg}(1-c), \overline{\text{sg}}(b))\rangle$
$\rightarrow_{98}$	$\langle \max(\overline{\text{sg}}(1-c), b), 1-\max(b, c)\rangle$
$\rightarrow_{99}$	$\langle \max(\overline{\text{sg}}(1-c), \overline{\text{sg}}(1-b)), \min(1-c, \overline{\text{sg}}(b))\rangle$
$\rightarrow_{100}$	$\langle \max(\min(b, \text{sg}(a)), c), min(a, \text{sg}(b), d)\rangle$
$\rightarrow_{101}$	$\langle \max(\min(b, \text{sg}(a)), \min(c, \text{sg}(d))), \min(a, \text{sg}(b), \text{sg}(c), d)\rangle$
$\rightarrow_{102}$	$\langle \max(b, \min(c, \text{sg}(d))), \min(a, \text{sg}(c), d)\rangle$
$\rightarrow_{103}$	$\langle \max(\min(1-a, \text{sg}(a)), 1-d), \min(a, \text{sg}(1-a), d)\rangle$
$\rightarrow_{104}$	$\langle \max(\min(1-a, \text{sg}(a)), \min(1-d, \text{sg}(d))), \min(a, \text{sg}(1-a), d, \text{sg}(1-d))\rangle$
$\rightarrow_{105}$	$\langle \max(1-a, \min(1-d, \text{sg}(d))), \min(a, d, \text{sg}(1-d))\rangle$
$\rightarrow_{106}$	$\langle \max(\min(b, \text{sg}(1-b)), c), \min(1-b, \text{sg}(b), 1-c)\rangle$
$\rightarrow_{107}$	$\langle \max(\min(b, \text{sg}(1-b)), \min(c, \text{sg}(1-c))), \min(1-b, \text{sg}(b), 1-c, \text{sg}(c))\rangle$
$\rightarrow_{108}$	$\langle \max(b, \min(c, \text{sg}(1-c))), \min(1-b, 1-c, \text{sg}(c))\rangle$
$\rightarrow_{109}$	$\langle b+\min(\overline{\text{sg}}(1-a), c), a.b+\min(\overline{\text{sg}}(1-a), d)\rangle$
$\rightarrow_{110}$	$\langle \max(b, c), \min(a.b+\overline{\text{sg}}(1-a), d)\rangle$
$\rightarrow_{111}$	$\langle \max(b, c.d+\overline{\text{sg}}(1-c)), \min(a.b+\overline{\text{sg}}(1-a), d.(c.d+\overline{\text{sg}}(1-c))+\overline{\text{sg}}(1-d))\rangle$

(continued)

Table 1 (continued)

$\rightarrow_{112}$	$\langle b+c-b.c, a.b+\overline{\text{sg}}(1-a).d\rangle$
$\rightarrow_{113}$	$\langle b+c.d-b.(c.d+\overline{\text{sg}}(1-c)), (a.b+\overline{\text{sg}}(1-a)).(d.(c.d+\overline{\text{sg}}(1-c))+\overline{\text{sg}}(1-d))\rangle$
$\rightarrow_{114}$	$\langle 1-a+\min(\overline{\text{sg}}(1-a), 1-d), a.(1-a)+\min(\overline{\text{sg}}(1-a), d)\rangle$
$\rightarrow_{115}$	$\langle 1-\min(a,d), \min(a.(1-a)+\overline{\text{sg}}(1-a), d)\rangle$
$\rightarrow_{116}$	$\langle \max(1-a, (1-d).d+\overline{\text{sg}}(d)), \min(a.(1-a)+\overline{\text{sg}}(1-a), d.((1-d).d$ $+\overline{\text{sg}}(d))+\overline{\text{sg}}(1-d))\rangle$
$\rightarrow_{117}$	$\langle 1-a-d+a.d, (a.(1-a)+\overline{\text{sg}}(1-a)).d\rangle$
$\rightarrow_{118}$	$\langle 1-a+(1-d).d-(1-a).((1-d).d+\overline{\text{sg}}(d)),\ (a.(1-a)+\overline{\text{sg}}(1-a))$ $.d.((1-d).d+\overline{\text{sg}}(d))+\overline{\text{sg}}(1-d)\rangle$
$\rightarrow_{119}$	$\langle b+\min(\overline{\text{sg}}(b), c), (1-b).b+\min(\overline{\text{sg}}(b), 1-c)\rangle$
$\rightarrow_{120}$	$\langle \max(b,c), \min((1-b).b+\overline{\text{sg}}(b), 1-c)\rangle$
$\rightarrow_{121}$	$\langle \max(b, c.(1-c)+\overline{\text{sg}}(1-c)), \min((1-b).b+\overline{\text{sg}}(b), (1-c).(c.(1-c)$ $+\overline{\text{sg}}(1-c)))+\overline{\text{sg}}(c)\rangle$
$\rightarrow_{122}$	$\langle b+c-b.c, ((1-c).b+\overline{\text{sg}}(b)).(1-c)\rangle$
$\rightarrow_{123}$	$\langle b+c.(1-c)-(b.(c.(1-c)+\overline{\text{sg}}(1-c))),$ $((1-b).b+\overline{\text{sg}}(b)).(((1-c).(c.(1-c)+\overline{\text{sg}}(1-c)))+\overline{\text{sg}}(c))\rangle$
$\rightarrow_{124}$	$\langle c+\min(\overline{\text{sg}}(1-d), b), c.d+\min(\overline{\text{sg}}(1-d), a)\rangle$
$\rightarrow_{125}$	$\langle \max(b,c), \min(c.d+\overline{\text{sg}}(1-d), a)\rangle$
$\rightarrow_{126}$	$\langle \max(c, a.b+\overline{\text{sg}}(1-b)), \min(c.d+\overline{\text{sg}}(1-d), a.(a.b$ $+\overline{\text{sg}}(1-b))+\overline{\text{sg}}(1-a))\rangle$
$\rightarrow_{127}$	$\langle b+c-b.c, (c.d+\overline{\text{sg}}(1-d)).a\rangle$
$\rightarrow_{128}$	$\langle c+a.b-c.(a.b+\overline{\text{sg}}(1-b)), (c.d+\overline{\text{sg}}(1-d)).(a.(a.b+\overline{\text{sg}}(1-b))$ $+\overline{\text{sg}}(1-a))\rangle$
$\rightarrow_{129}$	$\langle 1-d+\min(\overline{\text{sg}}(1-d), 1-a), d.(1-d)+\min(\overline{\text{sg}}(1-d), a)\rangle$
$\rightarrow_{130}$	$\langle 1-\min(d,a), \min(d.(1-d)+\overline{\text{sg}}(1-d), a)\rangle$
$\rightarrow_{131}$	$\langle \max(1-d, (1-a).a+\overline{\text{sg}}(a)), \min(d.(1-d)+\overline{\text{sg}}(1-d),$ $a.((1-a).a+\overline{\text{sg}}(a))+\overline{\text{sg}}(1-a))\rangle$
$\rightarrow_{132}$	$\langle 1-a.d, (d.(1-d)+\overline{\text{sg}}(1-d)).a\rangle$
$\rightarrow_{133}$	$\langle 1-d+(1-a).a-(1-d).((1-a).a+\overline{\text{sg}}(a)),$ $(d.(1-d)+\overline{\text{sg}}(1-d)).(a.((1-a).a+\overline{\text{sg}}(a))+\overline{\text{sg}}(1-a))\rangle$
$\rightarrow_{134}$	$\langle c+\min(\overline{\text{sg}}(c), b), (1-c).c+\min(\overline{\text{sg}}(c), (1-b))\rangle$
$\rightarrow_{135}$	$\langle \max(b,c), \min((1-c).c+\overline{\text{sg}}(c), 1-b)\rangle$
$\rightarrow_{136}$	$\langle \max(c, (b.(1-b)+\overline{\text{sg}}(1-b))), \min((1-c).c+\overline{\text{sg}}(c),$ $(1-b).(b.(1-b)+\overline{\text{sg}}(1-b))+\overline{\text{sg}}(b))\rangle$
$\rightarrow_{137}$	$\langle b+c-b.c, ((1-c).c+\overline{\text{sg}}(c)).(1-b)\rangle$
$\rightarrow_{138}$	$\langle c+b.(1-b)-c.(b.(1-b)+\overline{\text{sg}}(1-b)),$ $((1-c).c+\overline{\text{sg}}(c)).((1-b).(b.(1-b)+\overline{\text{sg}}(1-b))+\overline{\text{sg}}(b))\rangle$
$\rightarrow_{139}$	$\langle \frac{b+c}{2}, \frac{a+d}{2}\rangle$
$\rightarrow_{140}$	$\langle \frac{b+c+\min(b,c)}{3}, \frac{a+d+\max(a,d)}{3}\rangle$
$\rightarrow_{141}$	$\langle \frac{b+c+\max(b,c)}{3}, \frac{a+d+\min(a,d)}{3}\rangle$
$\rightarrow_{142}$	$\langle \frac{3-a-d-\max(a,d)}{3}, \frac{a+d+\max(a,d)}{3}\rangle$

(continued)

Table 1 (continued)

$\rightarrow_{143}$	$\langle \frac{1-a+c+\min(1-a,c)}{3}, \frac{2+a-c-\min(1-a,c)}{3} \rangle$
$\rightarrow_{144}$	$\langle \frac{1+b-d+\min(b,1-d)}{3}, \frac{2-b+d+\min(b,1-d)}{3} \rangle$
$\rightarrow_{145}$	$\langle \frac{b+c+\min(b,c)}{3}, \frac{3-b-c-\min(b,c)}{3} \rangle$
$\rightarrow_{146}$	$\langle \frac{3-a-d-\min(a,d)}{3}, \frac{a+d+\min(a,d)}{3} \rangle$
$\rightarrow_{147}$	$\langle \frac{1-a+c+\max(1-a,c)}{3}, \frac{2+a-c-\max(1-a,c)}{3} \rangle$
$\rightarrow_{148}$	$\langle \frac{1+b-d+\max(b,1-d)}{3}, \frac{2-b+d-\max(b,1-d)}{3} \rangle$
$\rightarrow_{149}$	$\langle \frac{b+c+\max(b,c)}{3}, \frac{3-b-c-\max(b,c)}{3}$
$\rightarrow_{150,\lambda}$	$\langle \frac{b+c+\lambda-1}{2\lambda}, \frac{a+d+\lambda-1}{2\lambda}$, where $\lambda \geq 1$
$\rightarrow_{151,\gamma}$	$\langle \frac{b+c+\gamma}{2\gamma+1}, \frac{a+d+\gamma-1}{2\gamma+1}$, where $\gamma \geq 1$
$\rightarrow_{152,\alpha,\beta}$	$\langle \frac{b+c+\alpha-1}{\alpha+\beta}, \frac{a+d+\beta-1}{\alpha+\beta}$, where $\alpha \geq 1, \beta \in [0,\alpha]$
$\rightarrow_{153,\varepsilon,\eta}$	$\langle \min(1, \max(\mu_B(x), \nu_A(x)+\varepsilon)), \max(0, \min(\nu_B(x), \mu_A(x)-\eta)) \rangle$ where $\varepsilon, \eta \in [0,1]$ and $\varepsilon \leq \eta$

Table 2 IF negations

$\neg_1$	$\langle b, a \rangle$
$\neg_2$	$\langle \overline{\text{sg}}(a), \text{sg}(a) \rangle$
$\neg_3$	$\langle b, a.b + a^2 \rangle$
$\neg_4$	$\langle b, 1-b \rangle$
$\neg_5$	$\langle \overline{\text{sg}}(1-b), \text{sg}(1-b) \rangle$
$\neg_6$	$\langle \overline{\text{sg}}(1-b), \text{sg}(a) \rangle$
$\neg_7$	$\langle \overline{\text{sg}}(1-b), a \rangle$
$\neg_8$	$\langle 1-a, a \rangle$
$\neg_9$	$\langle \overline{\text{sg}}(a), a \rangle$
$\neg_{10}$	$\langle \overline{\text{sg}}(1-b), 1-b \rangle$
$\neg_{11}$	$\langle \text{sg}(b), \overline{\text{sg}}(b) \rangle$
$\neg_{12}$	$\langle b.(b+a), \min(1, a.(b^2+a+b.a)) \rangle$
$\neg_{13}$	$\langle \text{sg}(1-a), \overline{\text{sg}}(1-a) \rangle$
$\neg_{14}$	$\langle \text{sg}(b), \overline{\text{sg}}(1-a) \rangle$
$\neg_{15}$	$\langle \overline{\text{sg}}(1-b), \overline{\text{sg}}(1-a) \rangle$
$\neg_{16}$	$\langle \overline{\text{sg}}(a), \overline{\text{sg}}(1-a) \rangle$
$\neg_{17}$	$\langle \overline{\text{sg}}(1-b), \overline{\text{sg}}(b) \rangle$
$\neg_{18}$	$\langle b.\text{sg}(a), a.\text{sg}(b) \rangle$
$\neg_{19}$	$\langle b.\text{sg}(a), 0 \rangle$
$\neg_{20}$	$\langle b, 0 \rangle$
$\neg_{21}$	$\langle \min(1-a, \text{sg}(a)), \min(a, \text{sg}(1-a)) \rangle$
$\neg_{22}$	$\langle \min(1-a, \text{sg}(a)), 0 \rangle$
$\neg_{23}$	$\langle 1-a, 0 \rangle$
$\neg_{24}$	$\langle \min(b, \text{sg}(1-b)), \min(1-b, \text{sg}(b)) \rangle$
$\neg_{25}$	$\langle \min(b, \text{sg}(1-b)), 0 \rangle$

(continued)

Table 2 (continued)

$\neg_{26}$	$\langle b, a.b + \overline{\text{sg}}(1-a)\rangle$
$\neg_{27}$	$\langle 1-a, a.(1-a) + \overline{\text{sg}}(1-a)\rangle$
$\neg_{28}$	$\langle b, (1-b).b + \overline{\text{sg}}(b)\rangle$
$\neg_{29}$	$\langle \max(0, b.a + \overline{\text{sg}}(1-b)), \min(1, a.(b.a + \overline{\text{sg}}(1-b)) + \overline{\text{sg}}(1-a))\rangle$
$\neg_{30}$	$\langle a.b,\ a.(a.b + \overline{\text{sg}}(1-b)) + \overline{\text{sg}}(1-a)\rangle$
$\neg_{31}$	$\langle \max(0, (1-a).a + \overline{\text{sg}}(a)), \min(1, a.((1-a).a + \overline{\text{sg}}(a)) + \overline{\text{sg}}(1-a))\rangle$
$\neg_{32}$	$\langle (1-a).a,\ a.((1-a).a + \overline{\text{sg}}(a)) + \overline{\text{sg}}(1-a)\rangle$
$\neg_{33}$	$\langle b.(1-b) + \overline{\text{sg}}(1-b), (1-b).(b.(1-b) + \overline{\text{sg}}(1-b)) + \overline{\text{sg}}(b))\rangle$
$\neg_{34}$	$\langle b.(1-b),\ (1-b).(b.(1-b) + \overline{\text{sg}}(1-b)) + \overline{\text{sg}}(b)\rangle$
$\neg_{35}$	$\langle \frac{b}{2}, \frac{1+a}{2}\rangle$
$\neg_{36}$	$\langle \frac{b}{3}, \frac{2+a}{3}\rangle$
$\neg_{37}$	$\langle \frac{2b}{3}, \frac{2a+1}{3}\rangle$
$\neg_{38}$	$\langle \frac{1-a}{3}, \frac{2+a}{3}\rangle$
$\neg_{39}$	$\langle \frac{b}{3}, \frac{3-b}{3}\rangle$
$\neg_{40}$	$\langle \frac{2-2a}{3}, \frac{1+2a}{3}\rangle$
$\neg_{41}$	$\langle \frac{2b}{3}, \frac{3-2b}{3}\rangle$
$\neg_{42,\lambda}$	$\langle \frac{b+\lambda-1}{2\lambda}, \frac{a+\lambda}{2\lambda}$, where $\lambda \geq 1$
$\neg_{43,\gamma}$	$\langle \frac{b+\gamma}{2\gamma+1}, \frac{a+\gamma}{2\gamma+1}$, where $\gamma \geq 1$
$\neg_{44,\alpha,\beta}$	$\langle \frac{b+\alpha-1}{\alpha+\beta}, \frac{a+\beta}{\alpha+\beta}$, where $\alpha \geq 1, \beta \in [0, \alpha]$
$\neg_{45,\varepsilon,\eta}$	$\langle \min(1, \nu_A(x) + \varepsilon), \max(0, \mu_A(x) - \eta)\rangle$ where $\varepsilon, \eta \in [0, 1]$ and $\varepsilon \leq \eta$

3 Intutionistic Fuzzy Implications and Klir and Yuan's Axioms

First, we mention that the last four from the above intutionistic fuzzy implications ($\rightarrow_{150,\lambda}, \rightarrow_{151,\gamma}, \rightarrow_{152,\alpha,\beta}, \rightarrow_{152,\alpha,\beta}$) contain parameters and by this reason the method used here is not applicable to them.

Now, we discuss the properties of the first 149 intuitionistic fuzzy implications taking into account the classic Georg Klir and Bo Yuan's book [6] that it is convenient for our purposes. However, a similar, if not practically identical, analyses can be performed in the new settings and views related to fuzzy implications, notably included in the Baczynksi and Jayaram's book [4].

Some variants of fuzzy implications (marked by $I(x, y)$) are described in [6] and the following nine axioms are discussed, where

$$I(x, y) \equiv x \rightarrow y.$$

Table 3 IF implications and the axioms

No	A_1	A_2	A_3	A_3^*	A_4	A_4^*	A_5	A_5^*	A_6	A_7	A_7^*	A_8	A_8^*
1		•	•	•	•		○						
2	•	•	•	•		•	•			•			
3	•	•	•	•	•	•	•	•		•			•
4	•	•	•	•	•		○	•			•	•	•
5	•	•	•	•	•		○	•			•	•	•
6		•	•	•	•		○						
7				○	•		○				•	•	•
8	•	•	•	•		•	•			•			
9		•	•	•	•			○					
10		•	•	•	•								
11	•	•	•	•	•	•	•	•		•			•
13	•	•	•	•	•		○	•			•	•	•
14	•	•	•	•	•	•	•	•	•	•			•
15	•	•	•	•		•	•		•	•			
16	•	•	•	•	•			•					•
17		•	•	•	•		○	•					
18	•	•	•	•	•		○	•					•
19	•	•	•	•	•			•					•
20	•	•	•	•		•	•	•		•	•	•	•
21			•	•			○						
22	•	•	•	•			○	•			•	•	•
23	•	•	•	•		•	•	•		•	•	•	•
24	•	•	•	•		•	•		•	•			
25	•	•	•	•				•					•
26	•	•	•	•	•			•					•
27	•	•	•	•			○	•			•	•	•
28	•	•	•	•	•		○	•					•
29	•	•	•	•			○						•
30		•	•	•			○						
31	•	•	•	•		•	•	•		•			
32	•	•	•	•		•	•	•		•			
33	•	•	•	•			○	•					
34	•	•	•	•		•	•	•		•			
35	•	•	•	•			○	•					
36				○			○	•					
37	•	•	•	•		•	•			•			
38		•	•	•			○						
39		•	•	•									
40	•	•	•	•		•	•			•			

(continued)

Table 3 (continued)

No	A_1	A_2	A_3	A_3^*	A_4	A_4^*	A_5	A_5^*	A_6	A_7	A_7^*	A_8	A_8^*
12	•	•	•	•				•					
41	•	•	•	•				•					
42	•	•	•	•		•	•	•		•			
43	•	•	•	•				•					
44	•	•	•	•			○						
45	•	•	•	•			○						
46		•	•	•									
47	•	•	•	•									
48	•	•	•	•				•					
49	•	•	•	•				•					
50	•	•	•	•				•					
51				○				•					
52	•	•	•	•									
53		•	•	•									
54		•	•	•									
55	•	•	•	•									
56	•	•	•	•				•					
57	•	•	•	•				•					
58	•	•	•	•									
59	•	•	•	•									
60	•	•	•	•									
61	•			○	•		○						
62	•	•	•	•		•	•			•			
63	•	•	•	•		•	•			•			
64	•			○	•		○						
65	•	•	•	•		•	•			•			
66	•			○			○						
67	•		•	•	•								
68	•	•	•	•		•	•			•			
69	•	•	•	•		•	•		•	•			
70	•	•	•	•									
71	•		•	•			○						
72	•	•	•	•			○						
73	•	•	•	•									
74	•	•	•	•		•	•	•		•	•	•	•
75			•	•			○						
76	•	•	•	•			○	•			•	•	•
77	•	•	•	•		•	•	•		•	•	•	•
78	•	•	•	•									

(continued)

Table 3 (continued)

No	A_1	A_2	A_3	A_3^*	A_4	A_4^*	A_5	A_5^*	A_6	A_7	A_7^*	A_8	A_8^*
79	•	•	•	•			∘	•			•	•	•
80	•	•	•	•			∘						
81	•	•	•	•			∘	•					•
82	•			∘			∘						
83	•	•	•	•		•	•			•			
84	•	•	•	•		•	•			•			
85	•			∘			∘						
86	•		•	•									
87	•	•	•	•									
88	•	•	•	•		•	•	•		•			
89	•	•	•	•			∘						
90	•	•	•	•			∘						
91	•			∘									
92	•	•	•	•									
93	•	•	•	•									
94	•			∘									
95	•		•	•									
96	•	•	•	•									
97	•	•	•	•				•					
98	•	•	•	•									
99	•	•	•	•									
100	∘	∘		∘			∘	•		∘			
101	∘	∘	∘	∘			∘			∘			•
102	∘	∘	•	•			∘			∘			•
103	∘	∘		∘			∘			∘			
104	∘	∘		∘			∘			∘			
105	∘	∘	•	•			∘			∘			
106	∘	∘		∘						∘			
107	∘	∘		∘						∘			•
108	∘	∘	•	•			∘			∘			•
109		•	•	•	•								
110		•	•	•	•		∘	•					
111			•	•			∘						
112		•	•	•	•		∘	•					
113				∘			∘						
114		•	•	•			∘						
115		•	•	•			∘						
116			•	•			∘						
117		•	•	•			∘						

(continued)

Table 3 (continued)

No	A_1	A_2	A_3	A_3^*	A_4	A_4^*	A_5	A_5^*	A_6	A_7	A_7^*	A_8	A_8^*
118				◦			◦						
119		•	•	•									
120		•	•	•									
121			•	•									
122		•	•	•									
123				◦									
124	•			◦									
125	•		•	•			◦						
126			•	•			◦						
127	•		•	•			◦						
128				◦			◦						
129	•			◦			◦						
130	•		•	•			◦						
131			•	•			◦						
132	•		•	•			◦						
133				◦			◦						
134	•			◦									
135	•		•	•									
136			•	•									
137	•		•	•									
138				◦									
139	◦	◦		◦			◦						
140													
141	◦	◦		◦			◦						
142													
143													
144													
145													
146	◦	◦		◦			◦						
147	◦	◦		◦			◦						
148	◦	◦		◦			◦						
149	◦	◦		◦									

Axiom 1 $(\forall x, y)(x \leq y \rightarrow (\forall z)(I(x, z) \geq I(y, z)))$,
Axiom 2 $(\forall x, y)(x \leq y \rightarrow (\forall z)(I(z, x) \leq I(z, y)))$,
Axiom 3 $(\forall y)(I(0, y) = 1)$,
Axiom 4 $(\forall y)(I(1, y) = y)$,

Axiom 5 $(\forall x)(I(x, x) = 1)$,
Axiom 6 $(\forall x, y, z)(I(x, I(y, z)) = I(y, I(x, z)))$,
Axiom 7 $(\forall x, y)(I(x, y) = 1 \text{ iff } x \leq y)$,
Axiom 8 $(\forall x, y)(I(x, y) = I(N(y), N(x)))$, where N is a negation operation,
Axiom 9 I is a continuous function.

For our research, having in mind the specific forms of the intuitionistic fuzzy implications, we modify five of these axioms, as follows.

Axiom 3* $(\forall y)(I(0, y)$ is an IFT),
Axiom 4* $(\forall y)(I(1, y) \leq y)$,
Axiom 5* $(\forall x)(I(x, x)$ is an IFT),
Axiom 7* $(\forall x, y)($ if $x \leq y$, then $I(x, y) = 1)$,
Axiom 8* $(\forall x, y)(I(x, y) = N(N(I(N(y), N(x)))))$.

Here, we ignore Axiom 9, because, obviously, it is valid for the implications that do not contain operations sg and $\overline{\text{sg}}$.

Theorem *The intuitionistic fuzzy implications that satisfy Klir and Yuan's axioms as (classical) tautologies, are marked in Table 3 by "•" and these that satisfy the same axioms (only) as IFTs—by "∘".*

The check of the validity of the Theorem was made by the software application IFSTool [3, 5], developed as a tool for automatic check of the properties of intuitionistic fuzzy implications and negations. The software has an option—to either check for intuitionistic fuzzy tautologies or only for fuzzy tautologies. For the needs of the present paper, each axiom was tested with all implications and their corresponding negations, and manual backup checks of some of the properties were made, as well.

4 Conclusion: Open Problems

We finish with the following interesting open problem.

Open problem To determine criteria that show the most suitable axioms that can have real applications.

In a next research, other properties of the implications will be studied.

Acknowledgments The authors are thankful for the support provided by the Bulgarian National Science Fund under Grant Ref. No. DFNI-I-02-5 "InterCriteria Analysis: A New Approach to Decision Making".

References

1. Atanassov, K.: On Intuitionistic Fuzzy Sets Theory. Springer, Berlin (2012)
2. Atanassov, K.: On intuitionistic fuzzy logics: results and problems. In: Atanassov, K., Baczynski, M., Drewniak, J., Kacprzyk, J., Krawczak, M., Szmidt, E., Wygralak, M., Zadrozny, S. (eds.) Modern Approaches in Fuzzy Sets, Intuitionistic Fuzzy Sets, Generalized Nets and Related Topics, Volume 1: Foundations, pp. 23–49. SRI-PAS, Warsaw (2014)
3. Dimitrov, D.: IFSTool—software for intuitionistic fuzzy sets. Issues Intuit. Fuzzy Sets Gen. Nets **9**, 61–69 (2011)
4. Baczynski, M., Jayaram, B.: Fuzzy Implications. Springer, Berlin (2008)
5. IFSTool. Ifigenia.net. http://www.ifigenia.org/wiki/IFSTool
6. Klir, G., Yuan, B.: Fuzzy Sets and Fuzzy Logic. Prentice Hall, New Jersey (1995)

On Separability of Intuitionistic Fuzzy Sets

Evgeniy Marinov, Peter Vassilev and Krassimir Atanassov

Abstract Intuitionistic fuzzy sets prove very useful in modelling uncertain and imprecise information when in the evaluations, concerned with a bipolar type of evidence, the "pro" and "contra" estimations do not sum to one (truth) but there is a degree of uncertainty. Relying on the concept of IF-neighbourhoods, introduced in Marinov et al. (On intuitionistic fuzzy metric neighbourhoods, 2015), we propose in this paper a few notions of separability between intuitionistic fuzzy sets and give some applications employing the extended modal operators.

Keywords Distances · Intuitionistic fuzzy sets · Operators · Separability

1 Introduction

An essential part of this work is founded on the concept of distances between Intuitionistic Fuzzy Sets (IFSs) (see [2, 3]), which are an extension of the concept of Zadeh's fuzzy sets (FSs) (cf. [18]). They are especially useful in modelling uncertain and imprecise information when the evaluations are concerned with a bipolar type of evidence. The generalization here is that the "pro" and "contra" estimations do not sum to one (truth) but there is a degree of uncertainty. Many distance measures have been proposed and studied in the context of IFSs, notably those which are counterparts of the respective distance measures proposed for fuzzy sets [6, 7, 14]. Our point of departure is the standard definition of neighbourhood for metric spaces

E. Marinov (✉) · P. Vassilev · K. Atanassov
Department of Bioinformatics and Mathematical Modelling, Institute of Biophysics and Biomedical Engineering, Bulgarian Academy of Sciences, 105 Acad. G. Bonchev Str., 1113 Sofia, Bulgaria
e-mail: evgeniy.marinov@biomed.bas.bg

P. Vassilev
e-mail: peter.vassilev@gmail.com

K. Atanassov
e-mail: krat@bas.bg

K.T. Atanassov et al. (eds.), *Novel Developments in Uncertainty Representation and Processing*, Advances in Intelligent Systems and Computing 401,
DOI 10.1007/978-3-319-26211-6_10

(taking the base of open balls to construct the metric topological spaces) in the framework of the already proposed metrics for IFSs. Further relying on the previously introduced IF-neighbourhood and IF-ball we explore the idea of separability of IFSs in Sect. 3. Normally, the two words "metric" and "distance" are interchangeable, as they will be used in this paper.

2 General Definitions and Notions About Metrics

Before continuing with our main idea and results we remind some facts and useful definitions from topology.

2.1 *Definitions and Notions About Metrics*

In what follows with $\mathcal{P}(X) = \{A \mid A \subset X\}$ or $2^X = \{\mu \mid \mu : X \to \{0,1\}\}$ we will denote the set of subsets of any set X. In fact the above defined concepts are bijective, as one can easily see by taking the mapping $A = \mu^{-1}(1)$. In what follows we will consider the set X as a general topological space, and then explain how we can generate such a topological space through the so called *basis elements* or a *base* for the topology. For more detailed information about general topology one can refer to [1, 9].

Definition 1 A topological structure, or just a *topology*, on a set X is given by a collection $\tau \subset \mathcal{P}(X)$ of subsets of X, each called *open set* and satisfying the following three axioms:

1. $\emptyset \in \tau$ and $X \in \tau$.
2. $(\forall \tau_0)(\tau_0 \subset \tau$ and τ_0 is finite $\Rightarrow \cap\tau_0 \in \tau)$.
3. $(\forall \tau_1)(\tau_1 \subset \tau \Rightarrow \cup\tau_1 \in \tau)$.

We call the pair (X, τ) a topology in X, and sometimes the underlying set X can be omitted. The complement of any open subset in regard to X is called *closed set*.

For any point $x \in X$ and any $A \subset X$, A is called a *neighborhood* of X if there exists a $A_0 \in \tau$, such that $x \in A_0 \subset A$.

Let us now take a collection $\mathcal{B} \subset \mathcal{P}(X)$ of subsets of X, where the following statements hold [9]:

- $\cup\mathcal{B} = X$, which exactly means that $(\forall x \in X)(\exists B \in \mathcal{B})(x \in B)$
- $(\forall B_1, B_2 \in \mathcal{B})(\forall x \in X)(x \in B_1 \cap B_2 \Rightarrow (\exists B_3 \in \mathcal{B})(x \in B_3))$

Definition 2 The above described collection $\mathcal{B}$ is called *base* (and its elements are called *basis elements*) for a special topology, named the topology $(X, \tau_{\mathcal{B}})$ generated by $\mathcal{B}$, which is obtained by defining the open sets to be the union of the empty set and arbitrary unions of basis elements:

$$\tau_{\mathcal{B}} = \{\cup\tau_0 \mid \tau_0 \subset \mathcal{B}\}.$$

In the above definition, when we use $\cup\tau_0$, by this denotation we mean the union of the elements of τ_0, which are, in our case elements of $\mathcal{B}$, i.e. subsets of the base set X.

We will also remind the following theorem

Theorem 1 (cf. [1, Theorem 1.4, p. 45]) *Let X be a set and $\mathcal{B}$ be a base for the topology $\tau_{\mathcal{B}}$. Then U is open in $\tau_{\mathcal{B}}$ iff for all $x \in U$ there exists a basis element $B_x \in \mathcal{B}$ such that $x \in B_x \subset U$.*

A metric (topological) space can be thought as a very basic space that satisfies a few axioms. The ability to measure and compare distances between elements of a set is often crucial, and it provides more structure than general topological space possesses. When we refer to the elements or "points" of the underlying set, we do not necessarily refer to geometrical points, although this is how most of us usually visualize them. They may be objects of any type, such as sequences, functions, images, sounds, signals, decisions, etc.

Definition 3 (*cf.* [1]) A **metric** on a set X is a function $d : X \times X \to \mathbb{R}$ with the following properties:

1. $d(x, y) \geq 0$ for all $x, y \in X$, with $d(x, y) = 0$ iff $x = y$.
2. $d(x, y) = d(y, x)$ for all $x, y \in X$ (symmetry).
3. $d(x, z) \leq d(x, y) + d(y, z)$ for all $x, y, z \in X$ (the triangle inequality).

We call $d(x, y)$ the **distance** between x and y, and the pair (X, d) a **metric space**.

It is evident that d has the properties we expect when we measure distance between points in rigid geometry. Let us now remind three of the most popular metrics in $\mathbb{R}^n$, for any positive integer n.

Definition 4 (*cf.* [1]) For any $x = (x_1, \dots, x_n), y = (y_1, \dots, y_n) \in \mathbb{R}^n$, let us define:

1. **Euclidean metric**:

$$d_2(x, y) = \sqrt{\sum_{i=1}^{n} (x_i - y_i)^2}$$

2. **Manhattan (Hamming) metric**:

$$d_1(x, y) = \sum_{i=1}^{n} |x_i - y_i|$$

3. **Chebychev (max) metric**:

$$d_\infty(x, y) = \max\{|x_i - y_i| : i \in \overline{1, n}\}$$

Definition 5 Let (X, d) be a metric space. Then the d-open ball (briefly, open ball) of radius ε, centered in $x \in X$, is defined by:

$$\mathcal{B}(x, d, \varepsilon) = \{y \mid y \in X \ \& \ d(x, y) < \varepsilon\}$$

and the corresponding closed ball is defined by:

$$\overline{\mathcal{B}}(x, d, \varepsilon) = \{y \mid y \in X \ \& \ d(x, y) \leq \varepsilon\}.$$

On Fig. 1 one can see a graphical representation of the balls in $\mathbb{R}^2$ with center the point $(0, 0)$ in respect to the three main metrics proposed in Definition 4.

Definition 6 (*cf.* [11]) Let us take $\mathbb{R}^2$ as universe and $\xi_1, \xi_2 > 0$, then the open (ξ_1, ξ_2)-**neighbourhood**, centered in $x = (x_1, x_2) \in \mathbb{R}^2$, is defined by:

$$\mathcal{B}(x, \xi) = \{y \mid y \in \mathbb{R}^2 \ \& \ |x_1 - y_1| < \xi_1 \ \& \ |x_2 - y_2| < \xi_2\},$$

and the corresponding closed (ξ_1, ξ_2)-**neighbourhood** is defined by:

$$\overline{\mathcal{B}}(x, \xi) = \{y \mid y \in \mathbb{R}^2 \ \& \ |x_1 - y_1| \leq \xi_1 \ \& \ |x_2 - y_2| \leq \xi_2\}.$$

2.2 Intuitionistic Fuzzy Sets and Metrics

We provide now some preliminary information about intuitionistic fuzzy sets (cf. [2, 3]) and the standard distance measures on them (cf. [6, 14, 17]).

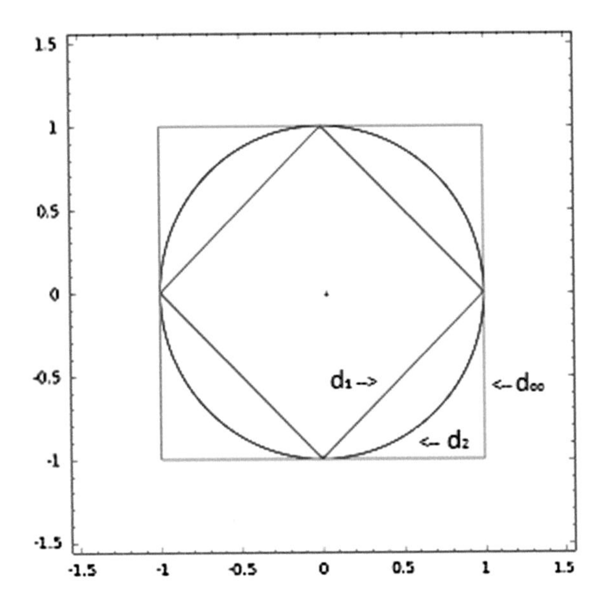

Fig. 1 The basis elements, i.e. the open balls of equal radius, in regard of the metrics $(\mathbb{R}^2, d_2)$, $(\mathbb{R}^2, d_1)$, $(\mathbb{R}^2, d_\infty)$, stated in Definition 4

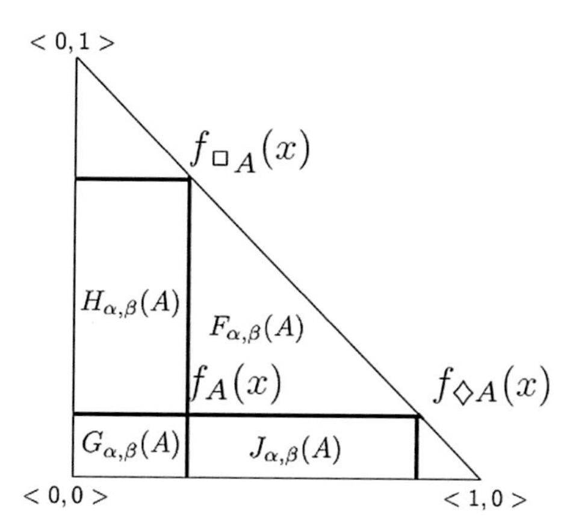

Fig. 2 Triangular representation of the intuitionistic fuzzy set $A \in \mathrm{IFS}(X)$ in a particular point $x \in X$, where $f_A(x)$ stands for the point on the *plane* with coordinates $(\mu_A(x), \nu_A(x))$. $\square A$ and $\Diamond A$ stand for the two modal operators "necessity" and "possibility" acting on A and the others are the extended modal operators

Let us consider a fixed set X as universe. An IFS A in the universe X is an object of the form (see, e.g., [2, 3]):

$$A = \{\langle x, \mu_A(x), \nu_A(x)\rangle | x \in X\}, \tag{1}$$

where the functions $\mu_A : X \to [0, 1]$ and $\nu_A : X \to [0, 1]$ define the degree of membership and the degree of non-membership of the element $x \in X$, respectively, and for every $x \in X$:

$$0 \leq \mu_A(x) + \nu_A(x) \leq 1. \tag{2}$$

Let us denote $f_A = (\mu_A, \nu_A)$ and $f_A(x) = (\mu_A(x), \nu_A(x))$, respectively. There are a few graphical representations of IFSs, we will employ the triangular representation shown on Fig. 2.

An additional concept for each IFS in X, that is an obvious result of (1) and (2), is called

$$\pi_A(x) = 1 - \mu_A(x) - \nu_A(x)$$

a *degree of uncertainty* of $x \in A$. It expresses a lack of knowledge of whether x belongs to A or not (cf. [2]). It is obvious that $0 \leq \pi_A(x) \leq 1$, for each $x \in X$. Uncertainty degree turns out to be relevant for both the applications and the development of theory of IFSs.

Distances between IFSs are calculated in the literature in two ways, using two parameters only (e.g., [2, 3, 6, 14]) or all three parameters ([8, 13–17]) describing the elements belonging to the sets. Both ways correctly define a metric space, that is, the three axioms for distance (non-negativity, symmetry and triangular inequality) are satisfied. Comparing the results obtained by the two different ways, one cannot say that both are equal. In [15–17], it is shown why in the calculation of distances between IFSs one should prefer all three parameters describing IFSs. Examples of the distances between any two IFSs A and B in $X = \{x_1, x_2, \ldots, x_n\}$ while using three parameter representation can be found in [15–17]. A *normalized distance* or *normalized metric* d in X is a metric such that $d: X \times X \rightarrow [0, 1] \subset \mathbb{R}_{\geq 0}$. Sometimes it is more convenient and easier to work with normalized metrics. Every metric can be normalized (cf. [1]). For a more generalized notion of distances between IFSs (the so called Modified weighted Hausdorff distance—MWHD) the reader may refer to [10].

Let us give the definition of the main metrics in IFSs.

- the **Hamming distance** (Szmidt and Kacprzyk form):

$$l_{3,\mathrm{IFS}}(A, B) = \sum_{i=1}^{n}(|\mu_A(x_i) - \mu_B(x_i)| + |\nu_A(x_i) - \nu_B(x_i)| + |\pi_A(x_i) - \pi_B(x_i)|)$$

- the **Euclidean distance** (Szmidt and Kacprzyk form):

$$e_{3,\mathrm{IFS}}(A, B) = (\sum_{i=1}^{n}(\mu_A(x_i) - \mu_B(x_i))^2 + (\nu_A(x_i) - \nu_B(x_i))^2 + (\pi_A(x_i) - \pi_B(x_i))^2)^{\frac{1}{2}}$$

Both distances, corresponding to the two former, are from the interval [0,1]. That is they are their corresponding normalized distances.

- the **normalized Hamming distance**:

$$L_{3,\mathrm{IFS}}(A, B) = \frac{1}{2n} l_{3,\mathrm{IFS}}(A, B)$$

- the **normalized Euclidean distance**:

$$E_{3,\mathrm{IFS}}(A, B) = \sqrt{\frac{1}{2n}} e_{3,\mathrm{IFS}}(A, B) \quad (3)$$

The counterparts of the above distances while using the first two parameter representation of IFSs are given, e.g., in [2]:

- the **Hamming distance**:

$$l_{2,\mathrm{IFS}}(A, B) = \sum_{i=1}^{n}(|\mu_A(x_i) - \mu_B(x_i)| + |\nu_A(x_i) - \nu_B(x_i)|)$$

- the **Euclidean distance**:

$$e_{2,\mathrm{IFS}}(A, B) = (\sum_{i=1}^{n}(\mu_A(x_i) - \mu_B(x_i))^2 + (\nu_A(x_i) - \nu_B(x_i))^2)^{\frac{1}{2}}$$

The normalized distances of the above two are stated as follows.

- the **normalized Hamming distance**:

$$L_{2,\mathrm{IFS}}(A, B) = \frac{1}{2n} l_{2,\mathrm{IFS}}(A, B)$$

- the **normalized Euclidean distance**:

$$E_{2,\mathrm{IFS}}(A, B) = \sqrt{\frac{1}{2n} e_{2,\mathrm{IFS}}(A, B)}$$

Further we will briefly remind the definition of intuitionistic fuzzy neighbourhood (IF-neighbourhood).

Definition 7 (*cf.* [11]) Let us be given a universe set X, $A, B \in \mathrm{IFS}(X)$ and a real number $\varepsilon \in [0, 1]$. We shall say that:

- $\mathcal{B}_{\mathrm{IF}}(A, d_j, \varepsilon) = \{B \mid B \in \mathrm{IFS}(X) \ \& \ (\forall x \in X)(d_j(f_A(x), f_B(x)) < \varepsilon)\}$ is **open IF-ball** with center A and radius ε (see Definition 5).
- $\overline{\mathcal{B}_{\mathrm{IF}}}(A, d_j, \varepsilon) = \{B \mid B \in \mathrm{IFS}(X) \ \& \ (\forall x \in X)(d_j(f_A(x), f_B(x)) \leq \varepsilon)\}$ is **closed IF-ball** with center A and radius ε (see Definition 5).

Analogically let us define the open and closed IF-neighbourhoods.

Definition 8 Let us be given a universe set X, $A, B \in \mathrm{IFS}(X)$ and $\xi, \eta \in [0, 1]$.

- $\mathcal{B}_{\mathrm{IF}}(A, \xi, \eta) = \{B \mid B \in \mathrm{IFS}(X) \ \& \ (\forall x \in X)(f_A(x) \in \mathcal{B}(f_B(x), \xi, \eta))\}$ is **open** (ξ, η)-**IF-neighbourhood** of A (see Definition 6)
- $\overline{\mathcal{B}_{\mathrm{IF}}}(A, \xi, \eta) = \{B \mid B \in \mathrm{IFS}(X) \ \& \ (\forall x \in X)(f_A(x) \in \overline{\mathcal{B}}(f_B(x), \xi, \eta))\}$ is **closed IF-**(ξ, η)**-neighbourhood** of A (see Definition 6)

Now having recalled all basic definitions and notions we are ready to introduce the new concept of separability.

3 Separability of Intuitionistic Fuzzy Sets

In general topology, one says that a and b, where Y is any topological space and a and b—elements or subsets of Y, are separable iff there are two open and disjoint subsets A and B of Y, containing a and b, respectively (cf. [9, 12]). Moreover, if we take the base of the metrical topology (Y, d) as the open d-balls, it is clear that every two elements a and b of Y are separable. If $d(a, b) > \varepsilon$ for a small enough $\varepsilon > 0$ it is enough to take $A = \mathcal{B}(a, d, \frac{\varepsilon}{2})$ and $B = \mathcal{B}(b, d, \frac{\varepsilon}{2})$, which provides that $A \cap B = \emptyset$. Therefore, one can say that a and b are (ε, d)-separable iff $b \notin \mathcal{B}(a, \varepsilon, d)$.

Let us now introduce a new concept of separability, which turns out to be more general than the standard definition in the theory of metric spaces.

Definition 9 For any $\varepsilon, \xi, \eta \in (0, 1]$ such that $\xi + \eta < 1$, we say that A and $B \in$ IFS(X) are:

1. (ε, d_j)**-IF-separable** ($j \in \{0, 1, \infty\}$) in $X_1 \subseteq X$ iff $B \in C_{\text{IF}}^{X_1}(A, d_j, \varepsilon)$, where

$$C_{\text{IF}}^{X_1}(A, d_j, \varepsilon) = \{D \mid D \in \text{IFS}(X) \; \& \; (\forall x \in X_1)$$

$(d_j(f_A(x), f_D(x)) \geq \varepsilon)\}$ is a subfamily of IFS(X).

2. $(\xi - \eta)$**-IF-separable** in $X_1 \subseteq X$ iff $B \in C_{\text{IF}}^{X_1}(A, \xi, \eta)$, where

$$C_{\text{IF}}^{X_1}(A, \xi, \eta) = \{D \mid D \in \text{IFS}(X) \; \& \; (\forall x \in X_1)$$

$(|\mu_A(x) - \mu_D(x)| \geq \xi \; \& \; |\nu_A(x) - \nu_D(x)| \geq \eta)\}$ is a subfamily of IFS(X).

Taking the standard concept of separability stated in the beginning of this section and the IF-neighbourhood (see Definition 7), one would say that A and $B \in$ IFS(X) are (ε, d_j)**-IF-separable** iff

$$B \notin \mathcal{B}_{\text{IF}}(A, \varepsilon, d_j).$$

From Definition 7 it follows that A and $B \in$ IFS(X) are (ε, d_j)**-IF-separable** exactly when there exists at least one point $x_0 \in X$ such that $d_j(f_A(x), f_B(x)) \geq \varepsilon$, i.e.

$$B \in C_{\text{IF}}^{\{x_0\}}(A, d_j, \varepsilon).$$

Analogically can be stated the $(\xi - \eta)$**-IF-separability**. On the other hand, we will say that,

Definition 10 A and $B \in$ IFS(X) are **completely** (ε, d_j)**-IF-separable** iff

$$B \in C_{\text{IF}}^{X}(A, d_j, \varepsilon). \tag{4}$$

Analogically, A and B are **completely** $(\xi - \eta)$**-IF-separable** iff

$$B \in C_{\text{IF}}^{X}(A, \xi, \eta). \tag{5}$$

For every IFS A a lot of operators are defined (see, e.g., [2, 3]), the most important of which, related to the applications of the present research, are ($\alpha, \beta \in [0, 1]$) the standard modal IF operators

$$\begin{array}{ll} A & = \{\langle x, \mu_A(x), 1 - \mu_A(x)\rangle | x \in X\}; \\ \Diamond A & = \{\langle x, 1 - \nu_A(x), \nu_A(x)\rangle | x \in X\}; \\ F_{\alpha,\beta}(A) & = \{\langle x, \mu_A(x) + \alpha\pi_A(x), \nu_A(x) + \beta\pi_A(x)\rangle \mid x \in X\}, \\ & \quad \text{where } \alpha + \beta \leq 1; \\ G_{\alpha,\beta}(A) & = \{\langle x, \alpha\mu_A(x), \beta\nu_A(x)\rangle \mid x \in X\}; \\ H_{\alpha,\beta}(A) & = \{\langle x, \alpha\mu_A(x), \nu_A(x) + \beta\pi_A(x)\rangle \mid x \in X\}; \end{array}$$

$$\begin{aligned}
H^*_{\alpha,\beta}(A) &= \{\langle x, \alpha\mu_A(x), \nu_A(x) + \beta(1 - \alpha\mu_A(x) - \nu_A(x))\rangle \mid x \in X\};\\
J_{\alpha,\beta}(A) &= \{\langle x, \mu_A(x) + \alpha\pi_A(x), \beta\nu_A(x)\rangle \mid x \in X\};\\
J^*_{\alpha,\beta}(A) &= \{\langle x, \mu_A(x) + \alpha(1 - \mu_A(x) - \beta\nu_A(x)), \beta\nu_A(x)\rangle \mid x \in X\};\\
C(A) &= \{\langle x, K_A, L_A\rangle | x \in X\};\\
I(A) &= \{\langle x, k_A, l_A\rangle | x \in X\}.
\end{aligned}$$

where

$$K_A = \sup_{y\in X} \mu_A(y), L_A = \inf_{y\in X} \nu_A(y)$$

$$k_A = \inf_{y\in X} \mu_A(y), l_A = \sup_{y\in X} \nu_A(y)$$

On Fig. 2 one may see the regions of operation of the extended modal operators.

Let us take,

$$P_A = \sup_{y\in X} \pi_A(y), p_A = \inf_{y\in X} \pi_A(y).$$

The following theorems about extended modal operators hold.

Theorem 2 *For every IFS A and for every $\xi, \eta \in [0, 1]$ we have that A is completely (ξ, η)-IF-separable with:*

1. *$F_{\alpha,\beta}(A)$ iff*

$$1 \geq \alpha \geq \frac{\xi}{p_A} \text{ and } 1 \geq \beta \geq \frac{\eta}{p_A},$$

where $\alpha + \beta \leq 1$;

2. *$G_{\alpha,\beta}(A)$ iff*

$$\max(0, 1 - \frac{\xi}{k_A}) \geq \alpha \geq 0 \text{ and } \max(0, 1 - \frac{\eta}{L_A}) \geq \beta \geq 0;$$

3. *$H_{\alpha,\beta}(A)$ iff*

$$\max(0, 1 - \frac{\xi}{k_A}) \geq \alpha \geq 0 \text{ and } 1 \geq \beta \geq \frac{\eta}{p_A};$$

4. *$J_{\alpha,\beta}(A)$ iff*

$$1 \geq \alpha \geq \frac{\xi}{p_A} \text{ and } \max(0, 1 - \frac{\eta}{L_A}) \geq \beta \geq 1;$$

5. *$H^*_{\alpha,\beta}(A)$ iff*

$$\max(0, 1 - \frac{\xi}{k_A}) \geq \alpha \geq 0 \text{ and } 0 \leq \beta \leq \min(1, \frac{\eta}{1 - \alpha\mu_A(x^*) - \nu_A(x^*)});$$

6. *$J^*_{\alpha,\beta}(A)$ iff*

$$0 \leq \alpha \leq \min(1, \frac{\xi}{1-\mu_A(x^*)-\beta\nu_A(x^*)}) \text{ and } \max(0, 1-\frac{\eta}{L_A}) \leq \beta \leq 1,$$

where $x^* \in X$ *satisfies the equality* $\mu_A(x^*) + \nu_A(x^*) = 1 - p_A$.

Proof Let us prove only (1) since the proofs of the next points are similar. Suppose that A and $F_{\alpha,\beta}(A)$ are completely (ξ, η)-IF-separable. From the definition of $F_{\alpha,\beta}(A)$ it follows that for every $x \in X$: $\alpha\pi_A(x) \geq \xi$ and $\beta\pi_A(x) \geq \eta$, which is satisfied iff $\alpha p_A \geq \xi$ and $\beta p_A \geq \eta$. Therefore, A and $F_{\alpha,\beta}(A)$ are completely (ξ, η)-IF-separable iff $1 \geq \alpha \geq \frac{\xi}{p_A}$ and $1 \geq \beta \geq \frac{\eta}{p_A}$, which completes the proof of (1).

Finally, let us introduce the concept of total separability with the following definition.

Definition 11 Let us take again A and $B \in \text{IFS}(X)$. We will call the two IFSs:

1. **totally** (ξ, η)**-IF-separable** or briefly $\overline{(\xi, \eta)}$*-separable* (see Fig. 3) iff

$$\begin{cases} K_A < k_B \text{ and } k_B - K_A \geq \xi \\ l_B < L_A \text{ and } L_A - l_B \geq \eta. \end{cases} \tag{6}$$

or

$$\begin{cases} K_B < k_A \text{ and } k_A - K_B \geq \xi \\ l_A < L_B \text{ and } L_B - l_A \geq \eta. \end{cases} \tag{7}$$

2. **totally** $\pi - (\xi, \eta)$**-IF-separable** or briefly $\overline{(\xi, \eta)} - \pi$**-separable** (see Fig. 4) iff

$$\begin{cases} K_A < k_B \text{ and } k_B - K_A \geq \xi \\ l_A < L_B \text{ and } L_B - l_A \geq \eta. \end{cases} \tag{8}$$

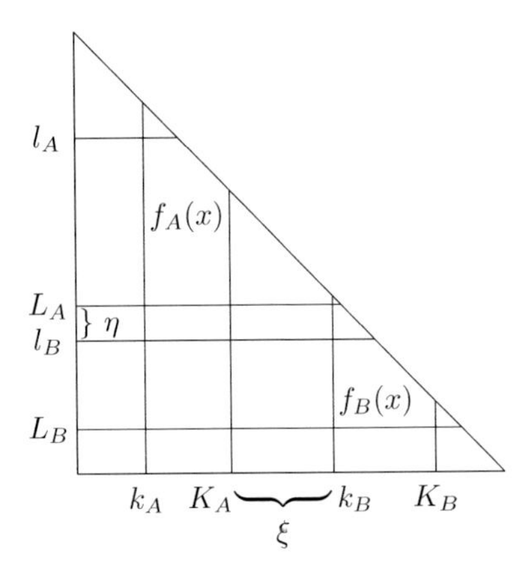

Fig. 3 The IFSs A and B are totally (ξ, η)-IF-separable

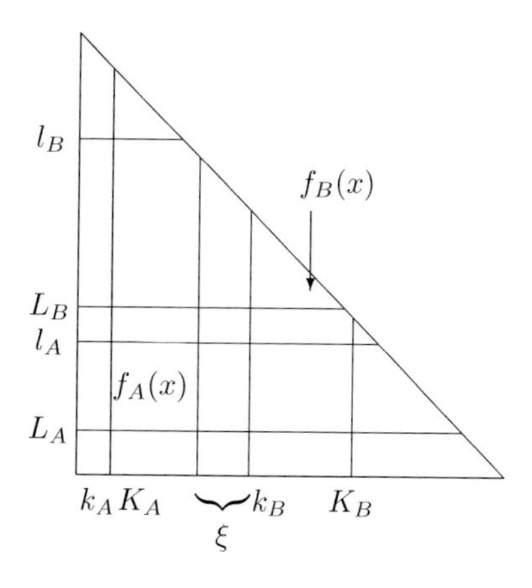

Fig. 4 The IFSs A and B are totally $\pi - (\xi, \eta)$-IF-separable

or

$$\begin{cases} K_B < k_A \text{ and } k_A - K_B \geq \xi \\ l_B < L_A \text{ and } L_A - l_B \geq \eta. \end{cases} \tag{9}$$

From Definition 11 easily follows the next theorem.

Theorem 3 *Let again A and B* $\in$ *IFS(X).*

1. *If A and B are* $\overline{(\xi, \eta)}$*-separable then they are completely* (ξ, η)*-IF-separable. That is (see (5))*
$$B \in C^X_{IF}(A, \xi, \eta).$$

2. *If A and B are* $\underline{(\xi, \eta)}$*-separable then they are completely* (ξ, η)*-IF-separable. That is (see (5))*
$$B \in C^X_{IF}(A, \xi, \eta).$$

Proof Let us prove only (1) since the proof of (2) is similar. If A and B are $\overline{(\xi, \eta)}$-separable let us take any point x from the universe X. From the definition of $\overline{K_A, k_A}$ and l_A, L_A and Definition 11 it follows that $|\mu_B(x) - \mu_A(x)| \geq k_B - K_A \geq \xi$ and $|\nu_B(x) - \nu_A(x)| \geq L_A - l_B \geq \eta$, which holds for any x form the universe. Therefore, A and B are completely (ξ, η)-IF-separable. The theorem is proved.

The proof of the next theorem is left as an exercise for the reader.

Theorem 4 *If A and B are* $\overline{(\xi, \eta)}$*-separable, then* $C(A) \lneq I(B)$ *or* $C(B) \lneq I(A)$*. On the other hand, if* $C(A) \lneq I(B)$*, then there are* $\xi, \eta \in (0, 1]$ *such that A and B are* $\underline{(\xi, \eta)}$*-separable.*

4 Conclusion

In summary, relaying on the notions of IF-neighbourhoods and IF-balls defined in [11], we have introduced the concepts of IF-separability in different points of the universe X. The weakest form of separability between A and $B \in \mathrm{IFS}(X)$ in the standard sense happens when we have separability in at least one point of the universe X. Whereas a stronger form, i.e. separability in all the points of X, corresponds to the newly introduced concept of complete IF-separability. Lastly we have defined the notion of total IF-separability which turns out to be the strongest form of separability proposed in this paper. All the forms of separability introduced in this paper help us to measure how far or "distinct" two IFSs from each other are in a more sensitive way compared to using only the standard properties of metric spaces (that is employing only the three axioms for distances). Many relations between them and the extended modal operators have also been stated as application to the theory proposed in this work.

In a next research of the authors, the introduced concept of IF-separabilities will be used for extending of some procedures related to InterCriteria Analysis (see [4, 5]). For example, the discussed in the paper distances will be used for determining of the degrees of consonance and dissonance between two criteria.

Acknowledgments The authors are grateful for the support provided by Grant DFNI-I-02-5 "InterCriteria Analysis—A New Approach to Decision Making" of the Bulgarian National Science Fund.

References

1. Adams, C., Franzosa, R.: Introduction to Topology—Pure and Applied. Pearson Prentice Hall, USA (2008)
2. Atanassov, K.: Intuitionistic Fuzzy Sets: Theory and Applications. Springer, Heidelberg (1999)
3. Atanassov, K.: On Intuitionistic Fuzzy Sets Theory. Springer, Berlin (2012)
4. Atanassov, K., Mavrov, D., Atanassova, V.: Intercriteria decision making: a new approach for multicriteria decision making, based on index matrices and intuitionistic fuzzy sets. In: Issues in Intuitionistic Fuzzy Sets and Generalized Nets, vol. 11, pp. 1–8 (2014)
5. Atanassova, V., Mavrov, D., Doukovska, L., Atanassov, K.: Discussion on the threshold values in the interCriteria decision making approach. Int. J. Notes Intuitionistic Fuzzy Sets **20**(2), 94–99 (2014)
6. Atanassov, K., Vassilev, P., Tcvetkov, R.: Intuitionistic Fuzzy Sets, Measures and Integrals. "Prof. M. Drinov" Academic Publishing House, Sofia (2013)
7. Ban, A.: Intuitionistic Fuzzy Measures. Theory and Applications. Nova Science Publishers, New York (2006)
8. Deng-Feng, L.: Multiattribute decision making models and methods using intuitionistic fuzzy sets. J. Comput. Syst. Sci. **70**, 73–85 (2005)
9. Kuratowski, K.: Topology, vol. 1. Academic Press, New York (1966)
10. Marinov, E., Szmidt, E., Kacprzyk, J., Tcvetkov, R.: A modified weighted Hausdorff distance between intuitionistic fuzzy sets. IEEE Int. Conf. Intell. Syst. IS'12. 138–141 (2012)
11. Marinov, E., Vassilev, P., Atanassov, K.: On intuitionistic fuzzy metric neighbourhoods, IFSA-EUSFLAT 2015 (submitted)
12. Munkres, J.: Topology, 2nd edn. Prentice Hall (2000)

13. Narukawa, Y., Torra, V.: Non-monotonic fuzzy measure and intuitionistic fuzzy set. LNAI **3885**, 150–160 (2006)
14. Szmidt, E.: Distances and similarities in intuitionistic fuzzy sets. In: Studies in Fuzziness and Soft Computing, vol. 307. Springer (2014)
15. Szmidt, E., Baldwin, J.: New similarity measure for intuitionistic fuzzy set theory and mass assignment theory. Notes Intuitionistic Fuzzy Sets **9**(3), 60–76 (2003)
16. Szmidt, E., Baldwin, J.: Entropy for intuitionistic fuzzy set theory and mass assignment theory. Notes Intuitionistic Fuzzy Sets **10**(3), 15–28 (2004)
17. Szmidt, E., Kacprzyk, J.: Distances between intuitionistic fuzzy sets. Fuzzy Sets Syst. **114**(3), 505–518 (2000)
18. Zadeh, L.: Fuzzy sets. Inf. Control **8**, 338–353 (1965)

On an Intuitionistic Fuzzy Probability Theory

Čunderlíková Katarína and Riečan Beloslav

Abstract We present some basic facts about a probability theory on IF-events. It is based on the Lukasiewicz operations and on the corresponding probability theory. We present a representation theorem originally published, as reported by Riečan (Soft Methodology and Random Information Systems, pp 243–248, [21]). We also show that the probability IF algebra can be embedded to a probability MV-algebra.

1 Probability Theory on a Family of IF-events

The first definition of probability on IF-events has been suggested in [7]. Consider a probability space $(\Omega, \mathcal{S}, P)$ and the family

$$\mathcal{F} = \{A = (\mu_A, \nu_A); \mu_A, \nu_A : \Omega \to [0,1] \; are \; \mathcal{S} - measurable, \mu_A + \nu_A \leq 1\}$$

Then they defined the probability $P(A)$ as a compact interval

$$P(A) = \left[\int_\Omega \mu_A \, dP, 1 - \int_\Omega \nu_A \, dP \right],$$

for $A \in \mathcal{F}$. Hence P is a mapping $P : \mathcal{F} \to \mathcal{J}$, where $\mathcal{J}$ is the family of all compact subintervals of the unit interval $[0, 1]$. They showed some properties of the mapping P, too.

In [20] an axiomatic definition of probability was suggested as a mapping $\mathcal{P} : \mathcal{F} \to \mathcal{J}$. We consider only the pair $(\Omega, \mathcal{S})$ without P, since the definition of $\mathcal{F}$ does

Č. Katarína
Faculty of Natural Sciences, Matej Bel University,
Tajovského 40, Banská Bystrica, Slovakia
e-mail: cunderlikova.lendelova@gmail.com

R. Beloslav (✉)
Mathematical Institute, Slovak Academy of Sciences,
Štefánikova 49, Bratislava, Slovakia
e-mail: beloslav.riecan@umb.sk

K.T. Atanassov et al. (eds.), *Novel Developments in Uncertainty Representation and Processing*, Advances in Intelligent Systems and Computing 401,
DOI 10.1007/978-3-319-26211-6_11

not depend on P. Recall that in $\mathcal{F}$ a partial ordering is used

$$A \leq B \Longleftrightarrow \mu_A \leq \mu_B, \nu_A \geq \nu_B,$$

i.e. A is less than B if the membership function of A is less than the membership function of B, and the nonmembership function of A is greater than the nonmembership function of B. With respect to this ordering the greatest element of $\mathcal{F}$ is $(\mathbf{1}, \mathbf{0})$ and the least element is $(\mathbf{0}, \mathbf{1})$, where $\mathbf{0}, \mathbf{1} : \Omega \to \mathcal{F}$ are the constant functions defined by $\mathbf{0}(\omega) = 0$, $\mathbf{1}(\omega) = 1$ for each $\omega \in \Omega$. Therefore in the axiomatic approach the probability of the maximal element is maximal

$$\mathcal{P}\big((\mathbf{1}, \mathbf{0})\big) = [1, 1] = \{1\}$$

and the probability of the minimal element is minimal

$$\mathcal{P}\big((\mathbf{0}, \mathbf{1})\big) = [0, 0] = \{0\}.$$

The classical additivity of the Kolmogorov probability measure

$$A \cap B = \emptyset \Longrightarrow P(A \cup B) = P(A) + P(B)$$

must be substituted by operations on IF-sets. In [20] the Lukasiewicz connectives have been used. If $a, b \in [0, 1]$ then

$$a \oplus_L b = \min(a + b, 1), a \odot_L b = \max(a + b - 1, 0).$$

Evidently these connectives corresponds to the usual union and intersection. If χ_A is the characteristic function of A (i.e. $\chi_A(\omega) = 1$, if $\omega \in A$ and $\chi_A(\omega) = 0$, if $\omega \notin A$), then

$$\chi_{A \cup B}(\omega) = \min(\chi_A(\omega) + \chi_B(\omega), 1),$$
$$\chi_{A \cap B}(\omega) = \max(\chi_A(\omega) + \chi_B(\omega) - 1, 0).$$

Extending the Lukasiewicz connectives to the IF-case, we can for each $A, B \in \mathcal{F}$ such that $A = (\mu_A, \nu_A), B = (\mu_B, \nu_B)$ define

$$A \oplus_L B = (\mu_A \oplus_L \mu_B, \nu_A \odot_L \nu_B),$$
$$A \odot_L B = (\mu_A \odot_L \mu_B, \nu_A \oplus_L \nu_B).$$

With respect to the operations the additivity means

$$A \odot_L B = (\mathbf{0}, \mathbf{1}) \Longrightarrow \mathcal{P}(A \oplus_L B) = \mathcal{P}(A) + \mathcal{P}(B)$$

Here $[a, b] + [c, d]$ is defined as $[a + c, b + d]$.

Finally continuity can be defined as follows:

$$A_n \nearrow A \Longrightarrow \mathcal{P}(A_n) \nearrow \mathcal{P}(A).$$

Of course, $A_n = (\mu_{A_n}, \nu_{A_n}) \nearrow A = (\mu_A, \nu_A)$ means (according to the ordering in $\mathcal{F}$) that

$$\mu_{A_n}(\omega) \nearrow \mu_A(\omega), \nu_{A_n}(\omega) \searrow \nu_A(\omega), \text{ for each } \omega \in \Omega.$$

On the other hand in $\mathcal{J}$

$$[a_n, b_n] \nearrow [a, b] \Longleftrightarrow a_n \nearrow a, b_n \nearrow b.$$

Let $\mathcal{P} : \mathcal{F} \to \mathcal{J}$ be a probability (i.e. an additive, a continuous mapping satisfying the boundary conditions). In [22] there was proved that there exists a probability $P : \mathcal{S} \to [0, 1]$ (in the Kolmogorov sense) and two numbers $\alpha, \beta \in [0, 1], \alpha \leq \beta$ such that for each $A = (\mu_A, \nu_A) \in \mathcal{F}$ there holds

$$\mathcal{P}(A) = \left[(1-\alpha)\int_\Omega \mu_A dP + \alpha\left(1 - \int_\Omega \nu_A dP\right), (1-\beta)\int_\Omega \mu_A dP + \beta\left(1 - \int_\Omega \nu_A dP\right)\right].$$

We can see that the Grzegorzewski—Mrówka definition [7] is a special case, when $\alpha = 0, \beta = 1$,

$$\mathcal{P}(A) = \left[\int_\Omega \mu_A dP, 1 - \int_\Omega \nu_A dP\right].$$

Similarly, if we put $\alpha = \beta = \frac{1}{2}$, then we obtain the probability

$$\mathcal{P}(A) = \left[\frac{1}{2}\int_\Omega \mu_A dP + \frac{1}{2}\left(1 - \int_\Omega \nu_A dP\right), \frac{1}{2}\int_\Omega \mu_A dP + \frac{1}{2}\left(1 - \int_\Omega \nu_A dP\right)\right] =$$
$$= \left\{\frac{1}{2}\int_\Omega \mu_A dP + \frac{1}{2}\left(1 - \int_\Omega \nu_A dP\right)\right\},$$

what is the definition by Gerstenkorn and Manko (see [6]).

Of course, in [9] another definition was suggested. Instead of the Lukasiewicz connectives to use the Gödel connectives

$$a \oplus_M b = \max(a, b) = a \vee b, a \odot_M b = \min(a, b) = a \wedge b,$$

hence for $A, B \in \mathcal{F}, A = (\mu_A, \nu_A), B = (\mu_B, \nu_B)$

$$A \oplus_M B = (\mu_A \vee \mu_B, \nu_A \wedge \nu_B),$$
$$A \odot_M B = (\mu_A \wedge \mu_B, \nu_A \vee \nu_B).$$

In this case the additivity of $\mathcal{P} : \mathcal{F} \to \mathcal{J}$ means that

$$A \odot_M B = (\mathbf{0}, \mathbf{1}) \Longrightarrow \mathcal{P}(A \oplus_M B) = \mathcal{P}(A) + \mathcal{P}(B).$$

Defining the continuity and the boundary conditions as before, we obtain again a good probability theory (see [25]).

Inspiring by these successes, two further definitions have been suggested in [2]:

$$A \oplus_Q B = \left(\sqrt{\mu_A^2 + \mu_B^2} \wedge 1, \left(1 - \sqrt{(1-\nu_A)^2 + (1-\nu_B)^2}\right) \wedge 1\right),$$
$$A \odot_Q B = \left((\mu_A + \mu_B - 1) \vee 0, (\nu_A + \nu_B) \wedge 1\right),$$

with the additivity

$$A \odot_Q B = (\mathbf{0}, \mathbf{1}) \Longrightarrow \mathcal{P}(A \oplus_Q B) = \mathcal{P}(A) + \mathcal{P}(B).$$

And

$$A \oplus_P B = (\mu_A + \mu_B - \mu_A \cdot \mu_B, \nu_A \cdot \nu_B),$$
$$A \odot_P B = (\mu_A \cdot \mu_B, \nu_A + \nu_B - \nu_A \cdot \nu_B),$$

with the additivity

$$A \odot_P B = (\mathbf{0}, \mathbf{1}) \Longrightarrow \mathcal{P}(A \oplus_P B) = \mathcal{P}(A) + \mathcal{P}(B).$$

Of course, in this paper we shall develop only the first theory.

2 Representation of L-Probability

Definition 1 By an L-probability on $\mathcal{F}$ we understand each function $\mathcal{P} : \mathcal{F} \to \mathcal{J}$ satisfying the following properties:

(i) $\mathcal{P}((\mathbf{1}, \mathbf{0})) = [1, 1] = 1$; $\mathcal{P}((\mathbf{0}, \mathbf{1})) = [0, 0] = 0$;
(ii) if $A \odot_L B = (\mathbf{0}, \mathbf{1})$ and $A, B \in \mathcal{F}$, then $\mathcal{P}(A \oplus_L B) = \mathcal{P}(A) + \mathcal{P}(B)$;
(iii) if $A_n \nearrow A$, then $\mathcal{P}(A_n) \nearrow \mathcal{P}(A)$.

Of course, each $\mathcal{P}(A)$ is an interval, denote it by $\mathcal{P}(A) = [\mathcal{P}^\flat(A), \mathcal{P}^\sharp(A)]$. By this way we obtain two real functions

$$\mathcal{P}^\flat : \mathcal{F} \to [0, 1], \mathcal{P}^\sharp : \mathcal{F} \to [0, 1]$$

and some properties of $\mathcal{P}$ can be characterized by some properties of $\mathcal{P}^\flat, \mathcal{P}^\sharp$.

Definition 2 By an L-state on $\mathcal{F}$ we understand each function $m : \mathcal{F} \to [0, 1]$ satisfying the following properties:

(i) $m((\mathbf{1}, \mathbf{0})) = 1$; $m((\mathbf{0}, \mathbf{1})) = 0$;
(ii) if $A \odot_L B = (\mathbf{0}, \mathbf{1})$ and $A, B \in \mathcal{F}$, then $m(A \oplus_L B) = m(A) + m(B)$;
(iii) if $A_n \nearrow A$, then $m(A_n) \nearrow m(A)$.

Theorem 1 Let $\mathcal{P} : \mathcal{F} \to \mathcal{J}$ and $\mathcal{P}(A) = [\mathcal{P}^\flat(A), \mathcal{P}^\sharp(A)]$ for each $A \in \mathcal{F}$. Then $\mathcal{P}$ is an L-probability if and only if $\mathcal{P}^\flat$ and $\mathcal{P}^\sharp$ are L-states.

Proof Let $\mathcal{P}$ be an L-probability, $A = (\mu_A, \nu_A), B = (\mu_B, \nu_B), A \odot_L B = (\mathbf{0}, \mathbf{1})$. Then

$$[\mathcal{P}^\flat(A \oplus_L B), \mathcal{P}^\sharp(A \oplus_L B)] = \mathcal{P}(A \oplus_L B) = \mathcal{P}(A) + \mathcal{P}(B) =$$
$$= [\mathcal{P}^\flat(A) + \mathcal{P}^\flat(B), \mathcal{P}^\sharp(A) + \mathcal{P}^\sharp(B)],$$

hence

$$\mathcal{P}^\flat(A \oplus_L B) = \mathcal{P}^\flat(A) + \mathcal{P}^\flat(B),$$
$$\mathcal{P}^\sharp(A \oplus_L B) = \mathcal{P}^\sharp(A) + \mathcal{P}^\sharp(B).$$

On the other hand, let $\mathcal{P}^\flat, \mathcal{P}^\sharp : \mathcal{F} \to [0, 1]$ be L-states, $A \odot_L B = (\mathbf{0}, \mathbf{1})$. Then

$$\mathcal{P}^\flat(A \oplus_L B) = \mathcal{P}^\flat(A) + \mathcal{P}^\flat(B),$$
$$\mathcal{P}^\sharp(A \oplus_L B) = \mathcal{P}^\sharp(A) + \mathcal{P}^\sharp(B).$$

hence also

$$\mathcal{P}(A \oplus_L B) = [\mathcal{P}^\flat(A \oplus_L B), \mathcal{P}^\sharp(A \oplus_L B)] = [\mathcal{P}^\flat(A) + \mathcal{P}^\flat(B), \mathcal{P}^\sharp(A) + \mathcal{P}^\sharp(B)] =$$
$$= [\mathcal{P}^\flat(A), \mathcal{P}^\sharp(A)] + [\mathcal{P}^\flat(B), \mathcal{P}^\sharp(B)] = \mathcal{P}(A) + \mathcal{P}(B),$$

hence $\mathcal{P}$ is L-additive.

Each L-probability has two special properties, whose we can use. First, we can represent it by integrals and secondly, we are able to embed L-probability theory to MV-algebra probability theory.

Theorem 2 To each L-state $m : \mathcal{F} \to [0, 1]$ there exists a probability measure $P : \mathcal{S} \to [0, 1]$ and a real number $\alpha \in [0, 1]$ such that for each $A = (\mu_A, \nu_A) \in \mathcal{F}$ there holds

$$m(A) = (1 - \alpha) \int_\Omega \mu_A dP + \alpha\left(1 - \int_\Omega \nu_A dP\right).$$

Proof With Respect to the Butnariu—Klement theorem [24] (Theorem 8.1.12) we shall find the value in the form $f(x, y)$, where $x = \int_\Omega \mu_A dP, y = \int_\Omega \nu_A dP$, and

$f : \triangle \to [0,1], \triangle = \{(u,v); u \geq 0, v \geq 0, u+v \leq 1\}$. From the boundary conditions we obtain

$$f(1,0) = f\left(\int_\Omega \mathbf{1}\, dP, \int_\Omega \mathbf{0}\, dP\right) = m(\mathbf{1},\mathbf{0}) = 1,$$
$$f(0,1) = f\left(\int_\Omega \mathbf{0}\, dP, \int_\Omega \mathbf{1}\, dP\right) = m(\mathbf{0},\mathbf{1}) = 0.$$

Now the additivity gives

$$\left(\frac{\mathbf{1}}{\mathbf{2}},\frac{\mathbf{1}}{\mathbf{2}}\right) \odot_L \left(\frac{\mathbf{1}}{\mathbf{2}},\frac{\mathbf{1}}{\mathbf{2}}\right) = (\mathbf{0},\mathbf{1}),$$
$$\left(\frac{\mathbf{1}}{\mathbf{2}},\frac{\mathbf{1}}{\mathbf{2}}\right) \oplus_L \left(\frac{\mathbf{1}}{\mathbf{2}},\frac{\mathbf{1}}{\mathbf{2}}\right) = (\mathbf{1},\mathbf{0}),$$

hence

$$f\left(\frac{1}{2},\frac{1}{2}\right) + f\left(\frac{1}{2},\frac{1}{2}\right) = f(1,0) = 1,$$
$$f\left(\frac{1}{2},\frac{1}{2}\right) = \frac{1}{2}.$$

Similarly

$$\left(\frac{\mathbf{1}}{\mathbf{4}},\frac{\mathbf{3}}{\mathbf{4}}\right) \odot_L \left(\frac{\mathbf{1}}{\mathbf{4}},\frac{\mathbf{3}}{\mathbf{4}}\right) = (\mathbf{0},\mathbf{1}),$$
$$\left(\frac{\mathbf{1}}{\mathbf{4}},\frac{\mathbf{3}}{\mathbf{4}}\right) \oplus_L \left(\frac{\mathbf{1}}{\mathbf{4}},\frac{\mathbf{3}}{\mathbf{4}}\right) = \left(\frac{\mathbf{1}}{\mathbf{2}},\frac{\mathbf{1}}{\mathbf{2}}\right),$$

hence

$$2f\left(\frac{1}{4},\frac{3}{4}\right) = \frac{1}{2},$$
$$f\left(\frac{1}{4},\frac{3}{4}\right) = \frac{1}{4}.$$

Analogously

$$\left(\frac{\mathbf{3}}{\mathbf{4}},\frac{\mathbf{1}}{\mathbf{4}}\right) \odot_L \left(\frac{\mathbf{1}}{\mathbf{4}},\frac{\mathbf{3}}{\mathbf{4}}\right) = (\mathbf{0},\mathbf{1}),$$
$$\left(\frac{\mathbf{3}}{\mathbf{4}},\frac{\mathbf{1}}{\mathbf{4}}\right) \oplus_L \left(\frac{\mathbf{1}}{\mathbf{4}},\frac{\mathbf{3}}{\mathbf{4}}\right) = (\mathbf{1},\mathbf{0}),$$

hence

$$f\left(\frac{3}{4},\frac{1}{4}\right)+f\left(\frac{1}{4},\frac{3}{4}\right)=1,$$
$$f\left(\frac{3}{4},\frac{1}{4}\right)=1-\frac{1}{4}=\frac{3}{4}.$$

Generally using additivity and continuity we can obtain the equality

$$f(x,1-x)=x, x\in[0,1].$$

Put $f(0,0)=\alpha$. By a similar procedure as before we can obtain

$$f(0,y)=\alpha(1-y), y\in[0,1].$$

Finally let $(x,y)\in\triangle$ be an arbitrary element, then

$$(x,1-x)\odot_L(0,x+y)=(0,1),$$
$$(x,1-x)\oplus_L(0,x+y)=(x,y).$$

Therefore

$$f(x,y)=f(x,1-x)+f(0,x+y)=x+\alpha(1-x-y)=$$
$$=(1-\alpha)x+\alpha(1-y),$$

hence

$$m(A)=(1-\alpha)\int_\Omega \mu_A\,dP+\alpha\left(1-\int_\Omega \nu_A\,dP\right).$$

Theorem 3 Let $\mathcal{P}:\mathcal{F}\to\mathcal{J}$ be an L-probability. Then there exists a probability measure $P:\mathcal{S}\to[0,1]$ and real numbers $\alpha,\beta\in[0,1]$ such that $\alpha\le\beta$ and for each $A=(\mu_A,\nu_A)\in\mathcal{F}$ there holds

$$\mathcal{P}(A)=\left[(1-\alpha)\int_\Omega \mu_A dP+\alpha\left(1-\int_\Omega \nu_A dP\right),(1-\beta)\int_\Omega \mu_A dP+\beta\left(1-\int_\Omega \nu_A dP\right)\right].$$

Proof Consider the boundary values $\mathcal{P}^\flat,\mathcal{P}^\sharp$ of $\mathcal{P}$. By Theorem 2

$$\mathcal{P}^\flat(A)=(1-\alpha)\int_\Omega \mu_A dP+\alpha\left(1-\int_\Omega \nu_A dP\right),$$
$$\mathcal{P}^\sharp(A)=(1-\beta)\int_\Omega \mu_A dP+\beta\left(1-\int_\Omega \nu_A dP\right).$$

Put $\int_\Omega \mu_A dP = u$, $\int_\Omega \nu_A dP = v$, evidently $u + v \leq 1$. Therefore

$$\begin{aligned}(\beta - \alpha)u &\leq (1 - \nu)(\beta - \alpha)\\(1 - \alpha)u - (1 - \beta)u &\leq (1 - \nu)(\beta - \alpha)\\(1 - \alpha)u + \alpha(1 - \nu) &\leq (1 - \beta)u + \beta(1 - \nu),\end{aligned}$$

hence $\mathcal{P}^\flat(A) \leq \mathcal{P}^\sharp(A)$, so that $\mathcal{P}(A) = [\mathcal{P}^\flat(A), \mathcal{P}^\sharp(A)]$ is actually an interval.

Theorem 4 Each L-probability $\mathcal{P}$ is strongly additive, i.e.

$$\mathcal{P}(A) + \mathcal{P}(B) = \mathcal{P}(A \oplus_L B) + \mathcal{P}(A \odot_L B)$$

for each $A, B \in \mathcal{F}$.

Proof It follows by Theorem 3 and the equality

$$a + b = a \oplus_L b + a \odot_L b$$

holding for each real numbers a, b.

3 Embedding

To embed $\mathcal{F}$ to a suitable MV-algebra it seems to be more convenient to use the definition of an MV-algebra by the help of an l-group.

Definition 3 By an ℓ-group we shall mean the structure $(G, +, \leq)$ such that the following properties are satisfied:

(i) $(G, +)$ is an Abelian group;
(ii) $(G, \leq)$ is a lattice;
(iii) $a \leq b \Longrightarrow a + c \leq b + c$.

For each ℓ-group G, an element $u \in G$ is said to be a strong unit of G, if for all $a \in G$ there is an integer $n \geq 1$ such that $nu \geq a$ (nu is the sum $u + \cdots + u$ with n) [23].

Example 1 Consider $G = R^2$,

$$\begin{aligned}(a, b) \hat{+} (c, d) &= (a + c, b + d - 1),\\(a, b) \leq (c, d) &\Longleftrightarrow a \leq c, b \geq d.\end{aligned}$$

then $(R^2, \hat{+}, \leq)$ is a lattice ordered group.

Evidently the operation $\hat{+}$ is commutative and associative, $(0, 1)$ is the neutral element, since

$$(0, 1) \hat{+} (a, b) = (a + 0, b + 1 - 1) = (a, b),$$

and $(-a, 2-b)$ is the inverse element, since

$$(a,b) \hat{+} (-a, 2-b) = (0,1).$$

Further $\leq$ is a partial order with

$$(a,b) \vee (c,d) = (\max(a,c), \min(b,d)),$$
$$(a,b) \wedge (c,d) = (\min(a,c), \max(b,d)).$$

Finally

$$(a,b) \leq (c,d) \Longrightarrow a \leq c, b \geq d,$$

hence

$$a+e \leq c+e,$$
$$b+f-1 \geq d+f-1,$$
$$(a,b) \hat{+} (e,f) = (a+e, b+f-1) \leq (c+e, d+f-1) = (c,d) \hat{+} (e,f).$$

Definition 4 ([23]) An MV-algebra is an algebraic system $(M, \oplus, \odot, \neg, 0, u)$, where $\oplus, \odot$ are binary operations, $\neg$ is a unary operation, $0, u$ are fixed elements, which can be obtained by the following way: there exists a lattice group

$$(G, +, \leq)$$

such that

$$M = \{x \in G; 0 \leq x \leq u\}$$

where 0 is the neutral element of G, u is a strong unit of G, and

$$a \oplus b = (a+b) \wedge u = \min(a+b, u),$$
$$a \odot b = (a+b-u) \vee 0 = \max(a+b-u, 0),$$
$$\neg a = u - a.$$

There $\vee, \wedge$ are the lattice operations with respect to the order and $-a$ is the opposite element of the element a with respect to the operation of the group.

Example 2 Let $([0,1], \oplus, \odot, \neg, 0, 1)$ be an MV-algebra, where $a \oplus b = \min(a+b, 1), a \odot b = \max(a+b-1, 0), \neg a = 1-a$. The corresponding group is $(R, +, \leq)$ where $+$ is usual sum, and $\leq$ is the usual ordering.

Example 3 Let $([0, 1]^2, \oplus, \odot, \neg, (0, 1), (1, 0))$ be an MV-algebra, where

$$\begin{aligned}(a, b) \oplus (c, d) &= (\min(a + c, 1), \max(b + d - 1, 0)),\\(a, b) \odot (c, d) &= (\max(a + c - 1, 0), \min(b + d, 1)),\\\neg(a, b) &= (1 - a, 1 - b).\end{aligned}$$

Here the corresponding group is $(R^2, \hat{+}, \leq)$ considered in Example 1.

Definition 5 Let $(M, \oplus, \odot, \neg, 0, u)$ is an MV-algebra. By an MV-probability on M we understand each function $P : M \to \mathcal{J}$ satisfying the following properties:

(i) $P(u) = [1, 1] = 1$; $P(0) = [0, 0] = 0$;
(ii) $a \odot b = 0 \Longrightarrow P(a \oplus b) = P(a) + P(b)$ for each $a, b \in M$;
(iii) if $a_n \nearrow a$, then $P(a_n) \nearrow P(a)$.

Definition 6 By an MV-state on M we consider each mapping $m : M \to [0, 1]$ satisfying the following conditions:

(i) $m(u) = 1, m(0) = 0$;
(ii) $a \odot b = 0 \Longrightarrow m(a \oplus b) = m(a) + m(b)$;
(iii) $a_n \nearrow a \Longrightarrow m(a_n) \nearrow m(a)$.

Theorem 5 If $(M, \oplus, \odot, \neg, 0, u)$ is an MV-algebra, $\mathcal{P} : M \to \mathcal{J}$ is a mapping, $\mathcal{P}(a) = [\mathcal{P}^\flat(a), \mathcal{P}^\sharp(a)]$, then $\mathcal{P}$ is an MV-probability if and only if $\mathcal{P}^\flat : M \to [0, 1]$, $\mathcal{P}^\sharp : M \to [0, 1]$ are MV-states.

Proof It is the same as in Theorem 1.

Theorem 6 Let $(\Omega, \mathcal{S})$ be a measurable space, $\mathcal{M}$ the family of all pairs $A = (\mu_A, \nu_A)$, where $\mu_A, \nu_A : \Omega \to [0, 1]$ are $\mathcal{S}$-measurable functions, $\mathbf{1} : \Omega \to 1$, $\mathbf{0} : \Omega \to 0$,

$$\begin{aligned}A \leq B &\Longleftrightarrow \mu_A \leq \mu_B, \nu_A \geq \nu_B,\\A \oplus_L B &= \big((\mu_A + \mu_B) \wedge \mathbf{1}, (\nu_A + \nu_B - 1) \vee \mathbf{0}\big),\\\neg A &= (1 - \mu_A, 1 - \nu_A).\end{aligned}$$

Then the system $(\mathcal{M}, \oplus_L, \odot_L, \neg, (\mathbf{0}, \mathbf{1}), (\mathbf{1}, \mathbf{0}))$ is an MV-algebra, $\mathcal{F} \subset \mathcal{M}$ and for each L-probability on $\mathcal{F}, \mathcal{P} : \mathcal{F} \to \mathcal{J}$ there exists exactly one probability $\bar{\mathcal{P}} : \mathcal{M} \to \mathcal{J}$ such that $\bar{\mathcal{P}} \mid \mathcal{F} = \mathcal{P}$.

Proof To see that $\mathcal{M}$ is an MV-algebra, consider the group $\mathcal{G}$ of all mappings from Ω to $(R^2, \hat{+}, \leq)$, $(\mu_A, \nu_A) \in \mathcal{G}$, μ_A, ν_A are $\mathcal{S}$-measurable (see *Example* 1), where

$$\begin{aligned}((\mu_A, \nu_A) \hat{+} (\mu_B, \nu_B))(\omega) &= (\mu_A(\omega), \nu_A(\omega)) \hat{+} (\mu_B(\omega), \nu_B(\omega)),\\(\mu_A, \nu_A) \leq (\mu_B, \nu_B) &\Longleftrightarrow \mu_A \leq \mu_B, \nu_A \geq \nu_B.\end{aligned}$$

Now

$$\mathcal{M} = \{(\mu_A, \nu_A); \mathbf{0} \le \mu_A \le \mathbf{1}, \mathbf{0} \le \nu_A \le \mathbf{1}\} =$$
$$= \{A; (\mathbf{0}, \mathbf{1}) \le A = (\mu_A, \nu_A) \le (\mathbf{1}, \mathbf{0})\}.$$

Evidently $\mathcal{F} \subset \mathcal{M}$. Let $\mathcal{P} : \mathcal{F} \to \mathcal{J}$ be a probability, $\mathcal{P}(A) = [\mathcal{P}^\flat(A), \mathcal{P}^\sharp(A)]$. Define

$$\bar{\mathcal{P}}^\flat\big((\mu_A, \nu_A)\big) = \mathcal{P}^\flat\big((\mu_A, 0)\big) - \mathcal{P}^\flat\big((0, 1 - \nu_A)\big),$$
$$\bar{\mathcal{P}}^\sharp\big((\mu_A, \nu_A)\big) = \mathcal{P}^\sharp\big((\mu_A, 0)\big) - \mathcal{P}^\sharp\big((0, 1 - \nu_A)\big),$$
$$\bar{\mathcal{P}}(A) = [\bar{\mathcal{P}}^\flat(A), \bar{\mathcal{P}}^\sharp(A)].$$

It is not difficult to prove that $\bar{\mathcal{P}}^\flat, \bar{\mathcal{P}}^\sharp : \mathcal{M} \to [0, 1]$ are MV-states, hence $\bar{\mathcal{P}} : \mathcal{M} \to \mathcal{J}$ is an MV-probability.

If $A = (\mu_A, \nu_A) \in \mathcal{F}$, then

$$(\mu_A, \nu_A) \odot_L (\mathbf{0}, \mathbf{1} - \nu_A) = (\mathbf{0}, \mathbf{1}),$$
$$(\mu_A, \nu_A) \oplus_L (\mathbf{0}, \mathbf{1} - \nu_A) = (\mu_A, \mathbf{0})$$

hence

$$\mathcal{P}\big((\mu_A, \mathbf{0})\big) = \mathcal{P}\big((\mu_A, \nu_A)\big) + \mathcal{P}\big((\mathbf{0}, \mathbf{1} - \nu_A)\big),$$
$$\mathcal{P}^\flat\big((\mu_A, \mathbf{0})\big) = \mathcal{P}^\flat\big((\mu_A, \nu_A)\big) + \mathcal{P}^\flat\big((\mathbf{0}, \mathbf{1} - \nu_A)\big),$$
$$\mathcal{P}^\sharp\big((\mu_A, \mathbf{0})\big) = \mathcal{P}^\sharp\big((\mu_A, \nu_A)\big) + \mathcal{P}^\sharp\big((\mathbf{0}, \mathbf{1} - \nu_A)\big).$$

Therefore

$$\mathcal{P}^\flat\big((\mu_A \nu_A)\big) = \mathcal{P}^\flat\big((\mu_A, \nu_A)\big) - \mathcal{P}^\flat\big((\mathbf{0}, \mathbf{1} - \nu_A)\big) = \bar{\mathcal{P}}^\flat\big((\mu_A, \nu_A)\big),$$
$$\mathcal{P}^\sharp\big((\mu_A \nu_A)\big) = \mathcal{P}^\sharp\big((\mu_A, \nu_A)\big) - \mathcal{P}^\sharp\big((\mathbf{0}, \mathbf{1} - \nu_A)\big) = \bar{\mathcal{P}}^\sharp\big((\mu_A, \nu_A)\big),$$

hence $\bar{\mathcal{P}}(A) = [\bar{\mathcal{P}}^\flat(A), \bar{\mathcal{P}}^\sharp(A)] = [\mathcal{P}^\flat(A), \mathcal{P}^\sharp(A)] = \mathcal{P}(A)$.

If $\mathcal{Q} : \mathcal{M} \to \mathcal{J}$ is a probability such that $\mathcal{Q}|\mathcal{F} = \mathcal{P}$ and $\mathcal{Q}(A) = (\mathcal{Q}^\flat(A), \mathcal{Q}^\sharp(A))$, then $\mathcal{P}^\flat, \mathcal{Q}^\flat : \mathcal{F} \to [0, 1]$ are L-states on $\mathcal{F}$. Therefore for each $A \in \mathcal{M}$

$$\mathcal{Q}^\flat((\mu_A, \nu_A)) = \mathcal{Q}^\flat((\mu_A, 0)) - \mathcal{Q}^\flat((0, 1 - \nu_A)) =$$
$$= \mathcal{P}^\flat((\mu_A, 0)) - \mathcal{P}^\flat((0, 1 - \nu_A)) = \bar{\mathcal{P}}^\flat(A)$$

hence $\mathcal{Q}^\flat = \bar{\mathcal{P}}^\flat$. Similarly $\mathcal{Q}^\sharp = \bar{\mathcal{P}}^\sharp$. Therefore

$$\mathcal{Q}(A) = (\mathcal{Q}^\flat(A), \mathcal{Q}^\sharp(A)) = (\bar{\mathcal{P}}^\flat(A), \bar{\mathcal{P}}^\sharp(A)) = \bar{\mathcal{P}}(A)$$

for each $A \in \mathcal{M}$.

Acknowledgments The support of the grant VEGA 1/0621/1 is kindly announced.

References

1. Atanassov, K.T.: Intuitionistic Fuzzy Sets: Theory and Applications. Studies in Fuzziness and Soft Computing. Physica-Verlag, Heidelberg (1999)
2. Atanassov, K.T., Riečan, B.: On two new types of probaility on IF-events. Notes IFS (2007)
3. Cignoli, L., D'Ottaviano, M., Mundici, D.: Algebraic Foundations of Many-valed Reasoning. Kluwer, Dordrecht (2000)
4. Foulis, D., Bennet, M.: Effect algebras and unsharp quantum logics. Found. Phys. **24**, 1325–1346 (1994)
5. Georgescu, G.: Bosbach states on fuzzy structures. Soft Comput. **8**, 217–230 (2004)
6. Gerstenkorn, T., Manko, J.: Probabilities of intuitionistic fuzzy events. In: Hryniewicz, O. et al. (ed.) Issues on Intelligent Systems Paradigms, pp. 63–68 (2005)
7. Grzegorzewski, P., Mrówka, E.: Probabilitty on intuitionistic fuzzy events. In: Grzegoruewsko, P. et al. (ed.) Soft Methods in Probability, Statistics and Data Analysis, pp. 105–115 (2002)
8. Kôpka, F., Chovanec, F.: D-posets. Math. Slovaca **44**, 21–34 (1994)
9. Krachounov M.: Intuitionistic probability and intuitionistic fuzzy sets. In: El-Darzi et al. (eds.) First International Workshop on IFS, pp. 714–717 (2006)
10. Lendelová, K.: Convergence of IF-observables. In: Issues in the Representation and Processing of Uncertain and Imprecise Information—Fuzzy Sets, Intuitionistic Fuzzy Sets, Generalzed Nets, and related Topics
11. Lendelová, K.: IF-probability on MV-algebras. Notes IFS **11**, 66–72 (2005)
12. Lendelová, K.: A note on invariant observables. Int. J. Theor. Physics **45**, 915–923 (2006)
13. Lendelová K.: Conditional IF-probability. In: Advances in Soft Computing Methods for Integrated Uncertainty Modeling, pp. 275–283 (2006)
14. Lendelová, K., Petrovičovā, J.: Representation of IF-probability for MV-algebras. Soft Comput. A **10**, 564–566 (2006)
15. Lendelová, K., Riečan, B.: Weak law of large numbers for IF-events. In: de Baets, B. et al. (ed.) Current Issues in Data and Knowledge Ebgineering, pp. 309–314 (2004)
16. Lendelová, K., Riečan, B.: Probability in triangle and square. In: Proceedings of the Eleventh International Conference IPMU, pp. 977–982. Paris (2006)
17. Lendelová, K., Riečan, B.: Strong law of large numbers for IF-events. In: Proceedings of the Eleventh International Conference IPMU, pp. 2363–2366. Paris (2006)
18. Mazúreková, Riečan B.: A measure ectension theorem. Notes on IFS **12**, 3–8 (2006)
19. Renčová, M., Riečan, B.: Probability on If sets: an elmentary approach. In: First International Workshop on IFS, Genralized Nets and Knowledge Systems, pp. 8–17 (2006)
20. Riečan, B.: A descriptive definition of probability on intutionistic fuzzy sets. In: Wagenecht, M., Hampet, R. (eds.) EUSFLAT'2003, pp. 263–266 (2003)
21. Riečan, B.: Representation of probabilities on IFS events. In: Dieaz, L. et al. (ed.) Soft Methodology and Random Information Systems, pp. 243–248 (2004)
22. Riečan, B.: On a problem of Radko Mesiar: general form of IF-probabilities. Fuzzy Sets Syst. **152**, 1485–1490 (2006)
23. Riečan, B., Mundici, D.: In: Pap, E. (ed.) Probability in MV-algebras. Handbook of Measure Thoery. Elsevier, Heidelberg (2002)
24. Riečan, B., Neubrunn, T.: Integral, Measure, and Ordering. Kluwer, Dordrecht (1997)
25. Riečan, B.: Probability theory on intuitionistic fuzzy events. A volume in Honour of Daniele Mundici's 60th Birthday Lecture Notes in Computer Science (2007)

Semi-properties of Atanassov Intuitionistic Fuzzy Relations

Urszula Bentkowska, Barbara Pękala, Humberto Bustince, Javier Fernandez and Edurne Barrenechea

Abstract In this paper properties of Atanassov intuitionistic fuzzy relations are examined, i.e.: semi-reflexivity, semi-irreflexivity, semi-symmetry, semi-connectedness, semi-asymmetry, semi-transitivity. The special attention is paid to the semi-transitivity property. Its characterization is given and connections with other transitivity properties are presented, i.e. transitivity itself and weak transitivity. Moreover, transformations of Atanassov intuitionistic fuzzy relations in the context of preservation of the given semi-properties of these relations are presented. The transformations that are considered: lattice operations, the converse, the complement, the composition of relations are the basic ones.

Keywords Atanassov intuitionistic fuzzy relations · Properties of interval-valued fuzzy relations · Weak transitivity · Semi-transitivity

---

U. Bentkowska (✉) · B. Pękala
Interdisciplinary Centre for Computational Modelling, University of Rzeszów,
Pigonia 1, 35-310 Rzeszów, Poland
e-mail: ududziak@ur.edu.pl

B. Pękala
e-mail: bpekala@ur.edu.pl

H. Bustince · J. Fernandez · E. Barrenechea
Departamento de Automatica y Computacion, Universidad Publica de Navarra,
Pamplona, Spain
e-mail: bustince@unavarra.es

J. Fernandez
e-mail: fcojavier.fernandez@unavarra.es

E. Barrenechea
e-mail: edurne.barrenechea@unavarra.es

K.T. Atanassov et al. (eds.), *Novel Developments in Uncertainty Representation and Processing*, Advances in Intelligent Systems and Computing 401,
DOI 10.1007/978-3-319-26211-6_12

1 Introduction

Atanassov intuitionistic fuzzy sets and relations (originally called intuitionistic fuzzy sets and relations, cf. [1]) are applied for example in group decision making [16], optimization problems, graph theory and neural networks (cf. [4]), medicine [9].

The concept of an Atanassov intuitionistic fuzzy set generalizes the concept of a fuzzy set introduced by Zadeh (cf. [18]). Namely, not only the degree of membership to a given set is considered but also the degree of non-membership to this set is taken into account in such way that the sum of both values is less than or equal to one. Atanassov intuitionistic fuzzy relations may have diverse types of properties (cf. [3, 5]) and there are many interesting problems to deal with in this area (cf. for example [6]).

In this paper we study the new class of properties, i.e. semi-reflexivity, semi-irreflexivity, semi-symmetry, semi-connectedness, semi-asymmetry, semi-transitivity and consider dependencies between semi-transitivity and other transitivity properties. Moreover, the problem of preservation of these properties by basic transformations is presented, especially we take into account the complement, the converse and composition of a relation and we also examine lattice operations. The regarded transformations are of the type $\mathfrak{F} : AIFR(X)^n \rightarrow AIFR(X), n \in \mathbb{N}$, where $AIFR(X)$ stands for the family of all Atanassov intuitionistic fuzzy relations described in a given set X. Semi-properties of Atanassov intuitionistic fuzzy relations may be important because of their possible applications for preference procedure which is of great interest nowadays (cf. for example [9, 14, 15, 17]).

2 Basic Notions

Now we recall some definitions which will be helpful in our investigations.

Definition 1 ([2]) Let $X \neq \emptyset$, $R, R^d : X \times X \rightarrow [0, 1]$ be fuzzy relations fulfilling condition

$$R(x, y) + R^d(x, y) \leqslant 1, \quad (x, y) \in (X \times X). \tag{1}$$

A pair $\rho = (R, R^d)$ is called an Atanassov intuitionistic fuzzy relation. The family of all Atanassov intuitionistic fuzzy relations in the set X is denoted by $AIFR(X)$.

The boundary elements in $AIFR(X)$ are $\mathbf{1} = (1, 0)$ and $\mathbf{0} = (0, 1)$, where $0, 1$ are the constant fuzzy relations. Basic operations for $\rho = (R, R^d), \sigma = (S, S^d) \in AIFR(X)$ are the union and the intersection, respectively

$$\rho \vee \sigma = (R \vee S, R^d \wedge S^d), \quad \rho \wedge \sigma = (R \wedge S, R^d \vee S^d). \tag{2}$$

Similarly, for arbitrary set $T \neq \emptyset$

$$(\bigvee_{t\in T} \rho_t)(x,y) = (\bigvee_{t\in T} R_t(x,y), \bigwedge_{t\in T} R_t^d(x,y)),$$

$$(\bigwedge_{t\in T} \rho_t)(x,y) = (\bigwedge_{t\in T} R_t(x,y), \bigvee_{t\in T} R_t^d(x,y)).$$

Moreover, the order is defined by

$$\rho \leqslant \sigma \Leftrightarrow (R \leqslant S,\ S^d \leqslant R^d). \tag{3}$$

If condition (3) will not be fulfilled we will write $\rho \nleq \sigma$ or $\rho > \sigma$. The pair $(AIFR(X), \leqslant)$ is a partially ordered set. Operations $\vee, \wedge$ are the binary supremum and infimum in the family $AIFR(X)$, respectively. Moreover, the family $(AIFR(X), \vee, \wedge)$ is a complete, distributive lattice.

Let us recall, useful in further considerations, definition of the composition and dual composition of fuzzy relations considered in the family $FR(X) = \{R | R : X \times X \to [0, 1]\}$ of all fuzzy relations in a given set $X \neq \emptyset$.

Definition 2 (*cf.* [18]) Let $R, S \in FR(X)$. The composition of fuzzy relations R and S is the fuzzy relation $(R \circ S) \in FR(X)$ such that

$$(R \circ S)(x,z) = \sup_{y\in X} \min(R(x,y), S(y,z)), \quad (x,z) \in X \times X. \tag{4}$$

The dual composition of fuzzy relations R and S is the fuzzy relation $(R \circ' S) \in FR(X)$ such that

$$(R \circ' S)(x,z) = \inf_{y\in X} \max(R(x,y), S(y,z)), \quad (x,z) \in X \times X. \tag{5}$$

Now, let us recall the notion of the composition and dual composition for Atanassov intuitionistic fuzzy relations

Definition 3 (*cf.* [4, 13]) Let $\rho = (R, R^d), \sigma = (S, S^d) \in AIFR(X)$. Then:

- the composition of relations ρ, σ is the relation

$$\rho \circ \sigma = (R \circ S, R^d \circ' S^d) \in AIFR(X),$$

- the dual composition of relations ρ, σ is the relation

$$\rho \circ' \sigma = (R \circ' S, R^d \circ S^d) \in AIFR(X),$$

where operations $\circ$ and $\circ'$ are described by the formulas (4) and (5).

For other operations on Atanassov intuitionistic fuzzy relations please see [8] and other types of compositions [5].

Definition 4 (*cf.* [2]) For arbitrary $\rho = (R, R^d) \in AIFR(X)$ the following relations are defined:

- the converse relation: $\rho^{-1} = (R^{-1}, (R^d)^{-1})$, where $R^{-1}(x, y) = R(y, x)$ and $(R^d)^{-1}(x, y) = R^d(y, x)$ for $x, y \in X$.
- the complement: $\rho' = (R^d, R)$.

3 Properties for Atanassov Intuitionistic Fuzzy Relations

3.1 Semi-properties

Now, we present semi-properties of Atanassov intuitionistic fuzzy relations. We follow the concept of such properties given by Drewniak (cf. [10]) for fuzzy relations but we restrict ourselves only to parameter $\alpha = 0.5$. This is why we will call these properties *semi-properties*.

Definition 5 ([11]) An Atanassov intuitionistic fuzzy relation $\rho = (R, R^d) \in AIFR(X)$ is called:

- semi-reflexive if

$$\forall_{x \in X} \ \rho(x, x) \geqslant (0.5, 0.5), \tag{6}$$

- semi-irreflexive if

$$\forall_{x \in X} \ \rho(x, x) \leqslant (0.5, 0.5), \tag{7}$$

- semi-symmetric if

$$\forall_{x,y \in X} \ \rho(x, y) \geqslant (0.5, 0.5) \Rightarrow \rho(y, x) = \rho(x, y), \tag{8}$$

- semi-asymmetric if

$$\forall_{x,y \in X} \ \rho(x, y) \wedge \rho(y, x) \leqslant (0.5, 0.5), \tag{9}$$

- semi-antisymmetric if

$$\forall_{x,y \in X, x \neq y} \ \rho(x, y) \wedge \rho(y, x) \leqslant (0.5, 0.5), \tag{10}$$

- totally semi-connected if

$$\forall_{x,y \in X} \ \rho(x, y) \vee \rho(y, x) \geqslant (0.5, 0.5), \tag{11}$$

- semi-connected if

$$\forall_{x,y\in X, x\neq y} \rho(x,y) \vee \rho(y,x) \geqslant (0.5, 0.5), \tag{12}$$

- semi-transitive if

$$\forall_{x,y,z\in X} \rho(x,y) \wedge \rho(y,z) \geqslant (0.5, 0.5) \Rightarrow \rho(x,z) \geqslant \rho(x,y) \wedge \rho(y,z). \tag{13}$$

Example 1 Let card $X = 3$, $\rho = (R, R^d), \sigma = (S, S^d) \in AIFR(X)$ be presented by matrices:

$$R = \begin{bmatrix} 0.5 & 0.2 & 0.9 \\ 0.7 & 0.5 & 0 \\ 0 & 0.9 & 0.5 \end{bmatrix}, \; R^d = \begin{bmatrix} 0.5 & 0.6 & 0 \\ 0.3 & 0.5 & 1 \\ 1 & 0 & 0.5 \end{bmatrix}, S = \begin{bmatrix} 0.6 & 0.7 & 0.3 \\ 0.7 & 0.5 & 0.3 \\ 0.1 & 0.4 & 1 \end{bmatrix}, \; S^d = \begin{bmatrix} 0.3 & 0.2 & 0.7 \\ 0.2 & 0.4 & 0.6 \\ 0.6 & 0.6 & 0 \end{bmatrix}.$$

ρ is semi-reflexive, semi-irreflexive, semi-asymmetric, semi-antisymmetric, totally semi-connected, semi-connected and σ is semi-symmetric. An example of a semi-transitive relation one may find in Example 2.

Theorem 1 ([11]) *Let $\rho = (R, R^d) \in AIFR(X)$ be an Atanassov intuitionistic fuzzy relation. Relation ρ is semi-transitive if and only if*

$$\forall_{x,z\in X} \rho^2(x,z) \geqslant (0.5, 0.5) \Rightarrow \rho(x,z) \geqslant \rho^2(x,z). \tag{14}$$

3.2 *Transitivity Properties for Atanassov Intuitionistic Fuzzy Relations*

Since transitivity is an important property for relations in decision making problems (it is one of the ways to guarantee the consistency of choices of decision makers [7]) we will compare presented in the previous section semi-transitivity with the other transitivity properties.

Definition 6 (*cf.* [17]) An Atanassov intuitionistic fuzzy relation $\rho = (R, R^d) \in AIFR(X)$ is:

- transitive, if

$$\forall_{x,y,z\in X} \rho(x,y) \wedge \rho(y,z) \leqslant \rho(x,z), \tag{15}$$

- weakly transitive, if

$$\forall_{x,y,z\in X} \rho(x,y) \geqslant (0.5, 0.5) \wedge \rho(y,z) \geqslant (0.5, 0.5) \Rightarrow \rho(x,z) \geqslant (0.5, 0.5). \tag{16}$$

Corollary 1 (cf. [3]) *An Atanassov intuitionistic fuzzy relation* $\rho = (R, R^d) \in AIFR(X)$ *is transitive, if and only if* $\rho \circ \rho \leqslant \rho$ ($\rho^2 \leqslant \rho$).

By definition of weak transitivity and that of composition, we can provide a characterization of weak transitivity.

Theorem 2 *Let* $\rho = (R, R^d) \in AIFR(X)$. ρ *is weakly transitive if and only if*

$$\forall_{x,z \in X} \; \rho^2(x,z) \geqslant (0.5, 0.5) \Rightarrow \rho(x,z) \geqslant (0.5, 0.5). \tag{17}$$

Proof If ρ is weakly-transitive, applying the tautologies for quantifiers we obtain

$$\forall_{x,y,z \in X} \; \min(R(x,y), R(y,z)) \geqslant 0.5 \Rightarrow R(x,z) \geqslant 0.5$$

and

$$\forall_{x,y,z \in X} \; \max(R^d(x,y), R^d(y,z)) \leqslant 0.5 \Rightarrow R^d(x,z) \leqslant 0.5.$$

As a result,

$$\forall_{x,z \in X} \; (\forall_{y \in X} \; \min(R(x,y), R(y,z)) \geqslant 0.5 \Rightarrow R(x,z) \geqslant 0.5)$$

and

$$\forall_{x,z \in X} \; (\forall_{y \in X} \; \max(R^d(x,y), R^d(y,z)) \leqslant 0.5 \Rightarrow R^d(x,z) \leqslant 0.5) .$$

This implies that

$$\forall_{x,z \in X} \; \sup_{y \in X} \min(R(x,y), R(y,z)) \geqslant 0.5 \Rightarrow R(x,z) \geqslant 0.5 \tag{18}$$

and

$$\forall_{x,z \in X} \; \inf_{y \in X} \max(R^d(x,y), R^d(y,z)) \leqslant 0.5 \Rightarrow R^d(x,z) \leqslant 0.5, \tag{19}$$

so by definition of compositions we get (17).

Let us assume that the condition in (17) is fulfilled, which is equivalent to those in (18) and (19). We will prove that ρ is weakly-transitive. Let $x, y, z \in X$ and $\rho(x,y) \geqslant (0.5, 0.5)$, $\rho(y,z) \geqslant (0.5, 0.5)$. As a result we have $R(x,y) \geqslant 0.5$, $R(y,z) \geqslant 0.5$ and $R^d(x,y) \leqslant 0.5$, $R^d(y,z) \leqslant 0.5$. We also have

$$\sup_{y \in X} \min(R(x,y), R(y,z)) \geqslant \min(R(x,y), R(y,z)) \geqslant 0.5$$

and

$$\inf_{y\in X} \max(R^d(x,y), R^d(y,z)) \leqslant \max(R^d(x,y), R^d(y,z)) \leqslant 0.5.$$

By (18), (19) we have $R(x,z) \geqslant 0.5$ and $R^d(x,z) \leqslant 0.5$, so $\rho(x,z) \geqslant (0.5, 0.5)$.

By the definition of order for Atanassov intuitionistic fuzzy relations the result follows. □

It is easy to see that if $\rho \in AIFR(X)$ is transitive, then it is weakly transitive and semi-transitive (it is enough to apply Corollary 1 and Theorems 1 and 2).

Example 2 Let card $X = 3$. The following relation $\rho = (R, R^d) \in AIFR(X)$

$$R = \begin{bmatrix} 0.6 & 0 & 0.5 \\ 0.4 & 0.3 & 0 \\ 0.2 & 0 & 0.4 \end{bmatrix}, \quad R^d = \begin{bmatrix} 0.4 & 1 & 0.5 \\ 0.6 & 0.7 & 1 \\ 0.8 & 1 & 0.6 \end{bmatrix},$$

is not transitive (see Corollary 1), but it is semi-transitive (see Theorem 1).

Note that semi-transitivity is a more restrictive (stronger) than weak transitivity.

Proposition 1 *Let $\rho \in AIFR(X)$. If ρ is semi-transitive, then it is weakly transitive.*

Example 3 Let card$(X) = 3$. The following $\rho = (R, R^d) \in AIFR(X)$ is weakly transitive (see Theorem 2), but it is not semi-transitive (so it is also not transitive), where $\sigma = \rho^2$, $\sigma = (S, S^d)$ and

$$R = \begin{bmatrix} 0.5 & 0.5 & 0.5 \\ 0.6 & 0.6 & 0.5 \\ 0.5 & 0.5 & 0.6 \end{bmatrix}, R^d = \begin{bmatrix} 0.4 & 0.5 & 0.3 \\ 0.2 & 0.4 & 0.5 \\ 0.5 & 0.5 & 0.4 \end{bmatrix}, S = \begin{bmatrix} 0.5 & 0.5 & 0.5 \\ 0.6 & 0.6 & 0.5 \\ 0.5 & 0.5 & 0.6 \end{bmatrix}, S^d = \begin{bmatrix} 0.4 & 0.5 & 0.4 \\ 0.4 & 0.4 & 0.3 \\ 0.5 & 0.5 & 0.4 \end{bmatrix}.$$

We see that $\sigma_{23} = (0.5, 0.3) \geqslant (0.5, 0.5)$ and $\rho_{23} = (0.5, 0.5) \ngeqslant \sigma_{23} = (0.5, 0.3)$ (cf. Theorem 1 and definition of order (3)).

4 Transformations and Semi-properties

Now, some transformations of Atanassov intuitionistic fuzzy relations having semi-properties will be considered (cf. [12]). Similar considerations for other properties of Atanassov intuitionistic fuzzy relations one may find in [3–5].

Proposition 2 *Let $\rho \in AIFR(X)$. ρ is semi-reflexive (semi-irreflexive) if and only if ρ' is semi-irreflexive (semi-reflexive). ρ is semi-asymmetric (semi-antisymmetric) if and only if ρ' is totally semi-connected (semi-connected). ρ is totally semi-connected (semi-connected) if and only if ρ' is semi-asymmetric (semi-antisymmetric).*

Example 4 Let card $X = 3$. We consider relation $\rho = (R, R^d) \in AIFR(X)$ from Example 2. This relation is semi-transitive but $\rho' = (W, W^d)$, is not semi-transitive because $s_{11} = 0.6 \geqslant 0.5$ and $w_{11} = 0.4 \ngeqslant s_{11}$ (see Theorem 1), where

$$W = \begin{bmatrix} 0.4 & 1 & 0.5 \\ 0.6 & 0.7 & 1 \\ 0.8 & 1 & 0.6 \end{bmatrix}, \quad W^d = \begin{bmatrix} 0.6 & 0 & 0.5 \\ 0.4 & 0.3 & 0 \\ 0.2 & 0 & 0.4 \end{bmatrix}, \quad S = W \circ W = \begin{bmatrix} 0.6 & 0.7 & 1 \\ 0.8 & 1 & 0.7 \\ 0.6 & 0.8 & 1 \end{bmatrix}.$$

Now, let card $X = 3$. We consider semi-symmetric relation σ from Example 1. Relation $\sigma' = (T, T^d)$, where

$$T = \begin{bmatrix} 0.3 & 0.2 & 0.7 \\ 0.2 & 0.4 & 0.6 \\ 0.6 & 0.6 & 0 \end{bmatrix}, T^d = \begin{bmatrix} 0.6 & 0.7 & 0.3 \\ 0.7 & 0.5 & 0.3 \\ 0.1 & 0.4 & 1 \end{bmatrix}$$

is not semi-symmetric because $t_{13} \geqslant 0.5$ but $t_{13} \neq t_{31}$.

Proposition 3 *Let $\rho \in AIFR(X)$. ρ is semi-reflexive (semi-irreflexive, semi-symmetric, semi-asymmetric, semi-antisymmetric, semi-connected, totally semi-connected, semi-transitive) if and only if ρ^{-1} is semi-reflexive (semi-irreflexive, semi-symmetric, semi-asymmetric, semi-antisymmetric, semi-connected, totally semi-connected, semi-transitive).*

Theorem 3 *Let $\rho, \sigma \in AIFR(X)$. If ρ is semi-irreflexive, then $\rho \wedge \sigma$ is semi-irreflexive. If ρ, σ are semi-reflexive (semi-symmetric, semi-asymmetric, semi-antisymmetric, semi-connected, totally semi-connected, semi-transitive), then $\rho \wedge \sigma$ is semi-reflexive (semi-symmetric, semi-asymmetric, semi-antisymmetric, semi-connected, totally semi-connected, semi-transitive).*

Proof Let $x, y \in X$, $\rho = (R, R^d), \sigma = (S, S^d) \in AIFR(X)$. If ρ is semi-irreflexive then $(R \wedge S)(x, x) = \min(R(x, x), S(x, x)) \leqslant \min(0.5, S(x, x)) \leqslant 0.5$ regardless of the value $S(x, x)$. Similarly $(R^d \vee S^d)(x, x) = \max(R(x, x), S(x, x)) \geqslant \max(0.5, S^d(x, x)) \geqslant 0.5$ for any value $S^d(x, x)$. As a result $(\rho \wedge \sigma)(x, x) \leqslant (0.5, 0.5)$, so $\rho \wedge \sigma$ is semi-irreflexive.

Now we will prove the property for semi-symmetry. If ρ and σ are semi-symmetric and $(\rho \wedge \sigma)(x, y) \geqslant (0.5, 0.5)$ then $\min(R(x, y), S(x, y)) \geqslant 0.5$ and $\max(R^d(x, y), S^d(x, y)) \leqslant 0.5$. Thus $R(x, y) \geqslant 0.5$, $S(x, y) \geqslant 0.5$, $R^d(x, y) \leqslant 0.5$ and $S^d(x, y) \leqslant 0.5$, so $\rho(x, y) \geqslant (0.5, 0.5)$ and $\sigma(x, y) \geqslant (0.5, 0.5)$. As a result $\rho(x, y) = \rho(y, x)$, $\sigma(x, y) = \sigma(y, x)$ and $(\rho \wedge \sigma)(x, y) = (\rho \wedge \sigma)(y, x)$, so $\rho \wedge \sigma$ is semi-symmetric. Other properties may be justified in a similar way. □

Example 5 Intersection of arbitrary two semi-connected and totally semi-connected Atanassov intuitionistic fuzzy relations need not be semi-connected, totally semi-connected, respectively. It follows from the fact that fuzzy relations representing the membership values of the given Atanassov intuitionistic fuzzy relation which is semi-connected, totally semi-connected need not be semi-connected, totally semi-connected, respectively (see [10], p. 78–79, where $\alpha = 0.5$).

Theorem 4 *Let $\rho, \sigma \in AIFR(X)$. If ρ is semi-reflexive, then $\rho \vee \sigma$ is semi-reflexive. If ρ, σ are semi-irreflexive (semi-symmetric, semi-asymmetric, semi-antisymmetric, semi-connected, totally semi-connected, semi-transitive), then $\rho \vee \sigma$ is semi-irreflexive (semi-symmetric, semi-asymmetric, semi-antisymmetric, semi-connected, totally semi-connected, semi-transitive).*

Proof We will prove only the property for semi-symmetry. Let $x, y \in X$, $\rho = (R, R^d)$, $\sigma = (S, S^d)$ be semi-symmetric and $(\rho \vee \sigma)(x, y) \geqslant (0.5, 0.5)$. Thus $\max(R(x, y), S(x, y)) \geqslant 0.5$ and $\min(R^d(x, y), S^d(x, y)) \leqslant 0.5$. There are four possible cases:

(1^0) $R(x, y) \geqslant S(x, y)$ and $R^d(x, y) \geqslant S^d(x, y)$,
(2^0) $R(x, y) \geqslant S(x, y)$ and $R^d(x, y) \leqslant S^d(x, y)$,
(3^0) $R(x, y) \leqslant S(x, y)$ and $R^d(x, y) \geqslant S^d(x, y)$,
(4^0) $R(x, y) \leqslant S(x, y)$ and $R^d(x, y) \leqslant S^d(x, y)$.

We will consider the first case and proof for the rest is analogous. >From semi-symmetry of ρ and σ it follows that $R(x, y) \geqslant 0.5 \Rightarrow R(x, y) = R(y, x)$, $R^d(x, y) \leqslant 0.5 \Rightarrow R^d(x, y) = R^d(y, x)$, $S(x, y) \geqslant 0.5 \Rightarrow S(x, y) = S(y, x)$, $S^d(x, y) \leqslant 0.5 \Rightarrow S^d(x, y) = S^d(y, x)$. Thus form the first case it follows that $R(x, y) \geqslant 0.5$ and $S^d(y, x) \leqslant 0.5$, so $R(x, y) = R(y, x)$ and $S^d(x, y) = S^d(y, x)$. We will show that $R(x, y) \geqslant S(y, x)$. Suppose that $R(x, y) < S(y, x)$. Then from assumptions of the first case we obtain $0.5 \leqslant R(x, y) < S(y, x)$ so from semi-symmetry of σ we have $S(x, y) = S(y, x)$. As a result $R(x, y) < S(x, y)$ which contradicts to assumptions of the first case. So $\max(R(x, y), S(x, y)) = R(x, y)$, $\max(R(y, x), S(y, x)) = \max(R(x, y), S(y, x)) = R(x, y)$ and this implies $(R \vee S)(x, y) = (R \vee S)(y, x)$. Similarly we can prove that $(R^d \wedge S^d)(x, y) = (R^d \wedge S^d)(y, x)$. As a result $(\rho \vee \sigma)(x, y) = (\rho \vee \sigma)(y, x)$, so $\rho \vee \sigma$ is symmetric. □

Example 6 Let card $X = 3$, $\rho = (R, R^d), \sigma = (S, S^d) \in AIFR(X)$ be presented by matrices

$$R = \begin{bmatrix} 0.7 & 0 & 0 \\ 0.8 & 0.9 & 0 \\ 0.8 & 0.9 & 0.8 \end{bmatrix}, \quad R^d = \begin{bmatrix} 0.2 & 1 & 1 \\ 0.1 & 0 & 1 \\ 0 & 0 & 0 \end{bmatrix}, \quad S = \begin{bmatrix} 0.7 & 0.8 & 0.8 \\ 0 & 0.9 & 0.9 \\ 0 & 0 & 0.8 \end{bmatrix}, \quad S^d = \begin{bmatrix} 0.2 & 0.1 & 0 \\ 1 & 0 & 0 \\ 1 & 1 & 0 \end{bmatrix}.$$

Relations ρ and σ are semi-transitive. Relation $\rho \vee \sigma = (T, T^d)$ is presented by the following matrices

$$T = R \vee S = \begin{bmatrix} 0.7 & 0.8 & 0.8 \\ 0.8 & 0.9 & 0.9 \\ 0.8 & 0.9 & 0.8 \end{bmatrix}, \quad T^d = R^d \wedge S^d = \begin{bmatrix} 0.2 & 0.1 & 0 \\ 0.1 & 0 & 0 \\ 0 & 0 & 0 \end{bmatrix}$$

and it is not semi-transitive. We can check it with the use of Theorem 1 where

$$W = T \circ T = \begin{bmatrix} 0.8 \; 0.8 \; 0.8 \\ 0.8 \; 0.9 \; 0.9 \\ 0.8 \; 0.8 \; 0.9 \end{bmatrix}, \quad T^d \circ' T^d = \begin{bmatrix} 0 \; 0 \; 0 \\ 0 \; 0 \; 0 \\ 0 \; 0 \; 0 \end{bmatrix}.$$

Relation $\rho \vee \sigma$ is not semi-transitive because $t_{11} \geqslant 0.5$ and $t_{11} \ngeqslant w_{11}$.

Example 7 Sum of any two semi-asymmetric and semi-antisymmetric Atanassov intuitionistic fuzzy relations need not be semi-asymmetric, semi-antisymmetric, respectively. It is the consequence of the fact that fuzzy relations representing the membership values of the given Atanassov intuitionistic fuzzy relation which is semi-asymmetric, semi-antisymmetric need not be semi-asymmetric, semi-antisymmetric, respectively ([10], p. 78–79, where $\alpha = 0.5$).

Theorem 5 *Let $\rho \in AIFR(X)$. If ρ is semi-reflexive (semi-irreflexive), then $\rho \circ \rho$ is semi-reflexive (semi-irreflexive).*

Proof Let $x \in X$, $\rho \in AIFR(X)$ be semi-reflexive. Thus $R(x,x) \geqslant 0.5$ and $R^d(x,x) \leqslant 0.5$, so $(R \circ R)(x,x) = \sup_{y \in X} \min(R(x,y), R(y,x)) \geqslant \sup_{y=x} \min(R(x,y), R(y,x)) \geqslant \min(0.5, 0.5) = 0.5$ and $(R^d \circ' R^d)(x,x) = \inf_{y \in X} \max(R^d(x,y), R^d(y,x)) \leqslant \inf_{y=x} \max (R^d(x,y), R^d(y,x)) \leqslant \max(0.5, 0.5)$. What is equal to 0.5, which means that $(\rho \circ \rho)(x,x) \geqslant (0.5, 0.5)$ and $\rho \circ \rho$ is semi-reflexive. Similarly we can prove the case of irreflexivity. □

The composition or dual composition of semi-asymmetric, semi-antisymmetric, semi-connected, totally semi-connected, semi-symmetric and semi-transitive relation $\rho \in AIFR(X)$ by itself is not semi-asymmetric, semi-antisymmetric, semi-connected, totally semi-connected, semi-symmetric, semi-transitive, respectively. We present one of the suitable examples.

Example 8 Let card $X = 3$, $\rho = (R, R^d) \in AIFR(X)$. Relation ρ is semi-asymmetric. However, the composition of this relation by itself is not semi-asymmetric because $\min(t_{13}, t_{31}) = 0.6 > 0.5$, where

$$\rho = \begin{bmatrix} (0.3, 0.6) \; (0.6, 0.4) \; (0.2, 0.5) \; (0.2, 0.7) \\ (0.3, 0.6) \; (0.2, 0.7) \; (0.6, 0.4) \; (0.2, 0.5) \\ (0.7, 0.2) \; (0.4, 0.5) \; (0.5, 0.5) \; (0.6, 0.3) \\ (0.8, 0.2) \; (0.7, 0.1) \; (0.4, 0.5) \; (0.4, 0.6) \end{bmatrix}, T = R \circ R = \begin{bmatrix} 0.3 \; 0.3 \; 0.6 \; 0.2 \\ 0.6 \; 0.4 \; 0.5 \; 0.6 \\ 0.6 \; 0.6 \; 0.5 \; 0.5 \\ 0.4 \; 0.6 \; 0.6 \; 0.4 \end{bmatrix}.$$

5 Conclusion

Semi-properties of Atanassov intuitionistic fuzzy relations were discussed. Especially, basic transformations $\mathfrak{F} : AIFR(X)^n \to AIFR(X)$, $n \in \mathbb{N}$, of this type of relations were considered. For the future work it may be interesting to consider other transformations, for example aggregations.

Acknowledgments This work was partially supported by the Centre for Innovation and Transfer of Natural Sciences and Engineering Knowledge in Rzeszów, through Project Number RPPK.01.03.00-18-001/10 and research project TIN2013-40765-P from the Spanish Government.

References

1. Atanassov, K.: Intuitionistic fuzzy relations. In: Proceedings of the Third International Symposium Automation and Science Instrumentation, Part II, Varna, 56–57 October 1984
2. Atanassov, K.: Intuitionistic fuzzy sets. Fuzzy Sets Syst. **20**, 87–96 (1986)
3. Burillo, P., Bustince, H.: Intuitionistic fuzzy relations (Part I). Mathw. Soft Comput. **2**, 5–38 (1995)
4. Burillo, P., Bustince, H.: Intuitionistic fuzzy relations (Part II). Effect of Atanassov's operators on the properties of the intuitionistic fuzzy relations. Mathw. Soft Comput. **2**, 117–148 (1995)
5. Burillo, P., Bustince, H.: Structures on intuitionistic fuzzy relations. Fuzzy Sets Syst. **78**, 293–303 (1996)
6. Bustince, H.: Construction of intuitionistic fuzzy relations with predetermined properties. Fuzzy Sets Syst. **109**, 379–403 (2000)
7. Chiclana, F., Herrera-Viedma, E., Alonso, S., Pereira, R.A.M.: Preferences and consistency issues in group decision making. In Bustince, H. et al. (eds.) Fuzzy Sets and Their Extensions: Representation, Aggregation and Models, pp. 219–237. Springer, Berlin (2008)
8. Deschrijver, G., Kerre, E.E.: A generalisation of operators on intuitionistic fuzzy sets using triangular norms and conorms. Notes IFS **8**, 19–27 (2002)
9. Deschrijver, G., Kerre, E.E.: On the composition of intuitionistic fuzzy relations. Fuzzy Sets Syst. **136**(3), 333–361 (2003)
10. Drewniak, J.: Fuzzy Relation Calculus. Silesian University, Katowice (1989)
11. Dudziak, U., Pękala, B.: Intuitionistic fuzzy preference relations. In: Galichet, S. et al. (eds.), Proceedings of the 7th conference of the European Society for Fuzzy Logic and Technology, pp. 529–536. Atlantis Press, Amsterdam (2011)
12. Dudziak, U.: Transformations of intuitionistic fuzzy relations. In: Atanassov, K.T. et al. (eds.) New Developments in Fuzzy Sets, Intuitionistic Fuzzy Sets, Generalized Nets and Related Topics, Foundations, vol. I, pp. 71–91. IBS PAN - SRI PAS, Warszawa (2012)
13. Goguen, A.: L-fuzzy sets. J. Math. Anal. Appl. **18**, 145–174 (1967)
14. Gong, Z.-W., Li, L.-S., Zhou, F.-X., Yao, T.-X.: Goal programming approaches to obtain the priority vectors from the intuitionistic fuzzy preference relations. Comput. Ind. Eng. **57**, 1187–1193 (2009)
15. Gong, Z.-W., Li, L.-S., Forrest, J., Zhao, Y.: The optimal priority models of the intuitionistic fuzzy preference relation and their application in selecting industries with higher meteorological sensitivity. Expert Syst. Appl. **38**, 4394–4402 (2001)
16. Li, D.-F.: Multiattribute decision making models and method using intuitionistic fuzzy sets. J. Comput. Syst. Sci. **70**, 73–85 (2005)
17. Xu, Z.: Intuitionistic preference relations and their application in group decision making. Inform. Sci. **177**, 2363–2379 (2007)
18. Zadeh, L.A.: Fuzzy sets. Inform. Control **8**, 338–353 (1965)

Intuitionistic Fuzzy Complete Lattices

Soheyb Milles, Ewa Rak and Lemnaouar Zedam

Abstract In this paper, the concept of intuitionistic complete lattices is introduced. Some characterizations of such intuitionistic complete lattices are given. The Tarski-Davis fixed point theorem for intuitionistic fuzzy complete lattices is proved, which establish an other criterion for completeness of intuitionistic fuzzy complete lattices in terms of fixed points of intuitionistic monotone maps.

Keywords Intuitionistic fuzzy set · Intuitionistic fuzzy order · Intuitionistic fuzzy complete lattice · Tarski-Davis fixed point theorem

1 Introduction

The notion of a fuzzy set was first introduced by Zadeh [25] by assuming the standard negation that the non-membership degree is equal to one minus membership degree and this makes the fuzzy sets compliment. In logical area, membership degree and non-membership degree can be interpreted as positive and negative. This means that if the membership is correct, then the non membership is wrong. Obviously it explains that the contraries relation exists.

In 1983 Atanassov [1] proposed a generalization of Zadeh non-membership degree and introduced the notion of intuitionistic fuzzy set (A-IFSs for short). The non-membership degree used for Atanassov's intuitionistic fuzzy set is a more-or-less independent degree: the only condition is that the non-membership degree is less

S. Milles · L. Zedam
Laboratory of Pure and Applied Mathematics, Department of Mathematics, Med Boudiaf University—Msila, Msila, Algeria
e-mail: milles.math@gmail.com

L. Zedam
e-mail: l.zedam@gmail.com

E. Rak (✉)
Faculty of Mathematics and Natural Sciences, University of Rzeszów, Rzeszów, Poland
e-mail: ewarak@ur.edu.pl

K.T. Atanassov et al. (eds.), *Novel Developments in Uncertainty Representation and Processing*, Advances in Intelligent Systems and Computing 401,
DOI 10.1007/978-3-319-26211-6_13

or equal to the one minus membership degree. Certainly fuzzy sets are intuitionistic fuzzy sets, but not conversely.

Inspired by the notion of intuitionistic fuzzy set, Burillo and Bustince [12, 13] introduced intuitionistic fuzzy relations as a natural generalizations of fuzzy relations. Intuitionistic fuzzy relations theory has been applied to many different fields, such as decision making, mathematical modelling, medical diagnosis, machine learning and market prediction, etc.

One of the important problems of fuzzy and intuitionistic fuzzy ordered set is to obtain an appropriate concepts of particular elements on a such structure like maximum, supremum, maximal elements and their duals, in particular, specific subclasses of fuzzy and intuitionistic fuzzy ordered sets. Several theoretical and applicational results connected with this problem can be found, e.g. in Bělohlávek [9], Bodenhofer and Klawonn [10], Bustince and Burillo [15, 16], Coppola et al. [17], Tripathy et al. [23], Zadeh [26], Zhang et al. [29].

In this paper, according to the intuitionistic fuzzy order introduced by Burillo and Bustince [12, 13] and based on the notions of supremum and infimum of subsets on a universe X with respect to an intuitionistic fuzzy order defined on it introduced by Tripathy et al. [23], we propose a notion of an intuitionistic fuzzy complete lattice which is a generalization of the crisp complete lattice notion. Some characterizations of such intuitionistic fuzzy complete lattice expressed in terms of supremum, infimum, chains and maximal chains are given.

One of the consequence of Tarski and Davis fixed point theorems in crisp lattices [18, 22] is that they established a criterion for completeness of lattices in terms of fixed points of monotone maps. In the last section we will focus on this criterion and Tarski-Davis fixed point theorem for intuitionistic fuzzy complete lattices will be proved.

2 Preliminaries

This section contains the basic definitions and properties of intuitionistic fuzzy sets, intuitionistic fuzzy relations, intuitionistic fuzzy lattices and some related notions that will be needed in the next sections of this paper. At first we recall some basic concepts of intuitionistic fuzzy sets. More details can be found in [1–8, 11, 20, 21, 27].

Let X be a universe, then a fuzzy set $A = \{\langle x, \mu_A(x)\rangle \;/\; x \in X\}$ defined by Zadeh [25] is characterized by a membership function $\mu_A : X \to [0, 1]$, where $\mu_A(x)$ is interpreted as the degree of a membership of the element x in the fuzzy subset A for each $x \in X$.

In [1] Atanassov introduced another fuzzy object, called intuitionistic fuzzy set (briefly IFS or A-IFS) as a generalization of the concept of fuzzy set, shown as follows

$$A = \{\langle x, \mu_A(x), \nu_A(x)\rangle \;/\; x \in X\},$$

which is characterized by a membership function $\mu_A : X \to [0, 1]$ and a non-membership function $\nu_A : X \to [0, 1]$ with the condition

$$0 \leq \mu_A(x) + \nu_A(x) \leq 1\,, \tag{1}$$

for any $x \in X$. The numbers $\mu_A(x)$ and $\nu_A(x)$ represent, respectively, the membership degree and the non-membership degree of the element x in the intuitionistic fuzzy set A for each $x \in X$.

In the fuzzy set theory, the non-membership degree of an element x of the universe is defined as $\nu_A(x) = 1 - \mu_A(x)$ (using the standard negation) and thus it is fixed. In intuitionistic fuzzy setting, the non-membership degree is a more-or-less independent degree: the only condition is that $\nu_A(x) \leq 1 - \mu_A(x)$. Certainly fuzzy sets are intuitionistic fuzzy sets by setting $\nu_A(x) = 1 - \mu_A(x)$, but not conversely.

Definition 1 Let A be an intuitionistic fuzzy set on universe X, the support of A is the crisp subset of X given by
$Supp(A) = \{x \in X \;/\; \mu_A(x) > 0 \text{ or } (\mu_A(x) = 0 \text{ and } \nu_A(x) < 1)\}$.

An intuitionistic fuzzy relation from a universe X to a universe Y is an intuitionistic fuzzy subset in $X \times Y$, i.e. is an expression R given by
$R = \{\langle (x, y), \mu_R(x, y), \nu_R(x, y)\rangle \;/\; (x, y) \in X \times Y\}$, where $\mu_R : X \times Y \to [0, 1]$, $\nu_A : X \times Y \to [0, 1]$ with the condition

$$0 \leq \mu_R(x, y) + \nu_R(x, y) \leq 1\,, \tag{2}$$

for any $(x, y) \in X \times Y$. The value $\mu_R(x, y)$ is called the degree of a membership of (x, y) in R and $\nu_R(x, y)$ is called the degree of a non-membership of (x, y) in R.

Next, we need the following definitions.

Let R be an intuitionistic fuzzy relation from a universe X to a universe Y. The transposition R^t of R is the intuitionistic fuzzy relation from the universe Y to the universe X defined by

$$R^t = \{\langle (x, y), \mu_{R^t}(x, y), \nu_{R^t}(x, y)\rangle \mid (x, y) \in X \times Y\},$$

where $\mu_{R^t}(x, y) = \mu_R(y, x)$ and $\nu_{R^t}(x, y) = \nu_R(y, x)$, for any $(x, y) \in X \times Y$.

Let R and P be two intuitionistic fuzzy relations from a universe X to a universe Y. R is said to be contained in P or we say that P contains R (notation $R \subseteq P$) if for all $(x, y) \in X \times Y$ $\mu_R(x, y) \leq \mu_P(x, y)$ and $\nu_R(x, y) \geq \nu_P(x, y)$.

The intersection (resp. the union) of two intuitionistic fuzzy relations R and P from a universe X to a universe Y is defined as

$$R \bigcap P = \{\langle (x, y), \mu_{R\cap P}(x, y), \nu_{R\cap P}(x, y)\rangle\},$$

where $\mu_{R\cap P}(x, y) = \min(\mu_R(x, y), \mu_P(x, y))$ and $\nu_{R\cap P}(x, y) = \max(\nu_R(x, y), \nu_P(x, y))$ for any $(x, y) \in X \times Y$.

The union of two intuitionistic fuzzy relations R and P from a universe X to a universe Y is defined as

$$R \bigcup P = \{\langle (x,y), \mu_{R \cup P}(x,y), \nu_{R \cup P}(x,y) \rangle\},$$

where $\mu_{R \cup P}(x,y) = \max(\mu_R(x,y), \mu_P(x,y))$ and $\nu_{R \cup P}(x,y) = \min(\nu_R(x,y), \nu_P(x,y))$ for any $(x,y) \in X \times Y$.

In general, if A is a set of intuitionistic fuzzy relations from a universe X to a universe Y, then

$$\bigcap_{R \in A} R = \{\langle (x,y), \mu_{\cap_{R \in A} R}(x,y), \nu_{\cap_{R \in A} R}(x,y) \rangle\},$$

where $\mu_{\cap_{R \in A} R}(x,y) = \inf_{R \in A} \mu_R(x,y)$ and $\nu_{\cap_{R \in A} R}(x,y) = \sup_{R \in A} \nu_R(x,y)$ for any $(x,y) \in X \times Y$;

$$\bigcup_{R \in A} R = \{\langle (x,y), \mu_{\cup_{R \in A} R}(x,y), \nu_{\cup_{R \in A} R}(x,y) \rangle\},$$

where $\mu_{\cup_{R \in A} R}(x,y) = \sup_{R \in A} \mu_R(x,y)$ and $\nu_{\cup_{R \in A} R}(x,y) = \inf_{R \in A} \nu_R(x,y)$ for any $(x,y) \in X \times Y$.

Let R be an intuitionistic fuzzy relation from a universe X to a universe X (intuitionistic fuzzy relation on a universe X, for short). The following properties are crucial in this paper (see e.g. [12, 16, 19, 23, 24, 28]):

(i) Reflexivity: $\mu_R(x,x) = 1$ for any $x \in X$. Just notice that $\nu_R(x,x) = 0$ for any $x \in X$.

(ii) Antisymmetry: if for any $x, y \in X$, $x \neq y$ then

$$\begin{cases} \mu_R(x,y) \neq \mu_R(y,x) \\ \nu_R(x,y) \neq \nu_R(y,x) \\ \pi_R(x,y) = \pi_R(y,x) \end{cases},$$

where $\pi_R(x,y) = 1 - \mu_R(x,y) - \nu_R(x,y)$.

(iii) Perfect antisymmetry: if for any $x, y \in X$ with $x \neq y$ $\mu_R(x,y) > 0$ or $(\mu_R(x,y) = 0$ and $\nu_R(x,y) < 1)$ then $\mu_R(y,x) = 0$ and $\nu_R(y,x) = 1$.

(iv) Transitivity: $R \supseteq R \circ^{\alpha,\beta}_{\lambda,\rho} R$.

Remark 1 The definition of perfect antisymmetry given in (iii) is equivalent to the following one for any $x, y \in X$, $(\mu_R(x,y) > 0$ and $\mu_R(y,x) > 0)$ or $(\nu_R(x,y) < 1$ and $\nu_R(y,x) < 1)$ implies that $x = y$.

The composition $R \circ^{\alpha,\beta}_{\lambda,\rho} R$ in the above definition of transitivity means that

$$R \circ^{\alpha,\beta}_{\lambda,\rho} R = \{\langle (x,z), \ \alpha_{y \in X}\{\beta[\mu_R(x,y), \mu_R(y,z)],$$

$$\lambda_{y\in X}\{\rho[\nu_R(x,y),\nu_R(y,z)]\rangle \mid x,z\in X\},$$

where α, β, λ and ρ are t-norms or t-conorms taken under the intuitionistic fuzzy condition

$$0\leq \alpha_{y\in X}\{\beta[\mu_R(x,y),\mu_R(y,z)]+\lambda_{y\in X}\{\rho[\nu_R(x,y),\nu_R(y,z)]\leq 1\,,$$

for any $x,z\in X$.

The properties of this composition and the choice of α, β, λ and ρ, for which this composition fulfills a maximal number of properties, are investigated in [12–16, 19]. If no other conditions are imposed, in the sequel we will take $\alpha=\sup$, $\beta=\min$, $\lambda=\inf$ and $\rho=\max$.

Notice that in [15], Bustince and Burillo mentioned that the definition of intuitionistic antisymmetry does not recover the fuzzy antisymmetry for the case in which the considered relation R is fuzzy. However, the definition of intuitionistic perfect antisymmetry does recover the definition of fuzzy antisymmetry given by Zadeh [26] when the considered relation is fuzzy. This note justifies the following definition of intuitionistic fuzzy order used in this paper.

Definition 2 ([12, 13]) Let X be a nonempty crisp set and $R=\{\langle(x,y),\mu_R(x,y),\nu_R(x,y)\rangle \mid x,y\in X\}$ be an intuitionistic fuzzy relation on X. R is called an intuitionistic fuzzy order or a partial intuitionistic fuzzy order if it is reflexive, transitive and perfect antisymmetric.

A nonempty set X with an intuitionistic fuzzy order R defined on it is called an intuitionistic fuzzy ordered set and we denote it by (X,μ_R,ν_R).

Notice that any partially ordered set $(X,\leq)$ and generally any fuzzy ordered set (X,R) can be regarded as intuitionistic fuzzy ordered sets.

Example 1 Let $m,n\in N$. Then, the intuitionistic fuzzy relation R defined for all $m,n\in N$ by

$$\mu_R(m,n)=\begin{cases}1\,,\ if\ m=n\\ 1-\frac{m}{n}\,,\ \ if\ m<n\\ 0\,,\ \ if\ m>n\end{cases},$$

and

$$\nu_R(m,n)=\begin{cases}0\,,\ if\ m=n\\ \frac{m}{2n}\,,\ \ if\ m<n\\ 1\,,\ if\ m>n\end{cases}$$

is an intuitionistic fuzzy order on N.

On the basis of the above definition of perfect antisymmetry we define linear or total intuitionistic fuzzy order as follows.

Definition 3 An intuitionistic fuzzy order R on a universe X is linear (or total) if for every $x, y \in X$ $[\mu_R(x,y) > 0$ or $(\mu_R(x,y) = 0$ and $\nu_R(x,y) < 1)]$
or
$[\mu_R(y,x) > 0$ or $(\mu_R(y,x) = 0$ and $\nu_R(y,x) < 1)]$.

Definition 4 An intuitionistic fuzzy ordered set (X, μ_R, ν_R) in which R is linear is called a linearly intuitionistic fuzzy ordered set or an intuitionistic fuzzy chain.

For an intuitionistic fuzzy ordered set (X, μ_R, ν_R) and $x \in X$, the intuitionistic fuzzy sets $R_{\geq[x]}$ and $R_{\leq[x]}$ are defined in X by
$R_{\geq[x]} = \{\langle y, \mu_{R_{\geq[x]}}(y), \nu_{R_{\geq[x]}}(y)\rangle \;/\; y \in X\}$, where $\mu_{R_{\geq[x]}}(y) = \mu_R(x,y)$ and $\nu_{R_{\geq[x]}}(y) = \nu_R(x,y)$.

$R_{\leq[x]} = \{\langle y, \mu_{R_{\leq[x]}}(y), \nu_{R_{\leq[x]}}(y)\rangle \,/\, y \in X\}$, where $\mu_{R_{\leq[x]}}(y) = \mu_R(y,x)$ and $\nu_{R_{\leq[x]}}(y) = \nu_R(y,x)$. $R_{\geq[x]}$ and $R_{\leq[x]}$ are called the dominating class of x and the class dominated by x, respectively.

Remark 2 The notions of the dominating class of x and the class dominated by x are generalizations of the classical notions $\uparrow x$ and $\downarrow x$ in a usual poset.

Next, we recall the definition of upper bounds, lower bounds, supremum and infimum on intuitionistic fuzzy ordered sets.

Definition 5 ([23]) Let (X, μ_R, ν_R) be an intuitionistic fuzzy ordered set and A be a subset of X.

(i) The set of upper bounds of A with respect to R is the intuitionistic fuzzy subset of X defined by

$$U(R,A)(y) = \bigcap_{x \in A} R_{\geq[x]}(y) \tag{3}$$

for any $y \in X$;

(ii) The set of lower bounds of A with respect to R is the intuitionistic fuzzy subset of X defined by

$$L(R,A)(y) = \bigcap_{x \in A} R_{\leq[x]}(y) \tag{4}$$

for any $y \in X$.

Definition 6 ([23]) Let (X, μ_R, ν_R) be an intuitionistic fuzzy ordered set and A be a subset of X. An element $x \in X$ is called the least upper bound (or a supremum) of A with respect to R if

(i) $x \in Supp(U(R,A))$ and
(ii) for all other $y \in Supp(U(R,A))$, $\mu_R(x,y) > 0$ or $(\mu_R(x,y) = 0$ and $\nu_R(x,y) < 1)$.

An element $x \in X$ is called the greatest lower bound (or an infimum) of A with respect to R if

(i) $x \in Supp(L(R, A))$ and
(ii) for all other $y \in Supp(L(R, A))$, $\mu_R(y, x) > 0$ or ($\mu_R(y, x) = 0$ and $\nu_R(y, x) < 1$).

Remark 3 Let (X, μ_R, ν_R) be an intuitionistic fuzzy ordered set and A be a subset of X. If the supremum and the infimum of A with respect to R exist, then from the perfect antisymmetry of R they are unique and denoted by $\sup_R(A)$ and $\inf_R(A)$, respectively.

Definition 7 An intuitionistic fuzzy ordered set (X, μ_R, ν_R) is called an intuitionistic fuzzy ordered lattice with respect to the intuitionistic fuzzy order R (or simply, intuitionistic fuzzy lattices) if each pair of elements $\{x, y\}$ of X has a supremum and an infimum.

Next, we introduce the notion of intuitionistic fuzzy complete lattices which is a natural generalization of the notion of crisp complete lattices.

Definition 8 An intuitionistic fuzzy ordered set (X, μ_R, ν_R) is called an intuitionistic fuzzy complete lattice if $\sup_R(A)$ and $\inf_R(A)$ exist for every nonempty subset $A \subseteq X$.

Definition 9 Let (X, μ_R, ν_R) be an intuitionistic fuzzy ordered set.

(i) An element $\top \in X$ is called the greatest element (the maximum) of X with respect to R or the intuitionistic fuzzy maximum of X if
$\mu_R(x, \top) > 0$ or ($\mu_R(x, \top) = 0$ and $\nu_R(x, \top) < 1$) for all $x \in X$.
(ii) An element $\perp \in X$ is called the least element (the minimum) of X with respect to R or the intuitionistic fuzzy minimum if
$\mu_R(\perp, x) > 0$ or ($\mu_R(\perp, x) = 0$ and $\nu_R(\perp, x) < 1$) for all $x \in X$.

Remark 4 Every intuitionistic fuzzy complete lattice must have a greatest element (or a maximum) and a least element (or a minimum). The greatest element will be denoted $\top_X$ and the least element $\perp_X$. It is easily follows that

$$\top_X = \sup_R(X) = \inf_R(\emptyset) \text{ and } \perp_X = \inf_R(X) = \sup_R(\emptyset).$$

In the last section we will need the following definitions.

Definition 10 Let (X, μ_R, ν_R) be an intuitionistic fuzzy ordered set. Then a map $f : X \to X$ is called intuitionistic fuzzy monotone if $\mu_R(f(x), f(y)) \geq \mu_R(x, y)$ and $\nu_R(f(x), f(y)) \leq \nu_R(x, y)$ for all $x, y \in X$.

Definition 11 An element $x \in X$ is called a fixed point of a map $f : X \to X$ if $f(x) = x$. The set of all fixed points of f will be denoted by $Fix(f)$.

3 Characterizations of Intuitionistic Fuzzy Complete Lattices

In this section we will provide an interesting characterization of intuitionistic fuzzy complete lattices in terms of supremum and infimum of its subsets, as well as in terms of its intuitionistic fuzzy chains and maximal intuitionistic fuzzy chains.

The following lemma is immediate.

Lemma 1 *Let (X, μ_R, ν_R) be an intuitionistic fuzzy ordered set, A be a subset of X and $x \in X$. Then, it holds that*

(i) $x = \sup_R(A)$ *with respect to R if and only if* $x = \inf_{R^t}(A)$ *with respect to* R^t;
(ii) $x = \inf_R(A)$ *with respect to R if and only if* $x = \sup_{R^t}(A)$ *with respect to* R^t;
(iii) (X, μ_R, ν_R) *is intuitionistic fuzzy complete lattice if and only if* $(X, \mu_{R^t}, \nu_{R^t})$ *is intuitionistic fuzzy complete lattice.*

Theorem 1 *Let (X, μ_R, ν_R) be an intuitionistic fuzzy ordered set. Then, it holds that*

(i) (X, μ_R, ν_R) *is an intuitionistic fuzzy complete lattice if and only if* $\sup_R(A)$ *exists for all* $A \subseteq X$;
(ii) (X, μ_R, ν_R) *is an intuitionistic fuzzy complete lattice if and only if* $\inf_R(A)$ *exists for all* $A \subseteq X$.

Proof Let (X, μ_R, ν_R) be an intuitionistic fuzzy ordered set and $A \subseteq X$.

(i) It is obvious that if (X, μ_R, ν_R) is an intuitionistic fuzzy complete lattice, then $\sup_R(A)$ exists for all $A \subseteq X$.
Conversely, suppose that $\sup_R(A)$ exists for all $A \subseteq X$ and we will show that every nonempty subset $A \subseteq X$ has an infimum. Let $L(R, A)$ be the intuitionistic fuzzy set of lower bounds of A with respect to R. Then, it holds that $\sup_R(Supp(L(R, A)))$ exists. Setting that $m = \sup_R(Supp(L(R, A)))$ and we show that $m = \inf_R(A)$. First, since $m \in Supp(U(R, Supp(L(R, A))))$, then it holds that $\mu_{U(R,Supp(L(R,A)))}(x) > 0$ or $[\mu_{U(R,Supp(L(R,A)))}(x) = 0$ and $\nu_{U(R,Supp(L(R,A)))}(x) < 1]$. By (3) we know that

$$U(R, Supp(L(R, A)))(y) = \bigcap_{x \in Supp(L(R,A))} R_{\geq[x]}(y).$$

Since $R_{\geq[x]} = \{\langle y, \mu_{R_{\geq[x]}}(y), \nu_{R_{\geq[x]}}(y)\rangle \ / \ y \in X\}$, where $\mu_{R_{\geq[x]}}(y) = \mu_R(x, y)$ and $\nu_{R_{\geq[x]}}(y) = \nu_R(x, y)$ and by the fact that $R_{\geq[x]} = R^t_{\leq[x]}$ and $U(R, A) = L(R^t, A)$, it follows that

$$U(R, Supp(L(R, A)) = L(R^t, Supp(L(R, A))) = L(R, A).$$

Hence,

$$m \in Supp(L(R, A)).$$

In the same way, for all $y \in Supp(L(R, A))$, it holds that $\mu_R(y, m) > 0$ or $(\mu_R(y, m) = 0$ and $\nu_R(y, m) < 1)$.
Thus $m = \inf_R(A)$, which implies that $\inf_R(A)$ exists. Therefore, (X, μ_R, ν_R) is an intuitionistic fuzzy complete lattice.

(ii) Follows from Lemma 1 and (i).

Remark 5 In the above Theorem 1 the existence of $\inf_R(\emptyset)$ guarantees the greatest element of (X, μ_R, ν_R), and in similar way, the existence of $\sup_R(\emptyset)$ guarantees the least element of (X, μ_R, ν_R). So an equivalent formulation of Theorem 1 can be written in the following way

(i) (X, μ_R, ν_R) is an intuitionistic fuzzy complete lattice if and only if it has the least element and $\sup_R(A)$ exists for all nonempty $A \subseteq X$;
(ii) (X, μ_R, ν_R) is an intuitionistic fuzzy complete lattice if and only if it has the greatest element and $\inf_R(A)$ exists for all nonempty $A \subseteq X$.

Theorem 2 *Let (X, μ_R, ν_R) be an intuitionistic fuzzy lattice. Then the following are equivalent:*

(i) (X, μ_R, ν_R) *is intuitionistic fuzzy complete lattice;*
(ii) (X, μ_R, ν_R) *is intuitionistic fuzzy chain-complete (i.e. every nonempty intuitionistic fuzzy chain in* (X, μ_R, ν_R) *has a supremum and an infimum);*
(iii) Every maximal intuitionistic fuzzy chain of X is an intuitionistic fuzzy complete lattice.

Proof (i)$\Rightarrow$ (ii) is obvious.

To prove (ii)$\Rightarrow$ (iii), let C be a maximal intuitionistic fuzzy chain (with respect to set inclusion) of X.

First, we will show that C has an intuitionistic fuzzy maximum and an intuitionistic fuzzy minimum. Since C is an intuitionistic fuzzy chain in (X, μ_R, ν_R) then it holds from (ii) that C has a supremum and an infimum. By using the fact that C is maximal (with respect to set inclusion) we obtain that $c_1 = \sup_R(C)$ is the maximum and $c_2 = \inf_R(C)$ is the minimum.

Second, let $A \subseteq C$. Since $A \subseteq C$ then it holds that A is an intuitionistic fuzzy chain. By (ii) we know that $\sup_R(A)$ exists in (X, μ_R, ν_R) and we denoted it by m. Now, it suffices to show that $m \in C$. Suppose that $m \notin C$, then it follows three cases:

(a) If $[\mu_R(x, m) > 0$ or $(\mu_R(x, m) = 0$ and $\nu_R(x, m) < 1]$ or $[\mu_R(m, x) > 0$ or $(\mu_R(m, x) = 0$ and $\nu_R(m, x) < 1]$ for all $x \in C$, then $C \cup \{m\}$ is intuitionistic fuzzy chain in (X, μ_R, ν_R). This is a contradiction with the fact that C is maximal.
(b) If there exist $x \in C$ such that $[\mu_R(x, m) = 0$ and $\nu_R(x, m) = 1]$, then it holds from the transitivity of R that

$$\mu_R(x, c_1) \wedge \mu_R(c_1, m) \leq \mu_R(x, m)$$

and

$$\nu_R(x, c_1) \vee \nu_R(c_1, m) \geq \nu_R(x, m).$$

Since $[\mu_R(x, m) = 0$ and $\nu_R(x, m) = 1]$, then it holds that $\mu_R(c_1, m) = 0$ and $\nu_R(c_1, m) = 1$. Hence $\sup_R\{c_1, m\} \notin C$. Thus, $C \bigcup \{\sup_R\{c_1, m\}\}$ is an intuitionistic fuzzy chain, which is a contradiction with maximality of C.

(c) If there exist $x \in C$ such that $[\mu_R(m,x) = 0$ and $\nu_R(m,x) = 1]$, then it follows similarly as (b).

As consequence of the above cases we get $m \in C$. Thus, A has a supremum in C. Therefore, C is an intuitionistic fuzzy complete lattice which follows from Theorem 1.

(iii)⇒ (i) Suppose that every maximal intuitionistic fuzzy chain of X is an intuitionistic fuzzy complete lattice and we will show that (X, μ_R, ν_R) is intuitionistic fuzzy complete lattice.

Let $A \subseteq X$ and $\mathcal{BFC}(Supp(U(R,A)))$ denote the set of all intuitionistic fuzzy fuzzy chains $C \subseteq Supp(U(R,A))$, ordered in classical way by $C_1 \sqsubseteq C_2$ if and only f C_1 is an intuitionistic fuzzy filter of C_2. This means that $C_1 \subseteq C_2$ or [if $x \in C_1$ and $y \in C_2$ with $\mu_R(x,y) > 0$ or $(\mu_R(x,y) = 0$ and $\nu_R(x,y) < 1)$ then $y \in C_1$].

Next, let $\{C_i : i \in I \subseteq N\}$ be a chain of $\mathcal{BFC}(Supp(U(R,A)))$ under the crisp order defined above. On the one hand, since C_i is an intuitionistic fuzzy chain of $Supp(U(R,A))$ and $C_i \subseteq C_{i+1}$ for all $i \in I$, then $\bigcup_{i\in I} C_i$ is an intuitionistic fuzzy chain of $Supp(U(R,A))$. Hence, $\bigcup_{i\in I} C_i \in \mathcal{BFC}(Supp(U(R,A)))$. On the other hand, $\bigcup_{i\in I} C_i$ is an upper bound of $\{C_i\}_{i\in I}$.

By Zorn's Lemma, we know that $\mathcal{BFC}(Supp(U(R,A)))$ has a maximal element denoted by C_m with respect to the above crisp order $\sqsubseteq$.

Let K be a maximal intuitionistic fuzzy chain such that $C_m \subseteq K$. By hypothesis, K is an intuitionistic fuzzy complete lattice, which implies that C_m has a an infimum denoted by c in (K, μ_R, ν_R).

Now, we will show that $c = \sup_R(A)$. Indeed, let $x \in A$. Since $C_m \subseteq Supp(U(R,A))$, then it holds that $\mu_R(x,y) > 0$ or $(\mu_R(x,y) = 0$ and $\nu_R(x,y) < 1)$ for all $y \in C_m$. Hence $\mu_R(x,c) > 0$ or $(\mu_R(x,c) = 0$ and $\nu_R(x,c) < 1)$. Thus, $c \in Supp(U(R,A))$. For all other $y \in Supp(U(R,A))$, it holds that $\mu_R(c,y) > 0$ or $(\mu_R(c,y) = 0$ and $\nu_R(c,y) < 1)$. Otherwise, we get a contradiction with the maximality of C_m. Thus, $c = \sup_R(A)$. Now, (X, μ_R, ν_R) is an intuitionistic fuzzy complete lattice follows from Theorem 1(i).

4 Tarski-Davis Fixed Point Theorem for Intuitionistic Fuzzy Complete Lattices

Tarski and Davis in their results established a criterion for completeness of lattices in terms of fixed points of monotone maps. In the last section, we will show that this criterion also stay valid for intuitionistic fuzzy complete lattices.

Additionally we need the following definition.

Definition 12 Let (X, μ_R, ν_R) be an intuitionistic fuzzy ordered set and $\{a_i\}_{i\in I\subseteq N}$ be a subset of elements of X. Then

(i) $\{a_i\}_{i \in I \subseteq N}$ is called an intuitionistic fuzzy ascending chain (or an ascending chain with respect to R) if $\mu_R(a_i, a_{i+1}) > 0$ or ($\mu_R(a_i, a_{i+1}) = 0$ and $\nu_R(a_i, a_{i+1}) < 1$), for all $i \in I$. Descending intuitionistic fuzzy chain (or an descending chain with respect to R) is defined dually.

(ii) (X, μ_R, ν_R) is said to be satisfy the intuitionistic fuzzy ascending chain condition (or the ACC_R, for short) if every intuitionistic fuzzy ascending chain $\{a_i\}_{i \in I \subseteq N}$ of elements of X is eventually stationary (i.e. there exist a positive integer $n \in I$ such that $a_m = a_n$ for all $m > n$). In other words, (X, R) contains no infinite intuitionistic fuzzy ascending chain.

(iii) Similarly, (X, μ_R, ν_R) is said to be satisfy the intuitionistic fuzzy descending chain condition (or the DCC_R, for short) if every intuitionistic fuzzy descending chain $\{a_i\}_{i \in I \subseteq N}$ of elements of X is ultimately stationary.

From Theorems 1 and 2, we derive the following results (unfortunately without very extensive proofs).

Proposition 1 *Let (X, μ_R, ν_R) be an intuitionistic fuzzy lattice. If (X, μ_R, ν_R) is not intuitionistic fuzzy complete lattice, then there exist an intuitionistic fuzzy chain C satisfying the ACC_R and having no infimum and an intuitionistic fuzzy chain D satisfying the DCC_R and having no supremum, such that*

(i) $\mu_R(d, c) > 0$ or ($\mu_R(d, c) = 0$ and $\nu_R(d, c) < 1$) for any $d \in D$ and $c \in C$;

(ii) For all $x \in X$, either there exists $c \in C$ with ($\mu_R(x, c) = 0$ and $\nu_R(x, c) = 1$) or there exists $d \in D$ with ($\mu_R(d, x) = 0$ and $\nu_R(d, x) = 1$), i.e. there is no element $x \in X$ such that $x \in Supp(L(R, C)) \bigcap Supp(U(R, D))$.

Theorem 3 *Let (X, μ_R, ν_R) be an intuitionistic fuzzy lattice. Then (X, μ_R, ν_R) is an intuitionistic fuzzy complete lattice if and only if every intuitionistic fuzzy monotone map $f : X \to X$ has a fixed point. Moreover, the set Fix(f) of all fixed points of f is an intuitionistic fuzzy complete lattice.*

5 Conclusion

In this paper we have introduced the notion of intuitionistic fuzzy complete lattice and investigated its most interesting properties. Some characterizations of intuitionistic fuzzy complete lattice expressed in terms of supremum, infimum, chains and maximal chains are given. Moreover, the Tarski-Davis fixed point theorem for intuitionistic fuzzy complete lattices is presented (without the extensive proof), which establish an other criterion for completeness of intuitionistic fuzzy complete lattices in terms of fixed points of intuitionistic monotone maps.

Acknowledgments This work is partially supported by the Centre for Innovation and Transfer of Natural Sciences and Engineering Knowledge No RPPK.01.03.00−18−001/10.

References

1. Atanassov, K.: Intuitionistic Fuzzy Sets. VII ITKRs Scientific Session, Sofia (1983)
2. Atanassov, K.: Intuitionistic fuzzy sets. Fuzzy Sets Syst. **20**, 87–96 (1986)
3. Atanassov, K.: Review and new results on intuitionistic fuzzy sets. IM-MFAIS **1** (1988)
4. Atanassov, K., Gargov, G.: Interval valued intuitionistic fuzzy sets. Fuzzy Sets Syst. **31**, 343–349 (1989)
5. Atanassov, K.: More on intuitionistic fuzzy sets. Fuzzy Sets Syst. **33**, 37–45 (1989)
6. Atanassov, K.: Remarks on the intuitionistic fuzzy sets. Fuzzy Sets Syst. **75**, 401–402 (1995)
7. Atanassov, K.: Intuitionistic Fuzzy Sets. Springer, New York (1999)
8. Biswas, R.: On fuzzy sets and intuitionistic fuzzy sets. NIFS **3**, 3–11 (1997)
9. Bělohlávek, R.: Concept lattices and order in fuzzy logic. Ann. Pure Appl. Log. **128**, 277–298 (2004)
10. Bodenhofer, U., Klawonn, F.: A formal study of linearity axioms for fuzzy orderings. Fuzzy Sets Syst. **145**, 323–354 (2004)
11. Burillo, P., Bustince, H.: Estructuras algebraicas en conjuntos IFS. II Congreso nacional de lógica y tecnologia fuzzy, pp. 135–147. Boadilla del Monte, Madrid (1992)
12. Burillo, P., Bustince, H.: Intuitionistic fuzzy relations. (Part I), Mathw. Soft Comput. Mag. **2**, 5–38 (1995)
13. Burillo, P., Bustince, H.: Intuitionistic fuzzy relations. (Part II), effect of Atanassov's operators on the properties of the intuitionistic fuzzy relations. Mathw. Soft Comput. Mag. **2**, 117–148 (1995)
14. Burillo, P., Bustince, H.: Antisymmetrical intuitionistic fuzzy relation. Order on the referential set induced by an bi fuzzy relation. Fuzzy Sets Syst. **62**, 17–22 (1995)
15. Burillo, P., Bustince, H.: Structures on intuitionistic fuzzy relations. Fuzzy Sets Syst. **3**, 293–303 (1996)
16. Bustince, H.: Construction of intuitionistic fuzzy relations with predetermined properties. Fuzzy Sets Syst. **3**, 79–403 (2003)
17. Coppola, C., Gerla, G., Pacelli, T.: Convergence and fixed points by fuzzy orders. Fuzzy Sets Syst. **159**, 1178–1190 (2008)
18. Davis, A.C.: A characterization of complete lattices. Pac. J. Math. **5**, 311–319 (1955)
19. Deschrijver, G., Kerre, E.E.: On the composition of intuitionistic fuzzy relations. Fuzzy Sets Syst. **136**, 333–361 (2003)
20. Gerstenkorn, T., Manko, J.: Intuitionistic fuzzy probabilistic sets. Fuzzy Sets Syst. **71**, 207–214 (1995)
21. Grzegorzewski, P., Mrówka, E.: Some notes on (Atanassov's) intuitionistic fuzzy sets. Fuzzy Sets Syst. **156**, 492–495 (2005)
22. Tarski, A.: A lattice-theoretical fixpoint theorem and its applications. Pac. J. Math. **5**, 285–309 (1955)
23. Tripathy, B.K., Satapathy, M.K., Choudhury, P.K.: Intuitionistic fuzzy lattices and intuitionistic fuzzy boolean algebras. Int. J. Eng. Technol. **5**, 2352–2361 (2013)
24. Xu, Z.S.: Intuitionistic fuzzy preference relations and their application in group decision making. Inform. Sci. **177**, 2363–2379 (2007)
25. Zadeh, L.A.: Fuzzy sets. Inf. Comput. **8**, 338–353 (1965)
26. Zadeh, L.A.: Similarity relations and fuzzy orderings. Inform. Sci. **3**, 177–200 (1971)
27. Zedam, L., Amroune, A.: On the representation of L-M algebra by intuitionistic fuzzy subsets. Arima J. **4**, 72–85 (2006)
28. Zedam, L., Amroune, A., Davvaz, B.: Szpilrajn theorem for intuitionistic fuzzy orderings. Annals of fuzzy mathematics and informatics, Article in press (2015)
29. Zhang, Q.Y., Xie, W., Fan, L.: Fuzzy complete lattices. Fuzzy Sets Syst. **160**, 2275–2291 (2009)

Traversing and Ranking of Elements of an Intuitionistic Fuzzy Set in the Intuitionistic Fuzzy Interpretation Triangle

Vassia Atanassova, Ivelina Vardeva, Evdokia Sotirova and Lyubka Doukovska

Abstract In this leg of research, we explore the question of traversing and ranking elements of an intuitionistic fuzzy set in the intuitionistic fuzzy interpretation triangle. This is necessary in the light of the new developments of the InterCriteria Analysis (ICA), a decision support approach based on intuitionistic fuzzy sets and index matrices. In the ICA, from the data about the evaluations or measurements of a set of objects against a set of criteria, we perform pairwise comparisons of any two objects against each pair of criteria, and perform computations that yield in result intuitionistic fuzzy pairs of numbers in the [0; 1]-interval that give the levels of correlation between any two of the evaluation criteria. In previous works, the correlations between the criteria (hence the term 'intercriteria') were analysed separately, by first setting priority on either the membership, or the non-membership component, and plotting them linearly; while currently the efforts are oriented to handling both IF components simultaneously by plotting them in the plane of the intuitionistic fuzzy interpretation triangle.

Keywords Intercriteria analysis • Intuitionistic fuzzy sets • Triangular geometrical interpretation of intuitionistic fuzzy sets • Closure • Interior

V. Atanassova (✉)
Bioinformatics and Mathematical Modelling Department, IBPhBME – Bulgarian Academy of Sciences, Acad. G. Bonchev Str., Bl. 105, 1113 Sofia, Bulgaria
e-mail: vassia.atanassova@gmail.com

I. Vardeva · E. Sotirova
Intelligent Systems Laboratory, "Prof. Dr. Asen Zlatarov" University, 1 "Prof. Yakimov" Blvd, 8010 Burgas, Bulgaria
e-mail: iveto@btu.bg

E. Sotirova
e-mail: esotirova@btu.bg

L. Doukovska
Intelligent Systems Department, IICT – Bulgarian Academy of Sciences, Acad. G. Bonchev Str., Bl. 2, 1113 Sofia, Bulgaria
e-mail: doukovska@iit.bas.bg

K.T. Atanassov et al. (eds.), *Novel Developments in Uncertainty Representation and Processing*, Advances in Intelligent Systems and Computing 401,
DOI 10.1007/978-3-319-26211-6_14

1 Introduction

Here we give a new line of research in the area of a novel decision support approach based on intuitionistic fuzzy sets and index matrices, titled InterCriteria Analysis.

In a previous paper [10] from this leg of the research, one of the authors discussed a new approach to represent the results of the InterCriteria Analysis, using the IF pairs as coordinates of points plotted onto the IF interpretational triangle with vertices (0; 0), (1; 0) and (0; 1), staying respectively for the complete uncertainty, complete truth and complete falsity.

Using this geometrical interpretation, we benefit from the possibility to see the exact shape of the intuitionistic fuzzy set of correlations in between a given set of criteria, but also from the possibility to have a new, and supposedly more precise, method of determining the top-correlating InterCriteria pairs and individual criteria. The proposed method is related to the use of two predefined thresholds α, β for the membership and the non-membership parts of the IF pair, forming the trapezium cut-out adjacent to the (1; 0) point and more generally to computing for each point in this cut-out its distance to the (1; 0) point, and then ranking the points according to the so calculated distances.

The proposed method requires that these threshold values α, β are known in advance and predefined, which for various reasons may not always be the case. In such cases, we need to find a method of defining these threshold values, and the present work aims to propose several such methods that may facilitate or objectify the process of decision making, and may further prompt new ideas of development of the InterCriteria Analysis approach.

2 Working Concepts

2.1 Intuitionistic Fuzzy Sets

Let X be a fixed universe. The intuitionistic fuzzy set over X has the form

$$A = \{\langle x, \mu_A(x), n_A(x)\rangle | x \in X\},$$

where $\mu_A(x)$, $\nu_A(x)$ are degrees of membership and non-membership of elements $x \in X$ to a fixed set $A \subset X$, with $0 \leq \mu_A(x), \nu_A(x) \leq 1$ and $0 \leq \mu_A(x) + \nu_A(x) \leq 1$.

Intuitionistic fuzzy sets were defined by Atanassov, originally in 1983 [1] and extensively presented in [3, 5, 7].

The IFSs have different geometrical interpretations (see, e.g., [5, 7]). One of them is linear, in analogy with the visualization of the ordinary fuzzy sets. In another, completely IFS-specific interpretation, first introduced in [4] (also, [5, 7]), the membership and non-membership values of each element x of the set A are interpreted as points into the an orthogonal triangle, with coordinates (0; 0), (1; 0) and (0; 1), where the hypotenuse is the graphical representation of the [0; 1], to which elements of the ordinary fuzzy set belong.

Over IFS, a number of operations, relations and operators have been defined. For the needs of the present research, the two major topological operators in use, defined by Atanassov in 1985 and later extended (see [2, 5, 6]).

2.2 InterCriteria Analysis

Here we will briefly repeat the theoretical framework of the proposed approach, firstly proposed in [9], while slightly improving in this part the notation from [11].

The approach of InterCriteria Analysis (see the ICA Research portal [14]) was originally devised in 2014 as an approach for IFS-based identification of correlations among a set of criteria involved in a decision making process. This is needed in problems, where measuring according to some of the criteria in the set is slower or more expensive, thus resulting in delay or raising the cost of the overall process of decision making. When such problems are being solved, the decision maker may deem appropriate to reasonably eliminate these criteria, in order to achieve economy and efficiency, while not compromising the overall level of accuracy. The ICA approach has been developed in order to address exactly such class of problems. Since then the approach has been extensively explored in both theoretical and applied direction. The aim is to have it approbated with real life examples with two major goals. The results may validate the approach by detecting patterns of correlations between the criteria, which are expected or known in advance through other methods, and furthermore the results may lead to discovery of new, previously unknown patterns that may.

The approach employs an index matrix (IM, see [8]) M of m rows $\{O_1, \ldots, O_m\}$ and n columns $\{C_1, \ldots, C_n\}$, where for every p, q $(1 \leq p \leq m, 1 \leq q \leq n)$, O_p in an evaluated object, C_q is a evaluation criterion, and $e_{O_pC_q}$ is the evaluation of the p-th object against the q-th criterion, defined as a real number or another object that is comparable according to relation R with all the rest elements of the index matrix M.

$$M = \begin{array}{c|ccccccc}
 & C_1 & \ldots & C_k & \ldots & C_l & \ldots & C_n \\
\hline
O_1 & e_{O_1,C_1} & \ldots & e_{O_1,C_k} & \ldots & e_{O_1,C_l} & \ldots & e_{O_1,C_n} \\
\vdots & \vdots & \ddots & \vdots & \ddots & \vdots & \ddots & \vdots \\
O_i & e_{O_i,C_1} & \ldots & e_{O_i,C_k} & \ldots & e_{O_i,C_l} & \ldots & e_{O_i,C_n} \\
\vdots & \vdots & \ddots & \vdots & \ddots & \vdots & \ddots & \vdots \\
O_j & e_{O_j,C_1} & \ldots & e_{O_j,C_k} & \ldots & e_{O_j,C_l} & \ldots & e_{O_j,C_n} \\
\vdots & \vdots & \ddots & \vdots & \ddots & \vdots & \ddots & \vdots \\
O_m & e_{O_m,C_1} & \ldots & e_{O_m,C_j} & \ldots & e_{O_m,C_l} & \ldots & e_{O_m,C_n}
\end{array},$$

From the requirement for comparability above, it follows that for each i, j, k it holds the relation $R(a_{OiCk}, a_{OjCk})$. The relation R has dual relation $\overline{R}$, which is true in the cases when relation R is false, and vice versa.

For the needs of our decision making method, pairwise comparisons between every two different criteria are made along all evaluated objects. During the comparison, it is maintained one counter of the number of times when the relation R holds, and another counter for the dual relation.

Let $S^{\mu}_{k,l}$ be the number of cases in which the relations $R(e_{OiCk}, e_{OjCk})$ and $R(e_{Oi Cl}, e_{OjCl})$ are simultaneously satisfied. Let also $S^{\nu}_{k,l}$ be the number of cases in which the relations $R(e_{OiCk}, e_{OjCk})$ and the dual relation $\overline{R}(e_{OiCl}, e_{OjCl})$ are simultaneously satisfied. As the total number of pairwise comparisons between the object is $m.n.(n-1)/2$, it is seen that there hold the inequalities:

$$0 \leq S^{\mu}_{k,l} + S^{\nu}_{k,l} \leq \frac{n(n-1)}{2}.$$

For every k, l, such that $1 \leq k \leq l \leq m$, and for $n \geq 2$ two numbers are defined:

$$\mu_{C_k,C_l} = 2\frac{S^{\mu}_{k,l}}{n(n-1)}, \quad \nu_{C_k,C_l} = 2\frac{S^{\nu}_{k,l}}{n(n-1)}.$$

The pair constructed from these two numbers plays the role of the intuitionistic fuzzy evaluation of the relations that can be established between any two criteria C_k and C_l. In this way the index matrix M that relates evaluated objects with evaluating criteria can be transformed to another index matrix M^* that gives the relations among the criteria:

$$M^* = \begin{array}{c|ccc} & C_1 & \cdots & C_m \\ \hline C_1 & \langle \mu_{C_1,C_1}, \nu_{C_1,C_1} \rangle & \cdots & \langle \mu_{C_1,C_m}, \nu_{C_1,C_m} \rangle \\ \vdots & \vdots & \ddots & \vdots \\ C_m & \langle \mu_{C_m,C_1}, \nu_{C_1,C_m} \rangle & \cdots & \langle \mu_{C_m,C_m}, \nu_{C_m,C_m} \rangle \end{array}.$$

From practical considerations, it has been more flexible to work with two index matrices M^{μ} and M^{ν}, rather than with the index matrix M^* of IF pairs.

The final step of the ICA algorithm is to determine the degrees of correlation between the criteria, depending on the user's choice of μ and ν. We call these correlations between the criteria: 'positive consonance', 'negative consonance' or 'dissonance'. Let $\alpha, \beta \in [0; 1]$ be the threshold values, against which we compare the values of $\mu_{Ck,Cl}$ and $\nu_{Ck,Cl}$. We call that criteria C_k and C_l are in:

- (α, β)-positive consonance, if $\mu_{Ck,Cl} > \alpha$ and $\nu_{Ck,Cl} < \beta$;
- (α, β)-negative consonance, if $\mu_{Ck,Cl} < \beta$ and $\nu_{Ck,Cl} > \alpha$;
- (α, β)-dissonance, otherwise.

The present work is a next step of the research, which was branched out in [10] in a new direction, different from the extensive studies related to the adequate methods to defining thresholds, among which [11–13].

3 Main Results

When we have an (arbitrary) IFS plotted on the IF interpretational triangle, we can apply to it the topological operators *Closure* and *Interior*, which are defined using the following formulas, and illustrated in Fig. 1. (see [2, 5, 7]).

$$C(A) = \{\langle x, \sup_{y \in E} \mu_A(y), \inf_{y \in E} \nu_A(y)\rangle | x \in E\}$$

$$I(A) = \{\langle x, \inf_{y \in E} \mu_A(y), \sup_{y \in E} \nu_A(y)\rangle | x \in E\}$$

We will note that since, in the context of InterCriteria Analysis we are only working with finite sets of m objects, of n criteria, and therefore with a resultant finite set of $n(n - 1)/2$ InterCriteria pairs, we can safely replace the functions 'supremum' and 'infimum', respectively by the functions 'maximum' and 'minimum'.

We will note that depending on the particular set, it may have the form of a triangle, in the case of fuzzy set, all of which elements are plotted onto the hypotenuse, or a trapezium, or a pentagon, see Fig. 2. Let us consider the most general case of a pentagon.

We are interested to find an appropriate procedure to rank the points corresponding to the InterCriteria pairs, and we have to make the stipulation that here we

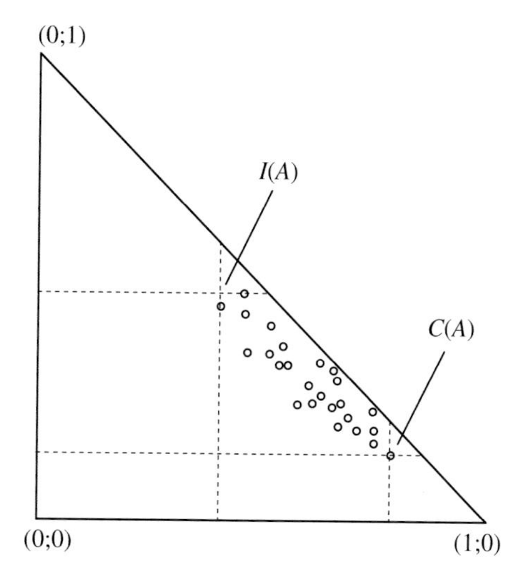

Fig. 1 An IFS, plotted onto the IF *Triangle*, with the indicated places of the topological operators Closure and Interior

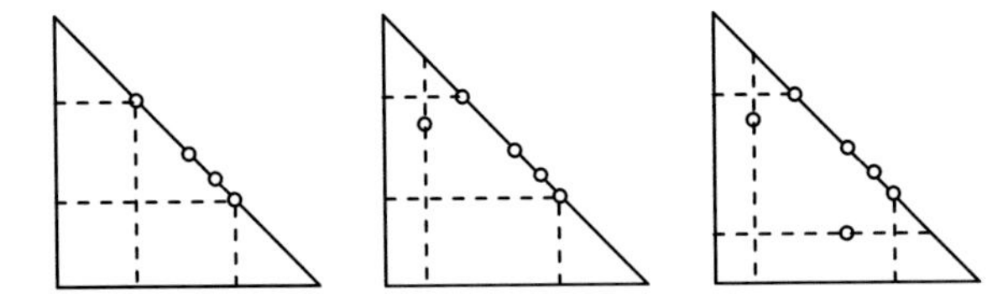

Fig. 2 *Triangle*, trapezium or pentagon are the possible shapes of the zone, enclosed by the topological operators Closure and Interior, and the hypotenuse of the IF *Triangle*

are considering a problem for which solution with the ICA method, we need to discover the highest possible consonances between the criteria, which back to the IF triangular interpretation means that we are looking for ranking them per their proximity to the point (1; 0).

The procedure for ranking the InterCriteria pairs, hence, comprises two phases: 1. To define the unit lengths of the rectangular grid that will divide the pentagon; 2. To define the consequence of traversing the so-defined subrectangles of the grid.

3.1 Defining the Rectangular Grid

A possible way to define the unit lengths a, b of the rectangular grid is given with the following two formulas:

$$a = \frac{\max\limits_{y \in E} \mu_A(y) - \min\limits_{y \in E} \mu_A(y)}{\frac{n(n-1)}{2}}, \; b = \frac{\max\limits_{y \in E} \nu_A(y) - \min\limits_{y \in E} \nu_A(y)}{\frac{n(n-1)}{2}}$$

The lengths PQ and QR are divided by the total number of points in the plotted set, and this is the finest possible division for the grid.

Another approach is to assign to a and b, respectively, the smallest possible positive, non-null difference in the first coordinates of any two points of the set, and the smallest possible positive, non-null difference in the second coordinates of any two points in the set, by the formulas:

$$a = \min_{i,j \in A} (|\mu_i - \mu_j|), \; b = \min_{k,l \in A} (|\nu_k - \nu_l|).$$

A simple check with arbitrary values can show that the results returned are not the identical.

For the sake of completeness, we can also note the most obvious way of defining the unit lengths of the rectangular grid by dividing PQ and QR into predefined number(s), not necessarily the same number of sections per side. Then, for predefined numbers u, w, the formulas will have the following forms:

$$a = \frac{\max_{y \in E} \mu_A(y) - \min_{y \in E} \mu_A(y)}{u}, \quad b = \frac{\max_{y \in E} \nu_A(y) - \min_{y \in E} \nu_A(y)}{w}$$

It is to be noted that the idea for the grid and its divisions is to a certain extent an analogue of the idea of setting a threshold.

3.2 Traversing the Subrectangles of the Grid

The question of the consequence of traversing the subrectangles of the grid, and thus determining the ordering of the InterCriteria pairs, is very interesting by itself, and reduces to the essential question of how we prioritize between the three intuitionistic fuzzy components of membership, non-membership and uncertainty. Different strategies, or scenarios, can be discussed here. Let us illustrate our discussions with the IFS from Fig. 3, and the grid in which the unit lengths have been determined in one of the possible ways, discussed above in Sect. 3.1.

(1) Strategy "Max μ First"

In response to this strategy, we start with the subrectangle with the maximal membership and minimal non-membership, i.e. the one containing the set's closure, and traverse through the grid in vertical direction (bottom-to-top), in a way that preserves the membership part as high as possible, while running through the gradually increasing non-membership parts, as illustrated in Fig. 4.

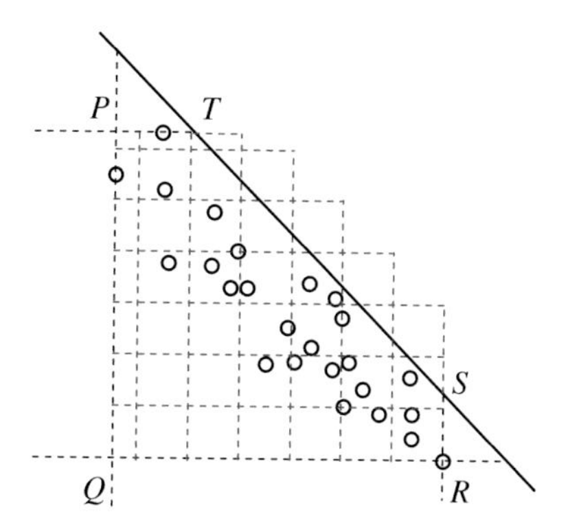

Fig. 3 The segment from the IF *Triangle*, containing the pentagonal zone, enclosed by the topological operators Closure and Interior (points R and P) and the hypotenuse, gridded in *red* with unit *rectangles*

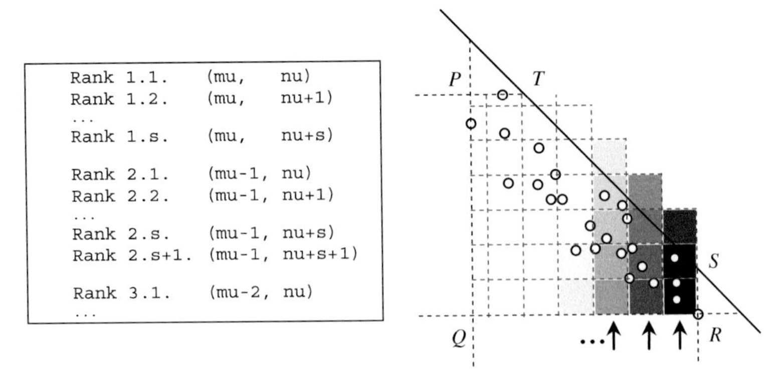

Fig. 4 Traversing the *subrectangles* of the grid, following the strategy "max μ first"

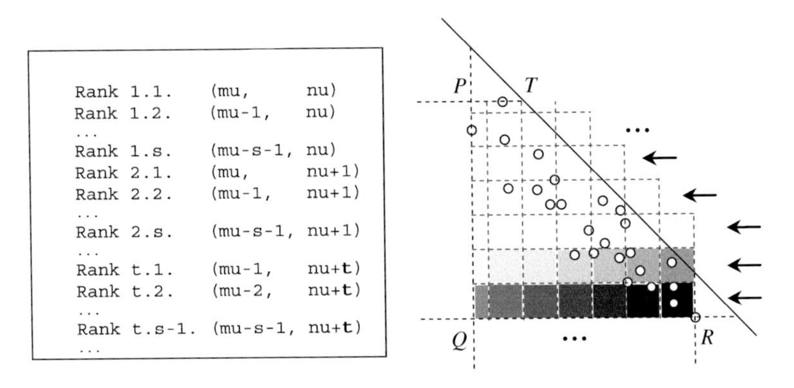

Fig. 5 Traversing the *subrectangles* of the grid, following the strategy "min ν first"

The following pseudocode gives it in a more formal way:

(2) Strategy "Min ν First"

In response to this strategy, we again start with the subrectangle with the maximal membership and minimal non-membership, but this time we traverse through the grid in horizontal direction (right-to-left), in a way that preserves the non-membership part as low as possible, while running through the gradually decreasing membership parts. The illustration of this strategy follows by analogy, Fig. 5.

The following pseudocode gives it in a more formal way:

(3) Diagonal Strategy

We start with the subrectangle with the maximal membership and minimal nonmembership. As a next step, we take the simultaneously the union of the subrectangles that are located one up and one left of the previous, and so forth, as illustrated in Fig. 4 and by the following pseudocode (Fig. 6).

```
Rank 1.    (mu, nu)
Rank 2.    (mu-1, nu), (mu+1, nu)
Rank 3.    (mu-2, nu), (mu-1,
nu+1), (mu, nu+2)
...
Rank r+1. (mu-r, nu), (mu-r+1,
nu+1),
(mu-r+2, nu+2), ..., (mu-2,
nu+r-2),
(mu-1, nu+r-1), (mu, nu+r)
...
```

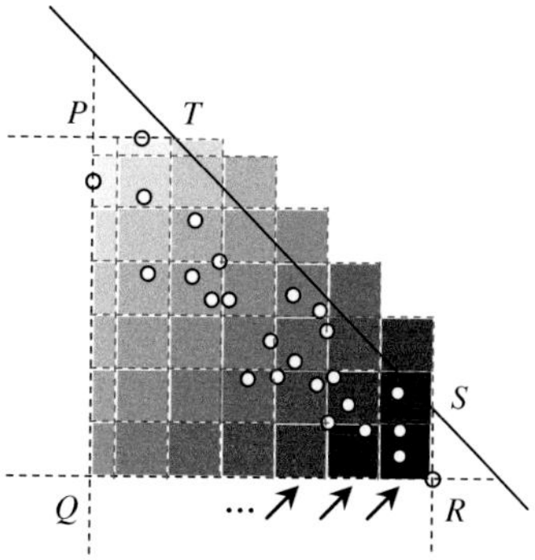

Fig. 6 Traversing the grid, following the diagonal strategy

Any of the strategies in A. can be combined with whichever strategy from B., per expert's decision, which makes the list of possible choices for the expert wide enough. New alternative options are also possible.

4 Example and Discussions

We will demonstrate the proposed strategies for traversing and ranking with an example. In a series of publications, we have illustrated these subsequent steps of elaborating the theory of ICA with a case study aimed at application of the ICA approach to data sourced from the World Economic Forum's annual Global Competitiveness Reports (GCRs), for the 28 EU member states from years 2008–2009 to 2013–2014, taking as a motivation the WEF's general address to policy makers to '*identify and strengthen the transformative forces that will drive future economic growth*' [16].

Here, we will present this new step of developing the theory of ICA with data from the GCR for the year 2014–2015. The data extracted from the GCR are the evaluations of the 28 EU Member States (in ICA: *objects*) according to the 12 pillars of competitiveness (in ICA: *criteria*), which by the adopted methodology of WEF are numbers between 1 and 7, with precision of one number after the decimal point. The 12 criteria are '1. Institutions'; '2. Infrastructure'; '3. Macroeconomic stability'; '4. Health and primary education'; '5. Higher education and training'; '6. Goods market efficiency'; '7. Labor market efficiency'; '8. Financial market sophistication'; '9. Technological readiness'; '10. Market size'; '11. Business sophistication'; '12. Innovation'. These evaluations are input in the form of an index matrix with dimensions 28×12.

A software application implementing the algorithm of ICA has performed the computations over this matrix and has produced the following two 12×12 matrices, which contain the membership and the non-membership components of the intuitionistic fuzzy pairs that represent for each pair of criteria the degree of either positive or negative consonance, or dissonance. The two matrices are symmetrical according to their main diagonal, along which all the IF pairs are all identical to the perfect Truth, $\mu = 1$, $\nu = 0$, since every criterion would perfectly correlated only with itself.

Taking the results produced by the ICA, with the IF pairs distributed in two index matrices M^{μ} and M^{ν}, collected respectively in Tables 1, 2, we plot them onto the IF interpretational triangle in Fig. 7.

Table 1 Discovered membership values with the application of ICA for 2014–2015

M^{μ}	1	2	3	4	5	6	7	8	9	10	11	12
1		0.696	0.585	0.690	0.796	0.839	0.772	0.706	0.833	0.497	0.783	0.817
2	0.696		0.460	0.659	0.714	0.651	0.563	0.545	0.751	0.648	0.775	0.765
3	0.585	0.460		0.407	0.529	0.595	0.664	0.704	0.585	0.439	0.540	0.548
4	0.690	0.659	0.407		0.751	0.667	0.558	0.492	0.661	0.495	0.698	0.698
5	0.796	0.714	0.529	0.751		0.743	0.653	0.598	0.759	0.577	0.762	0.812
6	0.839	0.651	0.595	0.667	0.743		0.780	0.675	0.770	0.476	0.751	0.738
7	0.772	0.563	0.664	0.558	0.653	0.780		0.722	0.728	0.418	0.656	0.683
8	0.706	0.545	0.704	0.492	0.598	0.675	0.722		0.677	0.516	0.656	0.664
9	0.833	0.751	0.585	0.661	0.759	0.770	0.728	0.677		0.542	0.786	0.786
10	0.497	0.648	0.439	0.495	0.577	0.476	0.418	0.516	0.542		0.614	0.603
11	0.783	0.775	0.540	0.698	0.762	0.751	0.656	0.656	0.786	0.614		0.857
12	0.817	0.765	0.548	0.698	0.812	0.738	0.683	0.664	0.786	0.603	0.857	

Table 2 Discovered non-membership values with the application of ICA for 2014–2015

M^{ν}	1	2	3	4	5	6	7	8	9	10	11	12
1		0.222	0.349	0.169	0.135	0.071	0.140	0.212	0.106	0.447	0.138	0.114
2	0.222		0.460	0.188	0.198	0.241	0.331	0.354	0.175	0.294	0.132	0.159
3	0.349	0.460		0.460	0.405	0.312	0.246	0.217	0.352	0.508	0.378	0.381
4	0.169	0.188	0.460		0.108	0.177	0.278	0.339	0.190	0.373	0.140	0.146
5	0.135	0.198	0.405	0.108		0.177	0.249	0.315	0.175	0.368	0.159	0.114
6	0.071	0.241	0.312	0.177	0.177		0.106	0.212	0.143	0.447	0.153	0.161
7	0.140	0.331	0.246	0.278	0.249	0.106		0.172	0.188	0.497	0.235	0.220
8	0.212	0.354	0.217	0.339	0.315	0.212	0.172		0.243	0.410	0.257	0.254
9	0.106	0.175	0.352	0.190	0.175	0.143	0.188	0.243		0.405	0.138	0.148
10	0.447	0.294	0.508	0.373	0.368	0.447	0.497	0.410	0.405		0.320	0.336
11	0.138	0.132	0.378	0.140	0.159	0.153	0.235	0.257	0.138	0.320		0.069
12	0.114	0.159	0.381	0.146	0.114	0.161	0.220	0.254	0.148	0.336	0.069	

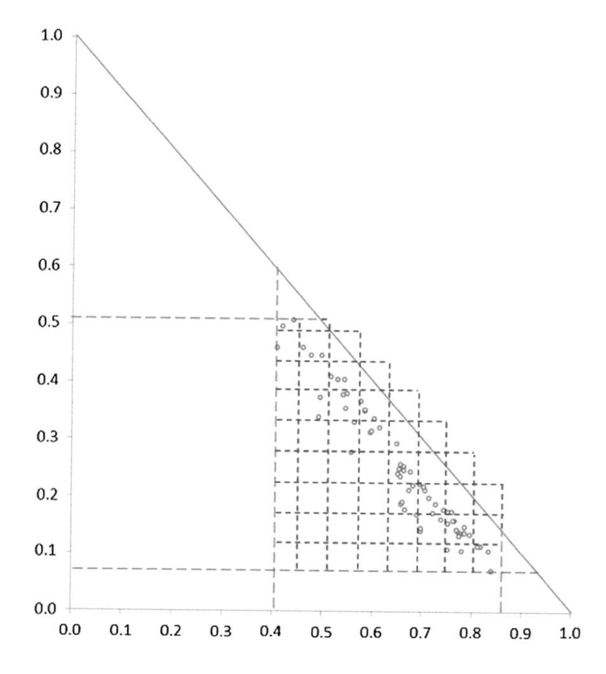

Fig. 7 Example with the ICA results for the year 2014–2015, plotted on the IF *triangle*

Table 3 Ranking of the InterCriteria consonance pairs

Rank	# Pairs	Criteria	μ	ν
1	4	11. Business sophistication—12. Innovation	0.857	0.069
		1. Institutions—6. Goods market efficiency	0.839	0.071
		1. Institutions—9. Technological readiness	0.833	0.106
		1. Institutions—12. Innovation	0.817	0.114
2	2	5. Higher education and training—12. Innovation	0.812	0.114
		1. Institutions—5. Higher education and training	0.796	0.135
3	10	9. Technological readiness—11. Business sophistication	0.786	0.138
		9. Technological readiness—12. Innovation	0.786	0.148
		1. Institutions—11. Business sophistication	0.783	0.138
		6. Goods market efficiency—7. Labor market efficiency	0.780	0.106
		2. Infrastructure—11. Business sophistication	0.775	0.132
		1. Institutions—7. Labor market efficiency	0.772	0.140
		6. Goods market efficiency—9. Technological readiness	0.770	0.143
		2. Infrastructure—12. Innovation	0.765	0.159
		5. Higher education and training—11. Business sophistication	0.762	0.159
		5. Higher education and training—9. Technological readiness	0.759	0.175

(continued)

Table 3 (continued)

Rank	# Pairs	Criteria	μ	ν
4	6	4. Health and primary education—5. Higher education and training	0.751	0.108
		6. Goods market efficiency—11. Business sophistication	0.751	0.153
		2. Infrastructure—9. Technological readiness	0.751	0.175
		5. Higher education and training—6. Goods market efficiency	0.743	0.177
		6. Goods market efficiency—12. Innovation	0.738	0.161
		7. Labor market efficiency—9. Technological readiness	0.728	0.188
5	8	7. Labor market efficiency—8. Financial market sophistication	0.722	0.172
		2. Infrastructure—5. Higher education and training	0.714	0.198
		1. Institutions—8. Financial market sophistication	0.706	0.212
		3. Macroeconomic stability—8. Financial market sophistication	0.704	0.217
		4. Health and primary education—11. Business sophistication	0.698	0.140
		4. Health and primary education—12. Innovation	0.698	0.146
		1. Institutions—2. Infrastructure	0.696	0.222
		1. Institutions—4. Health and primary education	0.690	0.169
6	5	7. Labor market efficiency—12. Innovation	0.683	0.220
		8. Financial market sophistication—9. Technological readiness	0.677	0.243
		6. Goods market efficiency—8. Financial market sophistication	0.675	0.212
		4. Health and primary education—6. Goods market efficiency	0.667	0.177
		3. Macroeconomic stability—7. Labor market efficiency	0.664	0.246
7	7	8. Financial market sophistication—12. Innovation	0.664	0.254
		4. Health and primary education—9. Technological readiness	0.661	0.190
		2. Infrastructure—4. Health and primary education	0.659	0.188
		7. Labor market efficiency—11. Business sophistication	0.656	0.235
		8. Financial market sophistication—11. Business sophistication	0.656	0.257
		5. Higher education and training—7. Labor market efficiency	0.653	0.249
		2. Infrastructure—6. Goods market efficiency	0.651	0.241
8	1	2. Infrastructure—10. Market size	0.648	0.294
9	4	10. Market size—11. Business sophistication	0.614	0.320
		10. Market size—12. Innovation	0.603	0.336
		5. Higher education and training—8. Financial market sophistication	0.598	0.315
		3. Macroeconomic stability—6. Goods market efficiency	0.595	0.312

(continued)

Table 3 (continued)

Rank	# Pairs	Criteria	μ	ν
10	5	1. Institutions—3. Macroeconomic stability	0.585	0.349
		3. Macroeconomic stability—9. Technological readiness	0.585	0.352
		5. Higher education and training—10. Market size	0.577	0.368
		2. Infrastructure—7. Labor market efficiency	0.563	0.331
		4. Health and primary education—7. Labor market efficiency	0.558	0.278
11	3	3. Macroeconomic stability—12. Innovation	0.548	0.381
		2. Infrastructure—8. Financial market sophistication	0.545	0.354
		9. Technological readiness—10. Market size	0.542	0.405
12	5	3. Macroeconomic stability—11. Business sophistication	0.540	0.378
		3. Macroeconomic stability—5. Higher education and training	0.529	0.405
		8. Financial market sophistication—10. Market size	0.516	0.410
		1. Institutions—10. Market size	0.497	0.447
		4. Health and primary education—10. Market size	0.495	0.373
13	3	4. Health and primary education—8. Financial market sophistication	0.492	0.339
		6. Goods market efficiency—10. Market size	0.476	0.447
		2. Infrastructure—3. Macroeconomic stability	0.460	0.460
14	1	3. Macroeconomic stability—10. Market size	0.439	0.508
15	2	7. Labor market efficiency—10. Market size	0.418	0.497
		3. Macroeconomic stability—4. Health and primary education	0.407	0.460

Let us assume that the unit length of the introduced rectangular grid is defined in the easiest possible way, with $a = b = 0.05$. Let us also assume that the traversing strategy adopted in the diagonal strategy discussed above. With respect to the listed steps, the ranking of the InterCriteria pairs is as follows in Table 3.

We will only note here that depending on the shape of IFS, obtained by its plotted elements, we can consider optimizing the traversing procedures, by skipping subrectangles of the grid that do not contain set elements. To do this, we should use the four topological operators $\Gamma_L(A)$, $\Gamma_R(A)$, $\Gamma_U(A)$, $\Gamma_D(A)$, which give the minimal convex region that completely contains the set A, per the definitions and formulas given in [15].

5 Conclusions

Despite the fact that the above proposed algorithms for traversing and ranking of the elements of an IFS have been developed for the needs of the InterCriteria Analysis, they can be used in any other relevant context, as well. The authors will be interested in their wider adoption and applicability.

This research is an essential development of previous works where the correlations between the criteria (hence the term 'InterCriteria') were analysed separately, by first setting priority on either the membership, or the non-membership component, and plotting them linearly. Starting from [10], and continuing here, however, we have started exploring a much more adequate approach for handling both IF components simultaneously by plotting them in the plane of the intuitionistic fuzzy interpretation triangle.

In future, it will be interesting to discuss the possibility of the usage of the six generalizations of the two topological operators *Closure* and *Interior*, as defined in [7].

Acknowledgments The authors are grateful for the support provided by the National Science Fund of Bulgaria under grant DFNI-I-02-5/2014.

References

1. Atanassov, K.: Intuitionistic fuzzy sets. Proceedings of VII ITKR's Session, Sofia (Bulgarian) (1983)
2. Atanassov, K.: Modal and topological operators, defined over intuitionistic fuzzy sets. In: Youth scientific contributions, vol. 1, pp. 18–21. Academic Publishing House, Sofia (1985)
3. Atanassov, K.: Intuitionistic fuzzy sets. Fuzzy Sets Syst. Elsevier **20**(1), 87–96 (1986)
4. Atanassov, K.: Geometrical interpretations of the elements of the intuitionistic fuzzy objects. Preprint IM-MFAIS-1-89, Sofia (1989)
5. Atanassov, K.: Intuitionistic Fuzzy Sets. Springer, Heidelberg (1999)
6. Atanassov, K.: On four intuitionistic fuzzy topological operators. Mathware Soft Comput. **8**, 65–70 (2001)
7. Atanassov, K.: On Intuitionistic Fuzzy Sets Theory. Springer, Berlin (2012)
8. Atanassov, K.: Index Matrices: Towards an Augmented Matrix Calculus. Springer, Cham (2014)
9. Atanassov, K., Mavrov, D., Atanassova, V.: InterCriteria decision making. A new approach for multicriteria decision making, based on index matrices and intuitionistic fuzzy sets. Issues Intuitionistic Fuzzy Sets Generalized Nets **11**, 1–8 (2014)
10. Atanassova, V.: Interpretation in the intuitionistic fuzzy triangle of the results, obtained by the intercriteria analysis. In: Proceedings of IFSA-EUSFLAT 2015, 30.06.2015–03.07.2015, pp. 1369–1374. Atlantic Press, Gijon, Spain (2015)
11. Atanassova, V., Doukovska, L., Atanassov, K., Mavrov, D.: InterCriteria decision making approach to EU member states competitiveness analysis. In: Proceedings of 4th International Symposium on Business Modeling and Software Design, pp. 289–294. 24–26 Jun 2014, Luxembourg (2014)
12. Atanassova, V., Mavrov, D., Doukovska, L., Atanassov, K.: Discussion on the threshold values in the InterCriteria decision making approach. Int. J. Notes Intuitionistic Fuzzy Sets **20** (2):94–99 (2014)
13. Atanassova, V., Vardeva, I.: Sum- and average-based approach to criteria shortlisting in the InterCriteria analysis. Int. J. Notes Intuitionistic Fuzzy Sets **20**(4):41–46 (2014)
14. InterCriteria Research Portal. http://www.intercriteria.net/publications
15. Rangasamy, P., Vassilev, P., Atanassov, K.: New topological operators over intuitionistic fuzzy sets. Adv. Stud. Contemp. Math. **18**(1), 49–57 (2009)
16. World Economic Forum. The Global Competitiveness Report 2014–2015. http://www.weforum.org/issues/global-competitiveness

A Novel Similarity Measure Between Intuitionistic Fuzzy Sets for Constructing Intuitionistic Fuzzy Tolerance

Janusz Kacprzyk, Dmitri A. Viattchenin, Stanislau Shyrai and Eulalia Szmidt

Abstract This paper deals with the problem of constructing intuitionistic fuzzy tolerance from a family of intuitionistic fuzzy sets. A method to calculate the intuitionistic fuzzy tolerance degrees between intuitionistic fuzzy sets on the basis of the Euclidean distance is proposed. An illustrative example used to compare the proposed similarity measure with other similarity measures and an application of the proposed similarity measure to clustering problem is considered. Preliminary conclusions are formulated.

Keywords Intuitionistic fuzzy set · Intuitionistic fuzzy tolerance · Similarity measure · Clustering

1 Introduction

Since the original Atanassov's [1] paper was published, intuitionistic fuzzy set theory has been applied to many areas and new concepts were introduced. Intuitionistic fuzzy set theory was developed by different researchers and monographs

J. Kacprzyk (✉) · E. Szmidt
Systems Research Institute Polish Academy of Sciences, Warsaw, Poland
e-mail: Janusz.Kacprzyk@ibspan.waw.pl

E. Szmidt
e-mail: szmidt@ibspan.waw.pl

D.A. Viattchenin · S. Shyrai
Department of Software Information Technology, Belarusian State University of Informatics and Radio-Electronics, Minsk, Belarus
e-mail: viattchenin@mail.ru

S. Shyrai
e-mail: ashaman410@gmail.com

J. Kacprzyk · E. Szmidt
WIT-Warsaw School of Information Technology Warsaw, Warsaw, Poland

K.T. Atanassov et al. (eds.), *Novel Developments in Uncertainty Representation and Processing*, Advances in Intelligent Systems and Computing 401,
DOI 10.1007/978-3-319-26211-6_15

[2, 3] are most informative bibliographical sources in this area. Intuitionistic fuzzy clustering procedures were also elaborated by different researchers. Different relational and prototype-based intuitionistic fuzzy clustering procedures are presented in literature. For example, intuitionistic fuzzy-set clustering methods are described by Xu in [4].

Distance measurement is a basis for clustering techniques. Different distances and similarity measures between intuitionistic fuzzy sets are considered in [5]. Moreover, similarity measures between intuitionistic fuzzy sets for constructing intuitionistic fuzzy tolerance relation were proposed in [6, 7].

In this paper, we propose a new method of construction of intuitionistic fuzzy tolerances based on a similarity measure between intuitionistic fuzzy sets. The proposed similarity measure is based on generalization of normalized Euclidean distance. So, the contents of this paper is as follow: in the second section some definitions of the intuitionistic fuzzy set theory are described, in the third section two similarity measures between intuitionistic fuzzy sets for constructing intuitionistic fuzzy tolerance are considered, in the fourth section a novel similarity measure is proposed, in the fifth section the new similarity measure is illustrated by a short example, in the sixth section an application of the proposed similarity measure to clustering is given in comparison with the similarity measure based on generalization of normalized Hamming distance, in the seventh section some preliminary conclusions are given.

2 Basic Definitions of the Intuitionistic Fuzzy Set Theory

Let us remind some basic definitions of the Atanassov's intuitionistic fuzzy set theory [1–3]. All concepts will be considered for a finite universe $X = \{x_1, \ldots, x_n\}$.

An intuitionistic fuzzy set IA in X is given by ordered triple $IA = \{\langle x_i, \mu_{IA}(x_i), \nu_{IA}(x_i)\rangle | x_i \in X\}$, where μ_{IA}, ν_{IA}: $X \to [0, 1]$ should satisfy a condition

$$0 \leq \mu_{IA}(x_i) + \nu_{IA}(x_i) \leq 1, \tag{1}$$

for all $x_i \in X$. The values $\mu_{IA}(x_i)$ and $\nu_{IA}(x_i)$ denote the degree of membership and the degree of non-membership of element $x_i \in X$ to IA, respectively. For each intuitionistic fuzzy set IA in X an intuitionistic fuzzy index [1] of an element $x_i \in X$ in IA can be defined as follows

$$\rho_{IA}(x_i) = 1 - (\mu_{IA}(x_i) + \nu_{IA}(x_i)). \tag{2}$$

The intuitionistic fuzzy index $\rho_{IA}(x_i)$ can be considered as a hesitancy degree of x_i to IA. It is seen that $0 \leq \rho_{IA}(x_i) \leq 1$ for all $x_i \in X$. Obviously, when $\nu_{IA}(x_i) = 1 - \mu_{IA}(x_i)$ for every $x_i \in X$, the intuitionistic fuzzy set IA is an ordinary fuzzy set in X. For each ordinary fuzzy set A in X, we have $\rho_A(x_i) = 0, \forall x_i \in X$.

Let $X = \{x_1, \ldots, x_n\}$ be an ordinary non-empty set. The binary intuitionistic fuzzy relation IR on X is an intuitionistic fuzzy subset IR of $X \times X$, which is given by the expression

$$IR = \{\langle (x_i, x_j), \mu_A(x_i, x_j), \nu_A(x_i, x_j) \rangle | x_i, x_j \in X\}, \tag{3}$$

where $\mu_{IR}\colon X \times X \to [0, 1]$ and $\nu_{IR}\colon X \times X \to [0, 1]$ satisfy the condition $0 \le \mu_{IR}(x_i, x_j) + \nu_{IR}(x_i, x_j) \le 1$ for every $(x_i, x_j) \in X \times X$ [8].

Let $\mathrm{IFR}(X)$ denote the set of all intuitionistic fuzzy relations on some universe X. An intuitionistic fuzzy relation $IR \in \mathrm{IFR}(X)$ is reflexive if for every $x_i \in X$, $\mu_{IR}(x_i, x_i) = 1$ and $\nu_{IR}(x_i, x_i) = 0$. An intuitionistic fuzzy relation $IR \in \mathrm{IFR}(X)$ is called symmetric if for all $(x_i, x_j) \in X \times X$, $\mu_{IR}(x_i, x_j) = \mu_{IR}(x_j, x_i)$ and $\nu_{IR}(x_i, x_j) = \nu_{IR}(x_j, x_i)$. An intuitionistic fuzzy relation IT in X is called an intuitionistic fuzzy tolerance if it is reflexive and symmetric. So, any intuitionistic fuzzy tolerance can be presented by a matrix $r_{n \times n} = [\mu_{IT}(x_i, x_j),\ \nu_{IT}(x_i, x_j)]$, $i, j = 1, \ldots, n$, where a tolerance coefficient $r(x_i, x_j) = (\mu_{IT}(x_i, x_j),\ \nu_{IT}(x_i, x_j))$, $i, j \in \{1, \ldots, n\}$ is called a closeness degree of x_i and x_j [6].

3 Constructing an Intuitionistic Fuzzy Tolerances Based on Measurement of Similarities Between Intuitionistic Fuzzy Sets

The method for constructing the intuitionistic fuzzy tolerance relation was proposed by Wang et al. in [6]. The similarity measure is based on the normalized Hamming distance and the similarity measure can be expressed by a formula

$$r(IA, IB) = \begin{cases} (1,\ 0), & IA = IB \\ \left(\begin{array}{l} 1 - \frac{1}{n}\sum\limits_{i=1}^{n} |\nu_{IA}(x_i) - \nu_{IB}(x_i)| - \frac{1}{n}\sum\limits_{i=1}^{n} |\rho_{IA}(x_i) - \rho_{IB}(x_i)|, \\ \frac{1}{n}\sum\limits_{i=1}^{n} |\nu_{IA}(x_i) - \nu_{IB}(x_i)| \end{array} \right), & IA \neq IB \end{cases}, \tag{4}$$

for all $i, j = 1, \ldots, n$. That is why the closeness degree $r(IA, IB) = (\mu_{IT}(IA, IB),\ \nu_{IT}(IA, IB))$ of intuitionistic fuzzy sets IA and IB can be constructed according to the formula (4).

On the other hand, the method for constructing the intuitionistic fuzzy tolerance which based on the normalized Hausdorff distance was proposed in [7]. The corresponding similarity measure can be expressed by a formula

$$h(IA,IB)=\begin{pmatrix} 1-\frac{1}{n}\sum_{i=1}^{n}\max\{|\nu_{IA}(x_i)-\nu_{IB}(x_i)|,|\rho_{IA}(x_i)-\rho_{IB}(x_i)|\}, \\ \frac{1}{n}\sum_{i=1}^{n}\max|\nu_{IA}(x_i)-\nu_{IB}(x_i)| \end{pmatrix}, \quad (5)$$

for all $i,j=1,\ldots,n$.

Corresponding intuitionistic fuzzy relations possesses the symmetry property and the reflexivity property. Moreover, the condition $0\leq\mu_{IT}(IA,IB)+\nu_{IT}(IA,IB)\leq 1$ is met for any intuitionistic fuzzy sets IA and IB. These facts were proved in [6, 7].

4 The Proposed Similarity Measure

The method for constructing the intuitionistic fuzzy tolerance can be developed for a case of the normalized Euclidean distance. So, the corresponding similarity measure can be written as a

$$e(IA,IB)=\begin{pmatrix} 1-\frac{1}{2n}\sum_{i=1}^{n}\sqrt{(\nu_{IA}(x_i)-\nu_{IB}(x_i))^2+(\rho_{IA}(x_i)-\rho_{IB}(x_i))^2}, \\ \frac{1}{2n}\sum_{i=1}^{n}\sqrt{(\nu_{IA}(x_i)-\nu_{IB}(x_i))^2} \end{pmatrix}, \quad (6)$$

for all $i,j=1,\ldots,n$.

Let us consider some basic properties of the proposed similarity measure (6). In the first place, we need to check whether $0\leq\mu_{IT}(IA,IB)+\nu_{IT}(IA,IB)\leq 1$ holds or not.

Lemma 1 Let IA and IB be two intuitionistic fuzzy sets on $X=\{x_1,\ldots,x_n\}$ and IT be a binary intuitionistic fuzzy relation on X. The condition $0\leq\mu_{IT}(IA,IB)+\nu_{IT}(IA,IB)\leq 1$ is met for the closeness degree of intuitionistic fuzzy sets IA and IB which is constructed according to the formula (6).

Proof Let $e(IA,IB)=(\mu_e(IA,IB),\ \nu_e(IA,IB))$. So,

$$\begin{aligned}\mu_e(IA,IB)+\nu_e(IA,IB)&=1-\frac{1}{2n}\sum_{i=1}^{n}\sqrt{(\nu_{IA}(x_i)-\nu_{IB}(x_i))^2+(\rho_{IA}(x_i)-\rho_{IB}(x_i))^2}\\&+\frac{1}{2n}\sum_{i=1}^{n}\sqrt{(\nu_{IA}(x_i)-\nu_{IB}(x_i))^2}\leq 1-\frac{1}{2n}\sum_{i=1}^{n}\sqrt{(\nu_{IA}(x_i)-\nu_{IB}(x_i))^2+(\rho_{IA}(x_i)-\rho_{IB}(x_i))^2}\\&+\frac{1}{2n}\sum_{i=1}^{n}\sqrt{(\nu_{IA}(x_i)-\nu_{IB}(x_i))^2+(\rho_{IA}(x_i)-\rho_{IB}(x_i))^2}=1.\end{aligned}$$

On the other hand, an equation $\sqrt{a^2+b^2}\leq|a|+|b|$ is met. That is why

$$\mu_e(IA, IB) + \nu_e(IA, IB) \geq 1 - \frac{1}{2n}\sum_{i=1}^{n} |\nu_{IA}(x_i) - \nu_{IB}(x_i)| - \frac{1}{2n}\sum_{i=1}^{n} |\rho_{IA}(x_i) - \rho_{IB}(x_i)|$$
$$+ \frac{1}{2n}\sum_{i=1}^{n} |\nu_{IA}(x_i) - \nu_{IB}(x_i)| = 1 - \frac{1}{2n}\sum_{i=1}^{n} |\rho_{IA}(x_i) - \rho_{IB}(x_i)| \geq 0,$$

which completes the proof. □

Lemma 2 The binary intuitionistic fuzzy relation IT on X which is constructed according to the formula (6) is the reflexive intuitionistic fuzzy relation on X.

Proof The proof is straightforward. □

Lemma 3 The binary intuitionistic fuzzy relation IT on X which is constructed according to the formula (6) is the symmetric intuitionistic fuzzy relation on X.

Proof The proof is straightforward. □

The corollary from these lemmas is the proposition that the intuitionistic fuzzy relation IT constructed according to the formula (6) is the intuitionistic fuzzy tolerance.

5 An Illustrative Example

Let us consider an example which was described by Wang, Xu, Liu, and Tang in [6]. Five different cars x_i, $i = 1, \ldots, 5$ must be classified into several kinds. Each car has six evaluation attributes which represent the oil consumption, coefficient of friction, price, comfortable degree, design and safety coefficient evaluated for five cars. Denote oil consumption by x^1, coefficient of friction by x^2, price by x^3, comfortable degree by x^4, design by x^5 and safety coefficient by x^6. The characteristics information of the cars is presented in Table 1.

So, each car can be considered as an intuitionistic fuzzy set x_i, $i = 1, \ldots, 5$, and $\mu_{x_i}(x^t) \in [0, 1]$, $i = 1, \ldots, 5$, $t = 1, \ldots, 6$ are their membership degrees and $\nu_{x_i}(x^t)$, $i = 1, \ldots, 5$, $t = 1, \ldots, 6$ are their non-membership degrees. In other words, each intuitionistic fuzzy set x_i, $i = 1, \ldots, 5$ is defined on the universe of attributes

Table 1 The initial data set, [4]

Objects	Attributes					
	x^1	x^2	x^3	x^4	x^5	x^6
x_1	(0.3, 0.5)	(0.6, 0.1)	(0.4, 0.3)	(0.8, 0.1)	(0.1, 0.6)	(0.5, 0.4)
x_2	(0.6, 0.3)	(0.5, 0.2)	(0.6, 0.1)	(0.7, 0.1)	(0.3, 0.6)	(0.4, 0.3)
x_3	(0.4, 0.4)	(0.8, 0.1)	(0.5, 0.1)	(0.6, 0.2)	(0.4, 0.5)	(0.3, 0.2)
x_4	(0.2, 0.4)	(0.4, 0.1)	(0.9, 0.0)	(0.8, 0.1)	(0.2, 0.5)	(0.7, 0.1)
x_5	(0.5, 0.2)	(0.3, 0.6)	(0.6, 0.3)	(0.7, 0.1)	(0.6, 0.2)	(0.5, 0.3)

$\{x^t | t=1, \ldots, 6\}$. That is why the membership degree $\mu_{x_i}(x^t)$ can be interpreted as the degree of expressiveness of some attribute x^t, $t \in \{1, \ldots, 6\}$ for the object x_i, $i \in \{1, \ldots, 5\}$ and the non-membership degree $\nu_{x_i}(x^t)$ can be considered as the degree of non-expressiveness of the attribute. Thus, if $X = \{x_1, \ldots, x_n\}$ is the set of objects which are defined on the universe of attributes $\{x^t | t=1, \ldots, m\}$ then the formula (4) can be rewritten as follows:

$$r(x_i, x_j) = \begin{cases} (1, 0), & x_i = x_j \\ \begin{pmatrix} 1 - \frac{1}{m} \sum_{t=1}^{m} \left| \nu_{x_i}(x^t) - \nu_{x_j}(x^t) \right| - \frac{1}{m} \sum_{t=1}^{m} \left| \rho_{x_i}(x^t) - \rho_{x_j}(x^t) \right|, \\ \frac{1}{m} \sum_{t=1}^{m} \left| \nu_{x_i}(x^t) - \nu_{x_j}(x^t) \right| \end{pmatrix}, & x_i \neq x_j \end{cases}, \tag{7}$$

for all $i, j = 1, \ldots, n$. In order of priority, the formula (5) can be rewritten as follows:

$$h(x_i, x_j) = \begin{pmatrix} 1 - \frac{1}{m} \sum_{t=1}^{m} \max\left\{ \left| \nu_{x_i}(x^t) - \nu_{x_j}(x^t) \right|, \left| \rho_{x_i}(x^t) - \rho_{x_j}(x^t) \right| \right\}, \\ \frac{1}{m} \sum_{t=1}^{m} \max \left| \nu_{x_i}(x^t) - \nu_{x_j}(x^t) \right| \end{pmatrix}, \tag{8}$$

for all $i, j = 1, \ldots, n$, and the formula (6) can be rewritten in similar manner as follows:

$$e(x_i, x_j) = \begin{pmatrix} 1 - \frac{1}{2m} \sum_{t=1}^{m} \sqrt{\left(\nu_{x_i}(x^t) - \nu_{x_j}(x^t) \right)^2 + \left(\rho_{x_i}(x^t) - \rho_{x_j}(x^t) \right)^2}, \\ \frac{1}{2m} \sum_{t=1}^{m} \sqrt{\left(\nu_{x_i}(x^t) - \nu_{x_j}(x^t) \right)^2} \end{pmatrix}, \tag{9}$$

for all $i, j = 1, \ldots, n$. By applying the formula (7), the matrix of intuitionistic fuzzy tolerance relation was obtained [6]. The matrix is presented in Table 2.

The matrix of intuitionistic fuzzy tolerance obtained by using formula (8) is equal to the matrix of intuitionistic fuzzy tolerance which was obtained by using formula (7). By applying the formula (9) to the initial data set, the matrix of intuitionistic fuzzy tolerance relation was also obtained. The matrix is presented in Table 3.

Thus, values $\mu_{IT}(x_i, x_j)$ are different and values $\nu_{IT}(x_i, x_j)$ are equal for all pairs (x_i, x_j) in Tables 2 and 3.

Table 2 The matrix of intuitionistic fuzzy tolerance obtained by using formula (7)

IT	x_1	x_2	x_3	x_4	x_5
x_1	(1.00, 0.00)				
x_2	(0.80, 0.10)	(1.00, 0.00)			
x_3	(0.72, 0.12)	(0.82, 0.08)	(1.00, 0.00)		
x_4	(0.75, 0.13)	(0.72, 0.10)	(0.70, 0.05)	(1.00, 0.00)	
x_5	(0.65, 0.22)	(0.68, 0.18)	(0.63, 0.23)	(0.63, 0.25)	(1.00, 0.00)

Table 3 The matrix of intuitionistic fuzzy tolerance obtained by using formula (9)

IT	x_1	x_2	x_3	x_4	x_5
x_1	(1.00, 0.00)				
x_2	(0.91, 0.10)	(1.00, 0.00)			
x_3	(0.88, 0.12)	(0.92, 0.08)	(1.00, 0.00)		
x_4	(0.90, 0.13)	(0.89, 0.10)	(0.86, 0.05)	(1.00, 0.00)	
x_5	(0.85, 0.22)	(0.87, 0.18)	(0.84, 0.23)	(0.85, 0.25)	(1.00, 0.00)

6 An Application to Clustering

Let us consider the results of processing the initial attributive intuitionistic fuzzy data by the D-PAIFC-TC-algorithm [9] for similarity measures (7) and (9).

In the first place, consider the result, obtained by using the similarity measure (7). By executing the D-PAIFC-TC-algorithm, the principal allotment $IR_P^*(X)$ among two intuitionistic fuzzy clusters was obtained. Membership functions and non-membership functions of two classes are presented in Fig. 1.

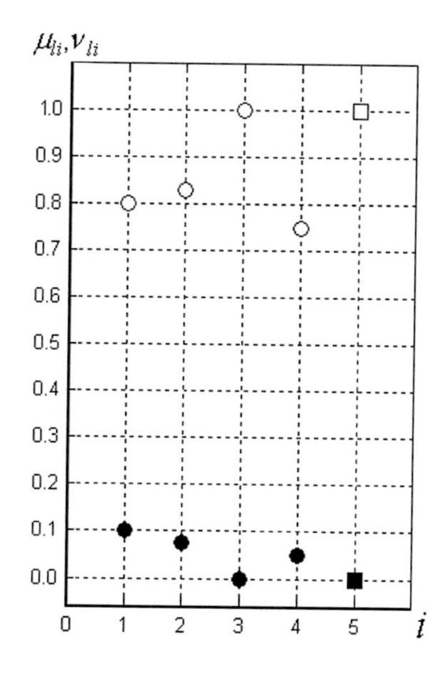

Fig. 1 The membership values and non-membership values of two intuitionistic fuzzy clusters obtained by using similarity measures (7) and (8)

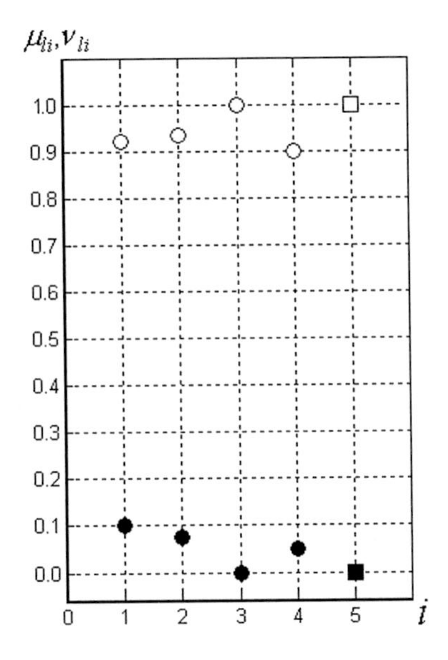

Fig. 2 The membership values and non-membership values of two intuitionistic fuzzy clusters obtained by using similarity measure (9)

In the second place, by executing the D-PAIFC-TC-algorithm for the formula (9), the allotment $IR_P^*(X)$ among two intuitionistic fuzzy clusters was obtained. Membership functions and non-membership functions of two classes are presented in Fig. 2.

Membership values of the first class are represented by ○, non-membership values of the first class are represented by ●, membership values of the second class are represented by □ and non-membership values of the second class are represented by ■ in Figs. 1 and 2. The third object is the typical point of the first intuitionistic fuzzy cluster and the fifth object is the typical point of the second intuitionistic fuzzy cluster in both experiments.

The illustrative example shows that membership values and non-membership values of elements depend on the selected similarity measure.

7 Conclusions

The new method for the similarity measurement between intuitionistic fuzzy sets is presented in the paper. The method is based on the normalized Euclidean distance. This distance was used to generate a new similarity measure to calculate the degree of similarity and degree of dissimilarity between intuitionistic fuzzy sets. Some properties of the proposed similarity measure are considered and results of application of the proposed similarity measure in comparison with the similarity measure which based on generalization of normalized Hamming distance to clustering the intuitionistic fuzzy data are discussed.

References

1. Atanassov, K.T.: Intuitionistic fuzzy sets. Fuzzy Sets Syst. **20**(1), 87–96 (1986)
2. Atanassov, K.T.: Intuitionistic Fuzzy Sets: Theory and Applications. Physica-Verlag, Heidelberg (1999)
3. Atanassov, K.T.: On Intuitionistic Fuzzy Sets Theory. Springer, Berlin (2012)
4. Xu, Z.: Intuitionistic Fuzzy Aggregation and Clustering. Springer, Berlin (2013)
5. Szmidt, E.: Distances and Similarities in Intuitionistic Fuzzy Sets. Springer, Berlin (2014)
6. Wang, Z., Xu, Z., Liu, S., Tang, J.: A netting clustering analysis method under intuitionistic fuzzy environment. Appl. Soft Comput. **11**(8), 5558–5564 (2011)
7. Viattchenin, D.A.: A method of construction of intuitionistic fuzzy tolerances based on a similarity measure between intuitionistic fuzzy sets. In: Atanassov K.T. et al. (Eds.) New Developments in Fuzzy Sets, Intuitionistic Fuzzy Sets, Generalized Nets and Related Topics, Vol. I: Foundations, pp. 191–202, IBS PAN– SRI PAS, Warsaw (2012)
8. Burillo, P., Bustince, H.: Intuitionistic fuzzy relations (Part I). Mathw. Soft Comput. **2**(1), 5–38 (1995)
9. Viattchenin, D.A., Shyrai, S.: Intuitionistic heuristic prototype-based algorithm of possibilistic clustering. Commun. Appl. Electron. **1**(8), 30–40 (2015)

A New Proposal of Defuzzification of Intuitionistic Fuzzy Quantities

Luca Anzilli and Gisella Facchinetti

Abstract In this paper we propose a method to defuzzify an intuitionistic fuzzy quantity that, depending on two parameters, recover previous methods and leaves freedom to the user.

Keywords Fuzzy sets · Fuzzy quantities · Intuitionistic fuzzy sets · Evaluation

1 Introduction

In many practical applications the available information corresponding to a fuzzy concept may be incomplete, that is the sum of the membership degree and the non-membership degree may be less than one. A possible solution is to use "Intuitionistic fuzzy sets" (IFSs) introduced by Atanassov [3–5].

Working in a fuzzy context, for different reasons, an optimization problem, a decision making problem and a control system need to transform the fuzzy result into a crisp value. In a fuzzy intuitionistic context the same problem occurs. This step is classically called "defuzzification". Many results are present in several papers for fuzzy numbers. There are in literature different approaches. One of the most recurring consists of the choice of a function that maps fuzzy numbers into the reals. This idea offers two opportunities. The first is to associate a real number to a fuzzy set, the second is to transfer the total order present into the reals to fuzzy numbers permitting to choose the better solution. One of the more used function is called the "centroid" that is the abscissa of the centre of gravity of output membership function hypograph. This method has the advantage to be useful even for a general fuzzy set. In the intuitionistic context the literature is not so wide. The problems in this context

L. Anzilli (✉) · G. Facchinetti
Department of Management, Economics, Mathematics and Statistics, University of Salento, Lecce, Italy
e-mail: luca.anzilli@unisalento.it

G. Facchinetti
e-mail: gisella.facchinetti@unisalento.it

K.T. Atanassov et al. (eds.), *Novel Developments in Uncertainty Representation and Processing*, Advances in Intelligent Systems and Computing 401,
DOI 10.1007/978-3-319-26211-6_16

are more than one. First of all an IFS is defined by two membership functions, the second one is that the fuzzy sets they identify an IFS may be non-normal and/or non-convex and so any result of defuzzification method introduced for fuzzy numbers is unusable. In [6, 7] the Authors present two methods, they call "de-i-fuzzification", to transform the IFS to evaluate in a fuzzy set. The solution to the problem they propose, justified by an optimization problem, is an average of the membership function (μ) and one minus the non membership function (ν). Starting from this idea, Yager in [9] propose a defuzzification way, in the discrete case, that matches the centroid.

In this paper we propose a different idea and approach for the second step. Even if the two membership functions that individuate an IFS are non-normal and non-convex we use an α-cut method and, solving an optimization problem, we introduce a defuzzification method that has its generality in the presence of two sets of weights. For a particular case of the weights we recover the centroid. But changing the weights we found other methods present in literature for non-normal and non-convex fuzzy sets. Our result is a sort of "generator" of defuzzification methods that, thanks to the presence of the two families of weights, leaves to the user a wide choice depending on his preferences and perceptions.

In Sect. 2 we give basic definitions and notations. In Sect. 3 we deal with defuzzification of intuitionistic fuzzy sets. In Sect. 4 we introduce intuitionistic fuzzy quantities. In Sect. 5 we propose an evaluation of intuitionistic fuzzy quantities.

2 Preliminaries and Notation

2.1 *Fuzzy Sets*

Let X denote a universe of discourse. A fuzzy set A in X is defined by a membership function $\mu_A : X \to [0, 1]$ which assigns to each element of X a grade of membership to the set A. The height of A is $h_A = height\, A = \sup_{x \in X} \mu_A(x)$. The support and the core of A are defined, respectively, as the crisp sets $supp(A) = \{x \in X; \mu_A(x) > 0\}$ and $core(A) = \{x \in X; \mu_A(x) = 1\}$. A fuzzy set A is normal if its core is nonempty. The union of two fuzzy sets A and B is the fuzzy set $A \cup B$ defined by the membership function $\mu_{A \cup B}(x) = \max\{\mu_A(x), \mu_B(x)\}$, $x \in X$. The intersection is the fuzzy set $A \cap B$ defined by $\mu_{A \cap B}(x) = \min\{\mu_A(x), \mu_B(x)\}$.

The α-cut of a fuzzy set A, with $0 \le \alpha \le 1$, is defined as the crisp set $A_\alpha = \left\{x \in X; \mu_A(x) \ge \alpha\right\}$ if $0 < \alpha \le 1$ and as the closure of the support if $\alpha = 0$.

We say that $A \subseteq B$ if $\mu_A(x) \le \mu_B(x)$ for each $x \in X$. Note that $A \subseteq B \iff A_\alpha \subseteq B_\alpha\ \forall \alpha$.

A fuzzy set is called convex if each α-cut is a closed interval $A_\alpha = [a_L(\alpha), a_R(\alpha)]$, where $a_L(\alpha) = \inf A_\alpha$ and $a_R(\alpha) = \sup A_\alpha$.

2.2 Intuitionistic Fuzzy Sets

An intuitionistic fuzzy set (IFS) A in X is defined as

$$A = \{\langle x, \mu_A(x), \nu_A(x)\rangle\,;\ x \in X\}$$

where $\mu_A : X \to [0,1]$ and $\nu_A : X \to [0,1]$ satisfy the condition

$$0 \le \mu_A(x) + \nu_A(x) \le 1\,.$$

The numbers $\mu_A(x), \nu_A(x) \in [0,1]$ denote the degree of membership and a degree of non-membership of x to A, respectively. For each IFS A in X, we denote

$$\pi_A(x) = 1 - \mu_A(x) - \nu_A(x)$$

the degree of the indeterminacy membership of the element x in A, that is the hesitation margin (or intuitionistic index) of $x \in A$ which expresses a lack of information of whether x belongs to A or not. We have $0 \le \pi_A(x) \le 1$ for all $x \in X$.

3 Defuzzification of IFSs

We now deal with the problem to defuzzificate an IFS A. A way to associate to an IFS A a real number may be described by the following procedure:

(i) transform the IFS A into a (standard) fuzzy set;
(ii) evaluate the standard fuzzy set by using a defuzzification method.

For step (i), in [7] the Authors called "de-i-fuzzification" a procedure to obtain a suitable fuzzy set starting from an IFS. Furthermore, they proposed to use the operator introduced in [4]

$$D_\lambda(A) = \{\langle x, \mu_A(x) + \lambda\pi_A(x), \nu_A(x) + (1-\lambda)\pi_A(x)\rangle;\ x \in X\}$$

with $\lambda \in [0,1]$. Note that $D_\lambda(A)$ is a standard fuzzy subset with membership function

$$\mu_\lambda(x) = \mu_A(x) + \lambda\pi_A(x)$$

In particular, they proposed $\lambda = 0.5$, as solution of the minimum problem

$$\min_{\lambda\in[0,1]} d(D_\lambda(A), A)$$

where d is the Euclidean distance. In this case the fuzzy set $D_{0.5}(A)$ is characterized by the membership function

$$\mu(x) = \frac{1}{2}(1 + \mu_A(x) - \nu_A(x)). \tag{1}$$

For step (ii), in agreement with the approach suggested in [9, Sect. 10], we may evaluate the IFS A by computing the center of gravity (COG) of the obtained fuzzy set, that is

$$Val_\lambda(A) = \frac{\int_{-\infty}^{+\infty} x\, \mu_\lambda(x)\, dx}{\int_{-\infty}^{+\infty} \mu_\lambda(x)\, dx} \tag{2}$$

with $\lambda = 0.5$.

Our aim is to propose a defuzzification method for an IFS A using α-cuts. In the following we will present a defuzzification procedure for intuitionistic fuzzy quantities.

4 Intuitionistic Fuzzy Quantities

4.1 Fuzzy Quantities

We now introduce the concept of fuzzy quantity as defined in [1, 2].

Definition 1 Let N be a positive integer and let $a_1, a_2, \dots, a_{4N}$ be real numbers with $a_1 < a_2 \le a_3 < a_4 \le a_5 < a_6 \le a_7 < a_8 \le a_9 < \dots < a_{4N-2} \le a_{4N-1} < a_{4N}$. We call fuzzy quantity

$$A = (a_1, a_2, \dots, a_{4N};\ h_1, h_2, \dots, h_N,\ h_{1,2}, h_{2,3}, \dots, h_{N-1,N}) \tag{3}$$

where $0 < h_j \le 1$ for $j = 1, \dots, N$ and $0 \le h_{j,j+1} < \min\{h_j, h_{j+1}\}$ for $j = 1, \dots, N-1$, the fuzzy set defined by a continuous membership function $\mu : \mathbb{R} \to [0, 1]$, with $\mu(x) = 0$ for $x \le a_1$ or $x \ge a_{4N}$, such that for $j = 1, 2, \dots, N$

(i) μ is strictly increasing in $[a_{4j-3}, a_{4j-2}]$, with $\mu(a_{4j-3}) = h_{j-1,j}$ and $\mu(a_{4j-2}) = h_j$,
(ii) μ is constant in $[a_{4j-2}, a_{4j-1}]$, with $\mu \equiv h_j$,
(iii) μ is strictly decreasing in $[a_{4j-1}, a_{4j}]$, with $\mu(a_{4j-1}) = h_j$ and $\mu(a_{4j}) = h_{j,j+1}$,

and for $j = 1, 2, \dots, N-1$

(iv) μ is constant in $[a_{4j}, a_{4j+1}]$, with $\mu \equiv h_{j,j+1}$,

where $h_{0,1} = h_{N,N+1} = 0$. Thus the height of A is $h_A = \max_{j=1,\dots,N} h_j$.

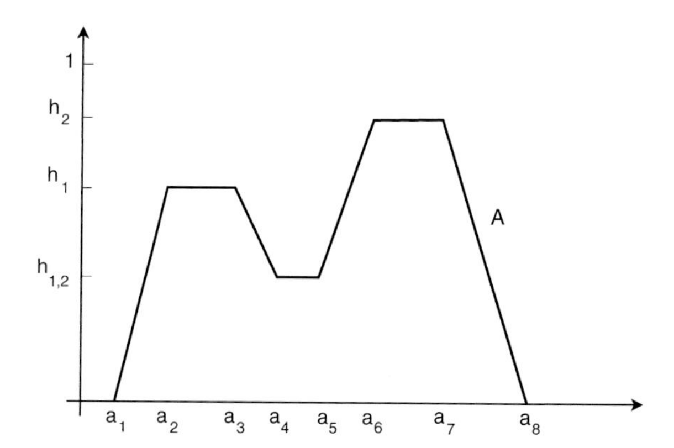

Fig. 1 Piecewise linear T1 FQ ($N = 2$)

We observe that in the case $N = 1$ the fuzzy quantity defined in (3) is fuzzy convex, that is every α-cut A_α is a closed interval. If $N \geq 2$ the fuzzy quantity defined in (3) is a non-convex fuzzy set with N humps and height $h_A = \max_{j=1,\ldots,N} h_j$.

Figure 1 shows an example of piecewise linear fuzzy quantity with $N = 2$.

Proposition 1 *Let A be the T1 FQ defined in (3) with height h_A. Then for each $\alpha \in [0, h_A]$ there exist an integer n^A_α, with $1 \leq n^A_\alpha \leq N$, and $A^\alpha_1, \ldots, A^\alpha_{n^A_\alpha}$ disjoint closed intervals such that*

$$A_\alpha = \bigcup_{i=1}^{n^A_\alpha} A^\alpha_i = \bigcup_{i=1}^{n_\alpha} [a^L_i(\alpha), a^R_i(\alpha)], \tag{4}$$

where we have denoted $A^\alpha_i = [a^L_i(\alpha), a^R_i(\alpha)]$, with $A^\alpha_i < A^\alpha_{i+1}$ (that is $a^R_i(\alpha) < a^L_{i+1}(\alpha)$). Thus n^A_α is the number of intervals producing the α-cut A_α.

From decomposition theorem for fuzzy sets and using previous result, we get the representation

$$A = \bigcup_{\alpha \in [0, h_A]} \alpha A_\alpha = \bigcup_{\alpha \in [0, h_A]} \alpha \bigcup_{i=1}^{n^A_\alpha} A^\alpha_i = \bigcup_{\alpha \in [0, h_A]} \bigcup_{i=1}^{n^A_\alpha} \alpha A^\alpha_i. \tag{5}$$

4.2 *Intuitionistic Fuzzy Quantities*

Definition 2 We call *intuitionistic fuzzy quantity* (IFQ) an IFS $A = \langle \mu_A, \nu_A \rangle$ of the real line such that μ_A and $1 - \nu_A$ are membership functions of fuzzy quantities.

Fig. 2 IFQ $A = (B, C)$

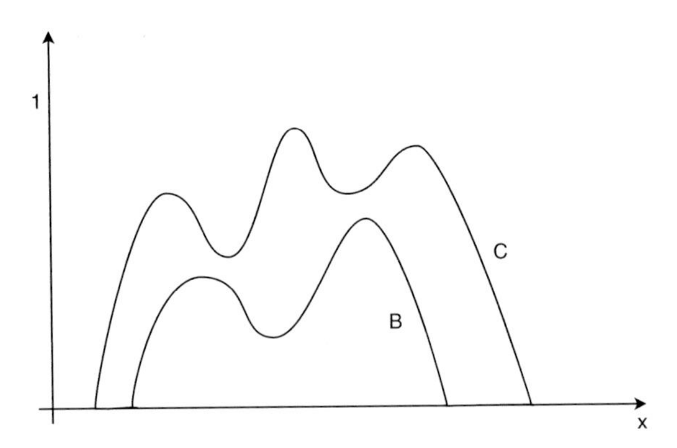

If A is an IFQ we denote by A^+ the fuzzy quantity with membership function $\mu_{A^+} = \mu_A$ and by A^- the fuzzy quantity with membership function $\mu_{A^-} = 1 - \nu_A$. An IFQ A may be indifferently denoted by $A = \langle \mu_A, \nu_A \rangle$ or $A = (A^+, A^-)$.

For the sake of notation simplicity, in the following an IFQ $A = (A^+, A^-)$ will be denoted by

$$A = (B, C).$$

Thus, B and C are fuzzy quantities with membership functions $\mu_B = \mu_{A^+} = \mu_A$ and $\mu_C = \mu_{A^-} = 1 - \nu_A$, respectively (Fig. 2).

5 Evaluation of Intuitionistic Fuzzy Quantities

A useful tool for dealing with fuzzy subsets are their α-cuts. In the case of an IFQ $A = (B, C)$ we have followed the procedure suggested in [8] and we call

$$B_\alpha = \left\{x \in X; \mu_A(x) \geq \alpha\right\}$$

and

$$C_\alpha = \left\{x \in X; 1 - \nu_A(x) \geq \alpha\right\} .$$

From (5) the α-cuts of fuzzy quantities B and C can be decomposed as

$$B_\alpha = \bigcup_{i=1}^{n_\alpha^B} B_i^\alpha, \qquad C_\alpha = \bigcup_{j=1}^{n_\alpha^C} C_j^\alpha.$$

Such decompositions enables us to introduce the family $\mathcal{A}$ of all the closed intervals B_i^α, C_j^α, that is

$$\mathcal{A} = \{B_1^\alpha, \ldots, B_{n_\alpha^B}^\alpha, C_1^\alpha, \ldots, C_{n_\alpha^C}^\alpha;\ 0 \leq \alpha \leq h\}$$

where

$$h = \max\{h_B, h_C\} = h_C.$$

For convenience, by defining

$$A_i^\alpha = \begin{cases} B_i^\alpha & i = 1, \dots, n_\alpha^B \\ C_{i-n_\alpha^B}^\alpha & i = n_\alpha^B + 1, \dots, n_\alpha^A \end{cases} \qquad \alpha \in [0, h] \tag{6}$$

where

$$n_\alpha^A = n_\alpha^B + n_\alpha^C, \tag{7}$$

we can represent $\mathcal{A}$ as the family of all the closed intervals $A_i^\alpha = [a_i^L(\alpha), a_i^R(\alpha)]$, that is

$$\mathcal{A} = \{A_i^\alpha;\ i = 1, \dots, n_\alpha^A,\ 0 \le \alpha \le h\}. \tag{8}$$

Our idea is to associate to IFQ A the nearest point to $\mathcal{A}$ respect to the Euclidean distance that depends on two parameters p and f. These two parameters will work as weights so we can say that we are looking for the real number $k^* = k^*(A; p, f)$ which minimizes the weighted mean of the squared distances.

Definition 3 We say that the real number k^* is an evaluation of the IFQ A with respect to (p, f) if it minimizes the weighted mean of the squared distances

$$\mathcal{D}_{p,f}^{(2)}(k; \mathcal{A}) = \int_0^{h_A} \sum_{i=1}^{n_\alpha^A} \left[(k - a_i^L(\alpha))^2 + (k - a_i^R(\alpha))^2\right] p_i(\alpha) f(\alpha)\, d\alpha \tag{9}$$

among all $k \in \mathbb{R}$, where, for each level α, the weights $p(\alpha) = (p_i(\alpha))_{i=1,\dots,\tilde{n}_\alpha}$ satisfy the properties

$$p_i(\alpha) \ge 0 \qquad \sum_{i=1}^{n_\alpha^A} p_i(\alpha) = 1, \tag{10}$$

the weight function $f : [0, 1] \to [0, +\infty[$ fulfil the condition

$$\int_0^h f(\alpha)\, d\alpha = 1. \tag{11}$$

The weights we have introduced work in a different manner: $p(\alpha)$ gives the possibility to evaluate in a different way the several intervals that produce anye α-cut, the weighting function f offers the possibility to give different importance to each α-level.

Theorem 1 *The real number k^* which minimizes (9) with respect to (p, f) is given by*

$$k^* = \int_0^h \sum_{i=1}^{n_\alpha^A} mid(A_i^\alpha)\, p_i(\alpha) f(\alpha)\, d\alpha, \tag{12}$$

where

$$mid(A_i^\alpha) = \frac{a_i^L(\alpha) + a_i^R(\alpha)}{2}$$

denotes the middle point of the interval $A_i^\alpha = [a_i^L(\alpha), a_i^R(\alpha)]$.

Proof We have to minimize the function $g : \mathbb{R} \to \mathbb{R}_+$ defined by

$$g(k) = \int_0^{h_A} \sum_{i=1}^{n_\alpha^A} \left[(k - a_i^L(\alpha))^2 + (k - a_i^R(\alpha))^2\right] p_i(\alpha) f(\alpha)\, d\alpha.$$

We have

$$g'(k) = 2 \int_0^{h_A} \sum_{i=1}^{n_\alpha^A} \left[2k - a_i^L(\alpha) - a_i^R(\alpha)\right] p_i(\alpha) f(\alpha)\, d\alpha.$$

By solving $g'(k) = 0$, taking into account that p and f satisfy conditions (10) and (11), respectively, we easily obtain that the solution k^* is given by (12). Moreover we get $g''(k) = 4 > 0$ and thus k^* minimizes g. □

In the following we indicate the evaluation of the IFQ $A = (B, C)$ as

$$V(A) = V(A; p, f) = \int_0^h \sum_{i=1}^{n_\alpha^A} mid(A_i^\alpha)\, p_i(\alpha) f(\alpha)\, d\alpha. \tag{13}$$

5.1 The Centroid as Particular Case

Let us consider an IFQ $A = (B, C)$. We show that defuzzification (2) with $\lambda = 0.5$, that is

$$Val(A) = \frac{\int_{-\infty}^{+\infty} x\, \mu(x)\, dx}{\int_{-\infty}^{+\infty} \mu(x)\, dx}$$

where μ is defined as $\mu(x) = (\mu_B(x) + \mu_C(x))/2$, may be obtained by the evaluation (13) we propose choosing particular values of the parameters involved.

In order to achieve this, we recall a previous result [2, Proposition 9.3] for fuzzy quantities.

Lemma 1 *Let A be a fuzzy quantity as defined in (3) with membership function μ_A, height h_A and α-cuts given by (4). Then for $t \geq 0$*

$$\int_{-\infty}^{+\infty} x^t \, \mu_A(x) \, dx = \frac{1}{t+1} \int_0^{h_A} \sum_{i=1}^{n_\alpha^A} \left(a_i^R(\alpha)^{t+1} - a_i^L(\alpha)^{t+1} \right) \, d\alpha.$$

In particular, for $t = 0$

$$\int_{-\infty}^{+\infty} \mu_A(x) \, dx = \int_0^{h_A} \sum_{i=1}^{n_\alpha^A} |A_i^\alpha| \, d\alpha = \int_0^{h_A} |A_\alpha| \, d\alpha \tag{14}$$

and, for $t = 1$

$$\int_{-\infty}^{+\infty} x \, \mu_A(x) \, dx = \int_0^{h_A} \sum_{i=1}^{n_\alpha^A} mid(A_i^\alpha) \, |A_i^\alpha| \, d\alpha, \tag{15}$$

where $|A_i^\alpha| = a_i^R(\alpha) - a_i^L(\alpha)$ is the length of interval A_i^α and $|A_\alpha|$ is the Lebesgue measure of A_α.

Proposition 2 *Let $A = (B, C)$ be an IFQ. Let A_i^α be the closed intervals defined in (6). If we choose*

$$p_i(\alpha) = \frac{|A_i^\alpha|}{\sum_{j=1}^{n_\alpha^A} |A_j^\alpha|}, \qquad f(\alpha) = \frac{\sum_{j=1}^{n_\alpha^A} |A_j^\alpha|}{\int_0^h \sum_{j=1}^{n_\alpha^A} |A_j^\alpha| \, d\alpha} \tag{16}$$

then we obtain

$$V(A) = Val(A).$$

Proof Substituting the weights (p, f) given in (16) in the expression of $V(A)$ (13) we obtain

$$V(A) = \int_0^h \sum_{i=1}^{n_\alpha^A} mid(A_i^\alpha) \, p_i(\alpha) f(\alpha) \, d\alpha = \frac{\int_0^h \sum_{i=1}^{n_\alpha^A} mid(A_i^\alpha) \, |A_i^\alpha| \, d\alpha}{\int_0^h \sum_{i=1}^{n_\alpha^A} |A_i^\alpha| \, d\alpha}.$$

Thus from (6) and (7) we get

$$V(A) = \frac{\int_0^{h_B} \sum_{i=1}^{n_\alpha^B} mid(B_i^\alpha)|B_i^\alpha|\, d\alpha + \int_0^{h_C} \sum_{j=1}^{n_\alpha^C} mid(C_j^\alpha)|C_j^\alpha|\, d\alpha}{\int_0^{h_B} \sum_{i=1}^{n_\alpha^B} |B_i^\alpha|\, d\alpha + \int_0^{h_C} \sum_{j=1}^{n_\alpha^C} |C_j^\alpha|\, d\alpha}$$

$$= \frac{\int_{-\infty}^{+\infty} x\, \mu_B(x)\, dx + \int_{-\infty}^{+\infty} x\, \mu_C(x)\, dx}{\int_{-\infty}^{+\infty} \mu_B(x)\, dx + \int_{-\infty}^{+\infty} \mu_C(x)\, dx} = Val(A)$$

where in the second equality we have applied (14) and (15) to the fuzzy quantities B and C. □

In a similar way we show the following result.

Proposition 3 *Let $A = (B, C)$ be an IFQ. Let A_i^α be the closed intervals defined in (6). If we choose*

$$p_i(\alpha) = \begin{cases} \dfrac{(1-\lambda)|B_i^\alpha|}{(1-\lambda)\sum_{i=1}^{n_\alpha^B} |B_i^\alpha| + \lambda \sum_{j=1}^{n_\alpha^C} |C_j^\alpha|} & i = 1, \dots, n_\alpha^B \\ \dfrac{\lambda |C_i^\alpha|}{(1-\lambda)\sum_{i=1}^{n_\alpha^B} |B_i^\alpha| + \lambda \sum_{j=1}^{n_\alpha^C} |C_j^\alpha|} & i = n_\alpha^B + 1, \dots, n_\alpha^A \end{cases}$$

and

$$f(\alpha) = \frac{(1-\lambda)\sum_{i=1}^{n_\alpha^B} |B_i^\alpha| + \lambda \sum_{j=1}^{n_\alpha^C} |C_j^\alpha|}{(1-\lambda)\int_0^h \sum_{i=1}^{n_\alpha^B} |B_i^\alpha|\, d\alpha + \lambda \int_0^h \sum_{j=1}^{n_\alpha^C} |C_j^\alpha|\, d\alpha}$$

then we obtain

$$V(A) = Val_\lambda(A),$$

where $Val_\lambda(A)$ is defined in (2).

6 Conclusion

Our proposal of defuzzification for IFSs is given depending of two groups of parameters that are real weights. As we have said the $p_i(\alpha)$ weights act on the several intervals of every α-cut, while the second weight f works along the vertical axis changing its importance for different level of α. These two actions give a wide opportunity of freedom to the operator taking into account his behaviour more pessimistic or optimistic. Our proposal has in itself even the centroid that works on x axis. In this case, as we have seen in (16), the weight $p_i(\alpha)$, with α fixed, is the width the ith interval that forms an α-cut, suitably normalized. The weight $f(\alpha)$ is, for every α fixed, the total length of that α cut suitably normalized. As in [7] the Authors conclude their work looking for an analogous study for other operators like $F_{\alpha,\beta}$, we are also working in this direction to see if our new general defuzzification method provides some new interesting results.

References

1. Anzilli, L., Facchinetti, G., Mastroleo, G.: Evaluation and interval approximation of fuzzy quantities. In: 8th Conference of the European Society for Fuzzy Logic and Technology (EUSFLAT-13), pp. 180–186 (2013)
2. Anzilli, L., Facchinetti, G., Mastroleo, G.: A parametric approach to evaluate fuzzy quantities. Fuzzy Sets Syst. **250**, 110–133 (2014)
3. Atanassov, K.T.: Intuitionistic fuzzy sets. Fuzzy sets Syst. **20**(1), 87–96 (1986)
4. Atanassov, K.T.: Intuitionistic Fuzzy Sets: Theory and Applications. Springer, Heidelberg (1999)
5. Atanassov, K.T.: Intuitionistic fuzzy sets: past, present and future. In: EUSFLAT Conference, pp. 12–19 (2003)
6. Atanassova, V., Sotirov, S.: A new formula for de-i-fuzzification of intuitionistic fuzzy sets. In: Notes on Intuitionistic Fuzzy Sets, 16th International Confrence on IFSs, Sofia, vol. 910, p. 4951 (2012)
7. Ban, A., Kacprzyk, J., Atanassov, K.: On de-i-fuzzification of intuitionistic fuzzy sets. Comptes Rendus de lAcademie bulgare des Sciences, Tome **61**(12), 1535–1540 (2008)
8. Grzegorzewski, P.: Distances and orderings in a family of intuitionistic fuzzy numbers. In: EUSFLAT Conference, pp. 223–227 (2003)
9. Yager, R.R.: Some aspects of intuitionistic fuzzy sets. Fuzzy Optim. Decis. Mak. **8**(1), 67–90 (2009)

Part III
Intuitionistic Fuzzy Sets: Applications

A New Heuristic Algorithm of Possibilistic Clustering Based on Intuitionistic Fuzzy Relations

Janusz Kacprzyk, Jan W. Owsiński, Dmitri A. Viattchenin and Stanislau Shyrai

Abstract This paper introduces a novel intuitionistic fuzzy set-based heuristic algorithm of possibilistic clustering. For the purpose, some remarks on the fuzzy approach to clustering are discussed and a brief review of intuitionistic fuzzy set-based clustering procedures is given, basic concepts of the intuitionistic fuzzy set theory and the intuitionistic fuzzy generalization of the heuristic approach to possibilistic clustering are considered, a general plan of the proposed clustering procedure is described in detail, two illustrative examples confirm good performance of the proposed algorithm, and some preliminary conclusions are formulated.

Keywords Intuitionistic fuzzy set · Intuitionistic fuzzy tolerance · Similarity measure · Clustering

1 Introduction

Cluster analysis aims at identifying groups of related objects and, hence, helps to discover assignment of objects and correlations in large data sets. The idea of data grouping is simple to use and in its nature is very near to the human method of thinking; whenever they are presented with a large amount of the data, humans are

J. Kacprzyk (✉) · J.W. Owsiński
Systems Research Institute Polish Academy of Sciences, Warsaw, Poland
e-mail: Janusz.Kacprzyk@ibspan.waw.pl

J.W. Owsiński
e-mail: jan.owsinski@ibspan.waw.pl

D.A. Viattchenin · S. Shyrai
Department of Software Information Technology, Belarusian State University of Informatics and Radio-Electronics, Minsk, Belarus
e-mail: viattchenin@mail.ru

S. Shyrai
e-mail: ashaman410@gmail.com

K.T. Atanassov et al. (eds.), *Novel Developments in Uncertainty Representation and Processing*, Advances in Intelligent Systems and Computing 401,
DOI 10.1007/978-3-319-26211-6_17

tend to summarize this huge number of the data into a small number of classes or categories in order to further facilitate its analysis.

A possibilistic approach to clustering was proposed by Krishnapuram and Keller [1]. Constraints in the possibilistic approach to clustering are less strong than constraints in the fuzzy objective function-based approach to clustering and values of the membership function of a possibilistic partition can be considered as typicality degrees. So, the possibilistic approach to clustering is more general and flexible approach to clustering than the fuzzy approach.

Since the fundamental Atanassov's [2] paper was published, intuitionistic fuzzy set theory has been developed in monographs [3, 4] and other publications. Moreover, this theory was applied to many areas. In particular, intuitionistic fuzzy clustering procedures were elaborated by different researchers. The intuitionistic fuzzy clustering methods are considered in [5].

Major clustering procedures are objective function-based algorithms of classification. However, heuristic clustering procedures display low level of complexity and high level of essential clarity. Some heuristic clustering algorithms are based on a definition of the cluster concept and the aim of these algorithms is cluster detection conform to a given definition. Such algorithms are called algorithms of direct classification or direct clustering algorithms.

A heuristic approach to possibilistic clustering is proposed in [6]. The essence of the heuristic approach to possibilistic clustering is that the sought clustering structure of the set of observations is formed based directly on the formal definition of fuzzy cluster and possibilistic memberships are determined also directly from the values of the pairwise similarity of observations. A concept of the allotment among fuzzy clusters is basic concept of the approach and the allotment among fuzzy clusters is a special case of the possibilistic partition [1]. The heuristic approach to possibilistic clustering is generalized for a case of intuitionistic fuzzy tolerance and the corresponding relational D-PAIFC-algorithm is described in [6]. Some intuitionistic fuzzy-set prototype-based heuristic algorithms of possibilistic clustering were also proposed in [7, 8].

The aim of the presented paper is a consideration of the novel intuitionistic fuzzy-set relational heuristic algorithm of possibilistic clustering. So, the contents of this paper are the following: in the second section basic definitions of the intuitionistic fuzzy set theory are described, in the third section basic concepts of the intuitionistic fuzzy generalization of the heuristic approach to possibilistic clustering are considered, in the fourth section a general plan of the D-AIFC(c)-algorithm is proposed, in the fifth section an illustrative example is given, in sixth section some preliminary conclusions are formulated.

2 Basic Notions of the Intuitionistic Fuzzy Set Theory

Let us remind some basic definitions of the Atanassov's intuitionistic fuzzy set theory [2–4] which will be used in further considerations.

An intuitionistic fuzzy set IA in X is given by ordered triple $IA = \{\langle x_i, \mu_{IA}(x_i), \nu_{IA}(x_i)\rangle | x_i \in X\}$, where $\mu_{IA}, \nu_{IA} : X \to [0, 1]$ should satisfy a condition

$$0 \leq \mu_{IA}(x_i) + \nu_{IA}(x_i) \leq 1, \tag{1}$$

for all $x_i \in X$. The values $\mu_{IA}(x_i)$ and $\nu_{IA}(x_i)$ denote the degree of membership and the degree of non-membership of element $x_i \in X$ to IA, respectively. For each intuitionistic fuzzy set IA in X an intuitionistic fuzzy index [2] of an element $x_i \in X$ in IA can be defined as follows

$$\rho_{IA}(x_i) = 1 - (\mu_{IA}(x_i) + \nu_{IA}(x_i)). \tag{2}$$

The intuitionistic fuzzy index $\rho_{IA}(x_i)$ can be considered as a hesitancy degree of x_i to IA. It is seen that $0 \leq \rho_{IA}(x_i) \leq 1$ for all $x_i \in X$. Obviously, when $\nu_{IA}(x_i) = 1 - \mu_{IA}(x_i)$ for every $x_i \in X$, the intuitionistic fuzzy set IA is an ordinary fuzzy set in X. For each fuzzy set A in X, we have $\rho_A(x_i) = 0$, for all $x_i \in X$.

Let $\mathrm{IFS}(X)$ denote the set of all intuitionistic fuzzy sets in X. Basic operations on intuitionistic fuzzy sets were defined by Atanassov in [2–4] and other publications. In particular, if $IA, IB \in \mathrm{IFS}(X)$ then

$$IA \cap IB = \{\langle x_i, \mu_{IA}(x_i) \wedge \mu_{IB}(x_i), \nu_{IA}(x_i) \vee \nu_{IB}(x_i)\rangle | x_i \in X\}, \tag{3}$$

and

$$IA \cup IB = \{\langle x_i, \mu_{IA}(x_i) \vee \mu_{IB}(x_i), \nu_{IA}(x_i) \wedge \nu_{IB}(x_i)\rangle | x_i \in X\}. \tag{4}$$

Moreover, some properties of intuitionistic fuzzy sets were given also in [9]. For example, if $IA, IB \in \mathrm{IFS}(X)$, then

$$IA \leq IB \Leftrightarrow \mu_{IA}(x_i) \leq \mu_{IB}(x_i) \text{ and } \nu_{IA}(x_i) \geq \nu_{IB}(x_i), \ \forall x_i \in X, \tag{5}$$

$$IA \preceq IB \Leftrightarrow \mu_{IA}(x_i) \leq \mu_{IB}(x_i) \text{ and } \nu_{IA}(x_i) \leq \nu_{IB}(x_i), \ \forall x_i \in X, \tag{6}$$

$$IA = IB \Leftrightarrow IA \leq IB \text{ and } IA \geq IB, \ \forall x_i \in X, \tag{7}$$

$$\bar{IA} = \{\langle x_i, \nu_{IA}(x_i), \mu_{IA}(x_i)\rangle | x_i \in X\}. \tag{8}$$

An α, β-level of an intuitionistic fuzzy set IA in X can be defined as

$$IA_{\alpha,\beta} = \{x_i \in X | \mu_{IA}(x_i) \geq \alpha, \ \nu_{IA}(x_i) \leq \beta\}, \tag{9}$$

where the condition

$$0 \leq \alpha + \beta \leq 1 \tag{10}$$

is met for any values α and β, $\alpha, \beta \in [0, 1]$.

The concept of the (α, β)-level intuitionistic fuzzy subset was defined in [6] as follows. The (α, β)-level intuitionistic fuzzy subset $IA_{(\alpha,\beta)}$ in X is given by the following expression:

$$IA_{(\alpha,\beta)} = \left\{ \langle x_i \in IA_{\alpha,\beta}, \mu_{IA_{(\alpha,\beta)}}(x_i) = \mu_{IA}(x_i), \nu_{IA_{(\alpha,\beta)}}(x_i) = \nu_{IA}(x_i) \rangle \right\}, \tag{11}$$

where $\alpha, \beta \in [0, 1]$ should satisfy the condition (10) and $IA_{\alpha,\beta}$ is the α, β-level of an intuitionistic fuzzy set IA which is satisfied the condition (9).

If IA is an intuitionistic fuzzy set in X, where X is the set of elements, then the (α, β)-level intuitionistic fuzzy subset $IA_{(\alpha,\beta)}$ in X, for which

$$\mu_{IA_{(\alpha,\beta)}}(x_i) = \begin{cases} \mu_{IA}(x_i), & \text{if } \mu_{IA}(x_i) \geqslant \alpha \\ 0, & \text{otherwise} \end{cases}, \tag{12}$$

and

$$\nu_{IA_{(\alpha,\beta)}}(x_i) = \begin{cases} \nu_{IA} \quad (x_i), & \text{if } \nu_{IA}(x_i) \leq \beta \\ 0, & \text{otherwise} \end{cases}. \tag{13}$$

is called an (α, β)-level intuitionistic fuzzy subset $IA_{(\alpha,\beta)}$ of the intuitionistic fuzzy set IA in X for some $\alpha, \beta \in [0, 1]$, $0 \leq \alpha + \beta \leq 1$.

Obviously, that the condition $IA_{(\alpha,\beta)} \preceq IA$ is met for any intuitionistic fuzzy set IA and its (α, β)-level intuitionistic fuzzy subset $IA_{(\alpha,\beta)}$ for any $\alpha, \beta \in [0, 1]$, $0 \leq \alpha + \beta \leq 1$. The important property will be very useful in further considerations.

Let us remind some basic definitions which were considered by Burillo and Bustince in [9, 10]. In cluster analysis, one is only interested in relations in a set X of classified objects.

Let $X = \{x_1, \ldots, x_n\}$ be an ordinary non-empty set. The binary intuitionistic fuzzy relation IR on X is an intuitionistic fuzzy subset IR of $X \times X$, which is given by the expression

$$IR = \{ \langle (x_i, x_j), \mu_A(x_i, x_j), \nu_A(x_i, x_j) \rangle | x_i, x_j \in X \}, \tag{14}$$

where $\mu_{IR} : X \times X \to [0, 1]$ and $\nu_{IR}: X \times X \to [0, 1]$ satisfy the condition $0 \leq \mu_{IR}(x_i, x_j) + \nu_{IR}(x_i, x_j) \leq 1$ for every $(x_i, x_j) \in X \times X$ [6].

Let $\text{IFR}(X)$ denote the set of all intuitionistic fuzzy relations on some universe X. An intuitionistic fuzzy relation $IR \in \text{IFR}(X)$ is reflexive if for every $x_i \in X$, $\mu_{IR}(x_i, x_i) = 1$ and $\nu_{IR}(x_i, x_i) = 0$. An intuitionistic fuzzy relation $IR \in \text{IFR}(X)$

is called symmetric if for all $(x_i, x_j) \in X \times X$, $\mu_{IR}(x_i, x_j) = \mu_{IR}(x_j, x_i)$ and $\nu_{IR}(x_i, x_j) = \nu_{IR}(x_j, x_i)$. An intuitionistic fuzzy relation *IT* in *X* is called an intuitionistic fuzzy tolerance if it is reflexive and symmetric.

An intuitionistic fuzzy relation *IT* in *X* is called an intuitionistic fuzzy tolerance if it is reflexive and symmetric. An intuitionistic fuzzy relation *IS* in *X* is called an intuitionistic fuzzy similarity relation if it is reflexive, symmetric and transitive.

An *n*-step procedure by using the composition of intuitionistic fuzzy relations beginning with an intuitionistic fuzzy tolerance can be used for construction of the transitive closure of an intuitionistic fuzzy tolerance *IT* and the transitive closure is an intuitionistic fuzzy similarity relation *IS*. The procedure is a basis of the clustering procedure which was proposed by Hung et al. in [11].

An α, β-level of an intuitionistic fuzzy relation *IR* in *X* was defined in [11] as

$$IR_{\alpha,\beta} = \{(x_i, x_j) | \mu_R(x_i, x_j) \geq \alpha,\ \nu_R(x_i, x_j) \leq \beta\}, \tag{15}$$

where the condition (10) is met for any values α and β, $\alpha, \beta \in [0, 1]$. So, if $0 \leq \alpha_1 \leq \alpha_2 \leq 1$ and $0 \leq \beta_2 \leq \beta_1 \leq 1$ with $0 \leq \alpha_1 + \beta_1 \leq 1$ and $0 \leq \alpha_2 + \beta_2 \leq 1$, then $IR_{\alpha_2,\beta_2} \subseteq IR_{\alpha_1,\beta_1}$. The proposition was formulated in [11].

The (α, β)-level intuitionistic fuzzy relation $IR_{(\alpha,\beta)}$ in *X* was defined in [6] as follows:

$$IR_{(\alpha,\beta)} = \left\{ \left\langle \begin{array}{c} (x_i, x_j) \in IR_{\alpha,\beta}, \mu_{IR_{(\alpha,\beta)}}(x_i, x_j) = \mu_{IR}(x_i, x_j), \\ \nu_{IR_{(\alpha,\beta)}}(x_i, x_j) = \nu_{IR}(x_i, x_j) \end{array} \right\rangle \right\}, \tag{16}$$

where $\alpha, \beta \in [0, 1]$ should satisfy the condition (10) and $IR_{\alpha,\beta}$ is the α, β-level of an intuitionistic fuzzy relation *IR* which is satisfied the condition (15).

The concept of the (α, β)-level intuitionistic fuzzy relation will be very useful in further considerations.

3 An Intuitionistic Fuzzy Generalization of the Heuristic Approach to Possibilistic Clustering

Let us consider intuitionistic extensions of basic concepts of the heuristic approach to possibilistic clustering, which were introduced in [6]. Let $X = \{x_1, \ldots, x_n\}$ be the initial set of elements and *IT* be some binary intuitionistic fuzzy tolerance on $X = \{x_1, \ldots, x_n\}$ with $\mu_{IT}(x_i, x_j) \in [0, 1]$ being its membership function and $\nu_{IT}(x_i, x_j) \in [0, 1]$ being its non-membership function. Let α and β be the α, β-level values of *IT*, $\alpha \in (0, 1]$, $\beta \in [0, 1)$, $0 \leq \alpha + \beta \leq 1$. Columns or lines of the intuitionistic fuzzy tolerance matrix are intuitionistic fuzzy sets $\{IA^1, \ldots, IA^n\}$.

Let $\{IA^1, \ldots, IA^n\}$ be intuitionistic fuzzy sets on *X*, which are generated by an intuitionistic fuzzy tolerance *IT*. The (α, β)-level intuitionistic fuzzy set $IA^l_{(\alpha,\beta)} = \{(x_i, \mu_{IA^l}(x_i), \nu_{IA^l}(x_i)) | \mu_{IA^l}(x_i) \geq \alpha, \nu_{IA^l}(x_i) \leq \beta, x_i \in X\}$ is intuitionistic fuzzy

(α,β)-cluster or, simply, intuitionistic fuzzy cluster. So $IA^l_{(\alpha,\beta)} \subseteq IA^l$, $\alpha \in (0,1]$, $\beta \in [0,1)$, $IA^l \in \{IA^1, \ldots, IA^n\}$ and μ_{li} is the membership degree of the element $x_i \in X$ for some intuitionistic fuzzy cluster $IA^l_{(\alpha,\beta)}$, $\alpha \in (0,1]$, $\beta \in [0,1)$, $l \in \{1, \ldots, n\}$. On the other hand, ν_{li} is the non-membership degree of the element $x_i \in X$ for the intuitionistic fuzzy cluster $IA^l_{(\alpha,\beta)}$.Value of α is the tolerance threshold of intuitionistic fuzzy clusters elements and value of β is the difference threshold of intuitionistic fuzzy clusters elements.

The membership degree of the element $x_i \in X$ for some intuitionistic fuzzy cluster $IA^l_{(\alpha,\beta)}$, $\alpha \in (0,1]$, $\beta \in [0,1)$, $0 \leq \alpha + \beta \leq 1$, $l \in \{1, \ldots, n\}$ can be defined as a

$$\mu_{li} = \begin{cases} \mu_{IA^l} & (x_i), \quad x_i \in IA^l_{\alpha,\beta} \\ 0, & \text{otherwise} \end{cases}, \tag{17}$$

where an α,β-level $IA^l_{\alpha,\beta}$ of an intuitionistic fuzzy set IA^l is the support of the intuitionistic fuzzy cluster $IA^l_{(\alpha,\beta)}$. So, condition $IA^l_{\alpha,\beta} = Supp(IA^l_{(\alpha,\beta)})$ is met for each intuitionistic fuzzy cluster $IA^l_{(\alpha,\beta)}$. The membership degree μ_{li} can be interpreted as a degree of typicality of an element to an intuitionistic fuzzy cluster.

On the other hand, the non-membership degree of the element $x_i \in X$ for an intuitionistic fuzzy cluster $IA^l_{(\alpha,\beta)}$, $\alpha \in (0,1]$, $\beta \in [0,1)$, $0 \leq \alpha + \beta \leq 1$, $l \in \{1, \ldots, n\}$ can be defined as a

$$\nu_{li} = \begin{cases} \nu_{IA^l} & (x_i), \quad x_i \in IA^l_{\alpha,\beta} \\ 0, & \text{otherwise} \end{cases} \tag{18}$$

So, the non-membership degree ν_{li} can be interpreted as a degree of non-typicality of an element to an intuitionistic fuzzy cluster. In other words, if columns or lines of intuitionistic fuzzy tolerance IT matrix are intuitionistic fuzzy sets $\{IA^1, \ldots, IA^n\}$ on X then intuitionistic fuzzy clusters $\{IA^1_{(\alpha,\beta)}, \ldots, IA^n_{(\alpha,\beta)}\}$ are intuitionistic fuzzy subsets of fuzzy sets $\{IA^1, \ldots, IA^n\}$ for some values $\alpha \in (0,1]$ and $\beta \in [0,1)$, $0 \leq \alpha + \beta \leq 1$. So, a condition $0 \leq \mu_{li} + \nu_{li} \leq 1$ is met for some intuitionistic fuzzy cluster $IA^l_{(\alpha,\beta)}$.

If conditions $\mu_{li} = 0$ and $\nu_{li} = 0$ are met for some element $x_i \in X$ and for some intuitionistic fuzzy cluster $IA^l_{(\alpha,\beta)}$, then the element will be called the residual element of the intuitionistic fuzzy cluster $IA^l_{(\alpha,\beta)}$. The value zero for a fuzzy set membership function is equivalent to non-belonging of an element to a fuzzy set. That is why values of tolerance threshold α are considered in the interval $(0,1]$. So, the value of a membership function of each element of the intuitionistic fuzzy cluster is the degree of similarity of the object to some typical object of fuzzy cluster. On the other hand, the value one for an intuitionistic fuzzy set non-membership function is equivalent to non-belonging of an element to an intuitionistic fuzzy set. That is why values of difference threshold β are considered in the interval $[0,1)$.

Let IT is an intuitionistic fuzzy tolerance on X, where X is the set of elements, and $\{IA^1_{(\alpha,\beta)}, \ldots, IA^n_{(\alpha,\beta)}\}$ is the family of intuitionistic fuzzy clusters for some $\alpha \in (0,1]$ and $\beta \in [0,1)$. The point $\tau^l_e \in IA^l_{\alpha,\beta}$, for which

$$\tau^l_e = \arg\max_{x_i} \mu_{li}, \ \forall x_i \in IA^l_{\alpha,\beta} \tag{19}$$

is called a typical point of the intuitionistic fuzzy cluster $IA^l_{(\alpha,\beta)}$. Obviously, the membership degree of a typical point of an intuitionistic fuzzy cluster is equal one because an intuitionistic fuzzy tolerance IT is the reflexive intuitionistic fuzzy relation. So, the non-membership degree of a typical point of an intuitionistic fuzzy cluster is equal zero. Moreover, a typical point of an intuitionistic fuzzy cluster does not depend on the value of tolerance threshold and an intuitionistic fuzzy cluster can have several typical points. That is why symbol e is the index of the typical point.

Let $IR^{\alpha,\beta}_{c(z)}(X) = \left\{IA^l_{(\alpha,\beta)} | l = \overline{1,c},\ c \le n, \alpha \in (0,1],\ \beta \in [0,1)\right\}$ be a family of intuitionistic fuzzy clusters for some value of tolerance threshold $\alpha \in (0,1]$ and some value of difference threshold $\beta \in [0,1)$, $0 \le \alpha + \beta \le 1$. These intuitionistic fuzzy clusters are generated by some intuitionistic fuzzy tolerance IT on the initial set of elements $X = \{x_1, \ldots, x_n\}$. If a condition

$$\sum_{l=1}^{c} \mu_{li} > 0, \ \forall x_i \in X, \tag{20}$$

and a condition

$$\sum_{l=1}^{c} \nu_{li} \ge 0, \ \forall x_i \in X \tag{21}$$

are met for all $IA^l_{(\alpha,\beta)}$, $l = \overline{1,c}$, $c \le n$ then the family is the allotment of elements of the set $X = \{x_1, \ldots, x_n\}$ among intuitionistic fuzzy clusters $\{IA^l_{(\alpha,\beta)}, l = \overline{1,c}, 2 \le c \le n\}$ for some value of the tolerance threshold $\alpha \in (0,1]$ and some value of the difference threshold $\beta \in [0,1)$. It should be noted that several allotments $IR^{\alpha,\beta}_{c(z)}(X)$ can exist for some pair of thresholds α and β. That is why symbol z is the index of an allotment.

The condition (20) requires that every object $x_i, i = \overline{1,n}$ must be assigned to at least one intuitionistic fuzzy cluster $IA^l_{(\alpha)}, l = \overline{1,c}, c \le n$ with the membership degree higher than zero and the condition is similar to the definition of the possibilistic partition [1]. The condition $2 \le c \le n$ requires that the number c of intuitionistic fuzzy clusters in $IR^{\alpha,\beta}_{c(z)}(X)$ must be more than two. Otherwise, the unique intuitionistic fuzzy cluster will contain all objects, possibly with different positive membership and non-membership degrees. The number c of fuzzy clusters can be equal the number of objects, n. This is taken into account in further considerations.

Allotment $IR_I^{\alpha,\beta}(X) = \left\{IA^l_{(\alpha,\beta)} | l = \overline{1,n}, \alpha \in (0,1], \ \beta \in [0,1)\right\}$ of the set of objects among n intuitionistic fuzzy clusters for some pair of thresholds α and β, $0 \le \alpha + \beta \le 1$, is the initial allotment of the set $X = \{x_1, \ldots, x_n\}$. In other words, if initial data are represented by a matrix of some intuitionistic fuzzy tolerance relation IT then lines or columns of the matrix are intuitionistic fuzzy sets $IA^l \alpha, l = \overline{1,n}$ and (α,β)-level fuzzy sets $IA^l_{(\alpha,\beta)}$, $l = \overline{1,n}$, $\alpha \in (0,1]$, $\beta \in [0,1)$ are intuitionistic fuzzy clusters. These intuitionistic fuzzy clusters constitute an initial allotment for some pair of thresholds α and β and they can be considered as clustering components.

If a condition

$$\bigcup_{l=1}^{c} IA^l_{\alpha,\beta} = X, \tag{22}$$

and a condition

$$card(IA^l_{\alpha,\beta} \cap IA^m_{\alpha,\beta)}) = 0, \ \forall IA^l_{(\alpha,\beta)}, IA^m_{(\alpha,\beta)}, l \ne m, \ \alpha, \beta \in (0,1] \tag{23}$$

are met for all intuitionistic fuzzy clusters $IA^l_{(\alpha,\beta)}$, $l = \overline{1,c}$ of some allotment $IR^{\alpha,\beta}_{c(z)}(X) = \left\{IA^l_{(\alpha,\beta)} | l = \overline{1,c}, \ c \le n, \alpha \in (0,1], \ \beta \in [0,1)\right\}$ then the allotment is the allotment among fully separate intuitionistic fuzzy clusters.

Intuitionistic fuzzy clusters in the sense of definition (17), (18) can have an intersection area. If the intersection area of any pair of different intuitionistic fuzzy clusters is an empty set, then conditions (22) and (23) are met and intuitionistic fuzzy clusters are called fully separate intuitionistic fuzzy clusters. Otherwise, intuitionistic fuzzy clusters are called partially separate intuitionistic fuzzy clusters and $w \in \{0, \ldots, n\}$ is the maximum number of elements in the intersection area of different intuitionistic fuzzy clusters. For $w = 0$ intuitionistic fuzzy clusters are fully separate intuitionistic fuzzy clusters. Thus, the conditions (22) and (23) can be generalized for a case of particularly separate intuitionistic fuzzy clusters. So, a condition

$$\sum_{l=1}^{c} card(A^l_{\alpha,\beta)}) \ge card(X), \forall A^l_{(\alpha,\beta)} \in IR^{\alpha,\beta}_{(z)}(X), \ \alpha \in (0,1], \beta \in [0,1), \tag{24}$$

and a condition

$$card(A^l_{\alpha,\beta} \cap A^m_{\alpha,\beta)}) \le w, \ \forall A^l_{(\alpha,\beta)}, A^m_{(\alpha,\beta)}, \ l \ne m, \ \alpha \in (0,1], \beta \in [0,1) \tag{25}$$

are generalizations of conditions (22) and (23). Obviously, if $w = 0$ in conditions (24) and (25) then conditions (22) and (23) are met. The adequate allotment $IR^{\alpha,\beta}_{c(z)}(X)$ for some value of tolerance threshold $\alpha \in (0,1]$ and some value of the difference threshold $\beta \in [0,1)$ is a family of fuzzy clusters which are elements of the initial allotment $IR_I^{\alpha,\beta}(X)$ for values of α and β and the family of fuzzy clusters should satisfy the conditions (24) and (25). So, the construction of adequate allotments $IR_{c(z)}^{\alpha,\beta}(X) = \{A_{\alpha,\beta}^l | l = \overline{1,c},\ c \le n\}$ for values α and β is a trivial combinatorial problem. Thus, the problem of cluster analysis can be defined in general as the problem of discovering the unique allotment $IR_c^*(X)$, resulting from the classification process, which corresponds to either most natural allocation of objects among intuitionistic fuzzy clusters or to the researcher's opinion about classification. In the first case, the number of intuitionistic fuzzy clusters c is not fixed. In the second case, the researcher's opinion determines the kind of the allotment sought and the number of intuitionistic fuzzy clusters c can be fixed. Detection of the a priori given number c of partially separated intuitionistic fuzzy clusters can be considered as the aim of classification. Several allotments among intuitionistic fuzzy clusters can exist for some pair of thresholds α and β. Thus, the problem consists in the selection of the unique principal allotment $IR_c^*(X)$ among c partially intuitionistic fuzzy clusters from the set $B(c)$ of allotments, $B(c) = \{IR_{c(z)}^{\alpha,\beta}(X)\}$, which is the class of possible solutions of the concrete classification problem. The symbol z is the index of the allotments. The selection of the unique allotment $IR_c^*(X)$ from the set $B(c) = \{IR_{c(z)}^{\alpha,\beta}(X)\}$ of allotments must be made on the basis of evaluation of allotments. The criterion

$$F(IR_{c(z)}^{\alpha,\beta}(X), \alpha, \beta) = \left(\sum_{l=1}^{c} \frac{1}{n_l} \sum_{i=1}^{n_l} \mu_{li} - \alpha \cdot c \right) - \left(\sum_{l=1}^{c} \frac{1}{n_l} \sum_{i=1}^{n_l} \nu_{li} - \beta \cdot c \right), \tag{26}$$

where c is the number of intuitionistic fuzzy clusters in the allotment $IR_{c(z)}^{\alpha,\beta}(X)$ and $n_l = card(IA_{\alpha,\beta}^l)$, $IA_{(\alpha,\beta)}^l \in IR_{c(z)}^{\alpha,\beta}(X)$ is the number of elements in the support of the intuitionistic fuzzy cluster $IA_{(\alpha,\beta)}^l$ can be used for evaluation of allotments.

Maximum of criterion (28) corresponds to the best allotment of objects among a priori given number c of intuitionistic fuzzy clusters. So, the classification problem can be characterized formally as determination of the solution $IR_c^*(X)$ satisfying

$$IR_c^*(X) = \arg \max_{IR_{c(z)}^{\alpha,\beta}(X) \in B(c)} F(IR_c^{\alpha,\beta}(X), \alpha, \beta), \tag{27}$$

where $B(c) = \{IR_{c(z)}^{\alpha,\beta}(X)\}$ is the set of allotments of objects among a priori given number c of intuitionistic fuzzy clusters corresponding to the pair of thresholds α and β.

A clustering procedure is based on the decomposition of initial intuitionistic fuzzy tolerance IT [6]. That is why basic concepts of the method of decomposition must be considered. Let IT be an intuitionistic fuzzy tolerance in X. Let $IT_{(\alpha,\beta)}$ be (α,β)-level intuitionistic fuzzy relation and the condition (10) is met for values α and β, $\alpha \in (0,1]$, $\beta \in [0,1)$. Let $IT_{\alpha,\beta}$ be a α,β-level of an intuitionistic fuzzy tolerance IT in X and $IT_{\alpha,\beta}$ be the support of $IT_{(\alpha,\beta)}$. The membership function $\mu_{IT_{(\alpha,\beta)}}(x_i,x_j)$ can be defined as

$$\mu_{IT_{(\alpha,\beta)}}(x_i,x_j)=\begin{cases}\mu_{IT}(x_i,x_j), & \text{if } \mu_{IT}(x_i,x_j)\geq\alpha\\ 0, & \text{otherwise}\end{cases}, \tag{28}$$

and the non-membership function $\nu_{IT_{(\alpha,\beta)}}(x_i,x_j)$ can be defined as

$$\nu_{IT_{(\alpha,\beta)}}(x_i,x_j)=\begin{cases}\nu_{IT}(x_i,x_j), & \text{if } \nu_{IT}(x_i,x_j)\leq\beta\\ 0, & \text{otherwise}\end{cases}. \tag{29}$$

Obviously, that the condition $IT_{(\alpha,\beta)}\preceq IT$ is met for any intuitionistic fuzzy tolerance IT and a (α,β)-level intuitionistic fuzzy relation $IT_{(\alpha,\beta)}$ for any $\alpha\in(0,1]$, $\beta\in[0,1)$, $0\leq\alpha+\beta\leq 1$. Thus we have the proposition that if $\alpha_{\ell(\alpha)}\leq\alpha_{\ell+1(\alpha)}$ and $\beta_{\ell+1(\beta)}\leq\beta_{\ell(\beta)}$ with $0\leq\alpha_{\ell(\alpha)}+\beta_{\ell(\beta)}\leq 1$, $0\leq\ \alpha_{\ell+1(\alpha)}+\beta_{\ell+1(\beta)}\leq 1$ then the condition $IT_{(\alpha_{\ell+1(\alpha)},\beta_{\ell+1(\beta)})}\preceq IT_{(\alpha_{\ell(\alpha)},\beta_{\ell(\beta)})}$ is met. So, the ordered sequences $0<\alpha_0\leq\cdots\leq\alpha_{\ell(\alpha)}\leq\cdots\leq\alpha_{Z(\alpha)}\leq 1$ and $0\leq\beta_{Z(\beta)}\leq\cdots\leq\beta_{\ell(\beta)}\leq\cdots\leq\beta_0<1$ must be constructed for the decomposition of an intuitionistic fuzzy tolerance *IT*. A method of construction of sequences was developed in [6].

4 A General Plan of the D-AIFC(c)-Algorithm

Constructing the allotment $IR_c^*(X)$ among a priori given number c of partially separate intuitionistic fuzzy clusters can be considered as the aim of classification. The corresponding D-AIFC(c)-algorithm for detecting the allotment $IR_c^*(X)$ is an eleven-step procedure of classification as given below.

1. Let $w := 0$
2. Construct ordered sequences $0<\alpha_0\leq\cdots\leq\alpha_{\ell(\alpha)}\leq\cdots\leq\alpha_{Z(\alpha)}\leq 1$ and $0\leq\beta_{Z(\beta)}\leq\cdots\leq\beta_{\ell(\beta)}\leq\cdots\leq\beta_0<1$ of thresholds values; let $\ell(\alpha)\ :=\ 0$ and $\ell(\beta)\ :=\ 0$;
3. The following condition is checked:

   if the condition $0\leq\alpha_{\ell(\alpha)}+\beta_{\ell(\beta)}\leq 1$ is met
   then construct the (α,β)-level intuitionistic fuzzy relation $IT_{(\alpha,\beta)}$ in the sense of definition (16) and go to step 3
   else the following condition is checked:

      if the condition $\ell(\beta)<Z(\beta)$ is met
      then let $\ell(\beta)\ :=\ \ell(\beta)+1$and go to step 3;

4. Construct the initial allotment $IR_I^{\alpha,\beta}(X)=\{IA^l_{(u,\beta)}|l=\overline{1,n},\ \alpha\in(0,1],\ \beta\in[0,1)\}$ for calculated values $\alpha_{\ell(\alpha)}$ and $\beta_{\ell(\beta)}$;

5. The following condition is checked:

   if for some intuitionistic fuzzy cluster $IA^l_{(\alpha,\beta)} \in IR^{\alpha,\beta}_I(X)$ the condition $card(IA^l_{\alpha,\beta}) = n$ is met
   then let $\ell(\beta) := \ell(\beta) + 1$ and go to step 3
   else go to step 6;

6. Construct allotments among a priori given number c of intuitionistic fuzzy clusters $IR^{\alpha,\beta}_{c(z)}(X) = \{IA^l_{(\alpha,\beta)} | l = \overline{1,c}, c \le n\}$, $\alpha = \alpha_{\ell(\alpha)}$, $\beta = \beta_{\ell(\beta)}$ which satisfy conditions (24) and (25) for the pair of values $\alpha_{\ell(\alpha)}$ and $\beta_{\ell(\beta)}$ from the sequences $0 < \alpha_0 \le \cdots \le \alpha_{\ell(\alpha)} \le \cdots \le \alpha_{Z(\alpha)} \le 1$ and $0 \le \beta_{Z(\beta)} \le \cdots \le \beta_{\ell(\beta)} \le \cdots \le \beta_0 < 1$;
7. The following condition is checked:

   if allotments among a priori given number c of intuitionistic fuzzy clusters $IR^{\alpha,\beta}_{c(z)}(X) = \{IA^l_{(\alpha,\beta)} | l = \overline{1,c}, c \le n\}$, $\alpha = \alpha_{\ell(\alpha)}$, $\beta = \beta_{\ell(\beta)}$ which satisfy conditions (24) and (25) are not constructed
   then the following condition is checked:

   if the condition $\ell(\beta) = Z(\beta)$ is met let $\ell(\alpha) := \ell(\alpha) + 1$ and $\ell(\beta) := 0$ and go to step 3
   else let $\ell(\beta) := \ell(\beta) + 1$ and go to step 3

   else go to step 8;

8. The following condition is checked:

   if allotments among a priori given number c of intuitionistic fuzzy clusters $IR^{\alpha,\beta}_{c(z)}(X) = \{IA^l_{(\alpha,\beta)} | l = \overline{1,c}, c \le n\}$, $\alpha = \alpha_{\ell(\alpha)}$, $\beta = \beta_{\ell(\beta)}$ which satisfy conditions (24) and (25) are not constructed
   then let $w := w + 1$ and go to step 2
   else go to step 9;

9. Construct the class of possible solutions of the classification problem $B(c) = \{IR^{\alpha,\beta}_{c(z)}(X)\}$, which satisfy conditions (24) and (25) for the calculated pair of values $\alpha_{\ell(\alpha)}$ and $\beta_{\ell(\beta)}$ and for the given number of fuzzy clusters c as follows:

   if for some allotment $IR^{\alpha,\beta}_{c(z)}(X)$ the condition $card(IR^{\alpha,\beta}_{c(z)}(X)) = c$ is met
   then $IR^{\alpha,\beta}_{c(z)}(X) \in B(c)$;

10. Calculate the value of the criterion (26) for each allotment $IR^{\alpha,\beta}_{c(z)}(X) \in B(c)$;
11. The result $IR^*_c(X)$ of classification is formed as follows:

   if for some unique allotment $IR^{\alpha,\beta}_{c(z)}(X) \in B(c)$ the condition (27) is met

then the allotment is the result of classification $IR_c^*(X)$
else the number c of classes is suboptimal.

The unique allotment $IR_c^*(X)$ among the a priori given number c of partially separate intuitionistic fuzzy clusters and corresponding values of the tolerance threshold α and the difference threshold β, $0 \le \alpha + \beta \le 1$ are results of classification.

5 Experimental Results

Firstly, let us consider the result of classification, which was presented by Hung et al. in [11]. The result was obtained by applying the clustering procedure to the relational intuitionistic fuzzy data. An original intuitionistic fuzzy tolerance relation matrix is given in Table 1.

An intuitionistic fuzzy similarity relation matrix $IS = [\mu_{IS}(x_i, x_j), \nu_{IS}(x_i, x_j)]$, i, $j = 1, \ldots, 10$ was obtained by the n-step $\max - T_3$ & $\min - S_3$ composition procedure [9]. A hard partition is the result of application of the clustering method to the intuitionistic fuzzy similarity relation IS. The partition $\{x_1, x_4, x_8\}$, $\{x_2, x_5, x_9, x_{10}\}$, $\{x_3\}$, $\{x_6\}$, $\{x_7\}$ was obtained for $\alpha = 0.55$ and $\beta = 0.35$.

Table 1 The matrix of intuitionistic fuzzy tolerance

IT	x_1	x_2	x_3	x_4	x_5	x_6	x_7	x_8	x_9	x_{10}
x_1	(1.0, 0.0)									
x_1	(0.2, 0.7)	(1.0, 0.0)								
x_3	(0.5, 0.5)	(0.3, 0.6)	(1.0, 0.0)							
x_4	(0.8, 0.1)	(0.6, 0.4)	(0.5, 0.4)	(1.0, 0.0)						
x_5	(0.6, 0.3)	(0.7, 0.2)	(0.3, 0.6)	(0.7, 0.2)	(1.0, 0.0)					
x_6	(0.2, 0.7)	(0.9, 0.1)	(0.4, 0.5)	(0.3, 0.6)	(0.2, 0.7)	(1.0, 0.0)				
x_7	(0.3, 0.7)	(0.2, 0.7)	(0.1, 0.9)	(0.5, 0.4)	(0.4, 0.5)	(0.1, 0.7)	(1.0, 0.0)			
x_8	(0.9, 0.1)	(0.8, 0.2)	(0.3, 0.6)	(0.4, 0.6)	(0.5, 0.5)	(0.3, 0.7)	(0.6, 0.3)	(1.0, 0.0)		
x_9	(0.4, 0.5)	(0.3, 0.7)	(0.7, 0.2)	(0.1, 0.8)	(0.8, 0.1)	(0.7, 0.2)	(0.1, 0.8)	(0.0, 0.9)	(1.0, 0.0)	
x_{10}	(0.3, 0.7)	(0.2, 0.7)	(0.6, 0.3)	(0.3, 0.7)	(0.9, 0.1)	(0.2, 0.7)	(0.3, 0.7)	(0.2, 0.8)	(0.1, 0.8)	(1.0, 0.0)

For comparison, the D-AIFC(c)-algorithm was applied to the matrix of the intuitionistic fuzzy tolerance IT directly for $c \in \{2, \ldots, 4\}$. Let us consider the result of the experiment.

By executing the D-AIFC(c)-algorithm for $c = 2$, the allotment $IR_c^*(X)$ among two partially separate intuitionistic fuzzy clusters, which corresponds to the result, is received for the value of tolerance threshold $\alpha = 0.1$ and the value of difference threshold $\beta = 0.5$.

Membership functions and non-membership functions of two classes are presented in Fig. 1, where membership values of the first class are represented by symbol ○, non-membership values of the first class are represented by symbol ●, membership values of the second class are represented by symbol □ and non-membership values of the second class are represented by symbol ■.

The third object is the typical point of the first intuitionistic fuzzy cluster and the eighth object is the typical point of the second intuitionistic fuzzy cluster.

By executing the D-AIFC(c)-algorithm for $c = 3$, the allotment $IR_c^*(X)$ among three partially separate intuitionistic fuzzy clusters, which corresponds to the result, is received for the value of tolerance threshold $\alpha = 0.4$ and the value of difference threshold $\beta = 0.6$. Figure 2 shows membership functions and non-membership functions of three classes.

In Fig. 2 membership values of the first class are represented by ○, non-membership values of the first class are represented by ●, membership values of the second class are represented by □, non-membership values of the second class are represented by ■, membership values of the third class are represented by ▽ and non-membership values of the third class are represented by ▼.

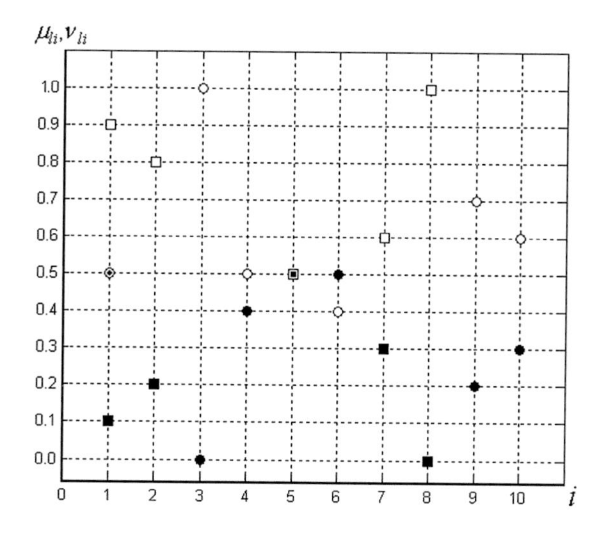

Fig. 1 Membership values and non-membership values of two intuitionistic fuzzy clusters

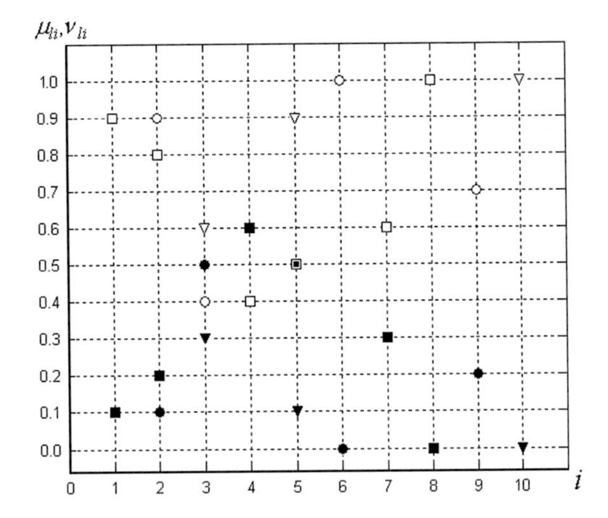

Fig. 2 Membership values and non-membership values of three intuitionistic fuzzy clusters

The sixth object is the typical point of the first intuitionistic fuzzy cluster, the eighth object is the typical point of the second intuitionistic fuzzy cluster and the tenth object is the typical point of the third intuitionistic fuzzy cluster.

By executing the D-AIFC(c)-algorithm for $c = 4$, the allotment $IR_c^*(X)$ among four partially separate intuitionistic fuzzy clusters is obtained for the value of tolerance threshold $\alpha = 0.1$ and the value of difference threshold $\beta = 0.3$. Membership functions and non-membership functions of four classes are presented in Fig. 3.

Membership values of the first class are represented by ○ in Fig. 3, non-membership values of the first class are represented by •, membership values of the second class are represented by □, non-membership values of the second class are

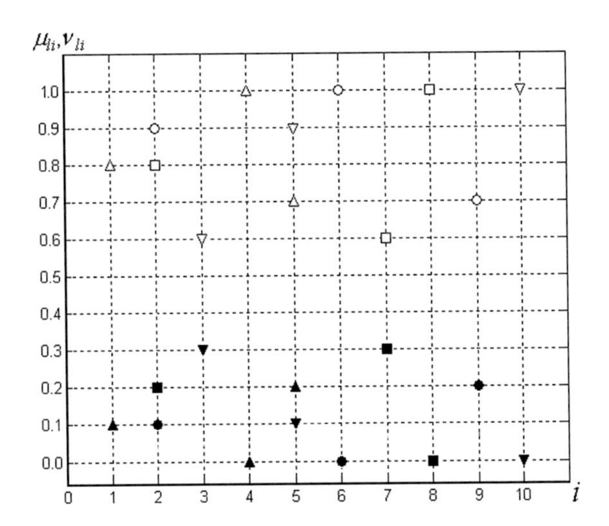

Fig. 3 Membership values and non-membership values of four intuitionistic fuzzy clusters

represented by ■, membership values of the third class are represented by ▽, non-membership values of the third class are represented by ▼, membership values of the fourth class are represented by Δ and non-membership values of the fourth class are represented by ▲.

The sixth object is the typical point of the first intuitionistic fuzzy cluster, the eighth object is the typical point of the second intuitionistic fuzzy cluster, the tenth object is the typical point of the third intuitionistic fuzzy cluster and the fourth object is the typical point of the fourth intuitionistic fuzzy cluster.

6 Conclusions

The novel relational heuristic D-AIFC(c)-algorithm of possibilistic clustering based on intuitionistic fuzzy set theory is proposed in the paper. Constructing the allotment $IR_c^*(X)$ among given number c of partially separate intuitionistic fuzzy clusters is the aim of classification.

The D-AIFC(c)-algorithm was tested on the illustrative data set for different number of intuitionistic fuzzy clusters in the sought allotment. The results of application of the proposed algorithm to the data show that the proposed algorithm is the effective tool for solving the classification problem under ambiguity of the initial data.

Acknowledgements The authors are grateful to Prof. Eulalia Szmidt for her useful remarks and fruitful discussions during the paper preparation.

References

1. Krishnapuram, R., Keller, J.M.: A possibilistic approach to clustering. IEEE Trans. Fuzzy Syst. **1**(2), 98–110 (1993)
2. Atanassov, K.T.: Intuitionistic fuzzy sets. Fuzzy Sets Syst. **20**(1), 87–96 (1986)
3. Atanassov, K.T.: Intuitionistic Fuzzy Sets: Theory and Applications. Physica-Verlag, Heidelberg (1999)
4. Atanassov, K.T.: On Intuitionistic Fuzzy Sets Theory. Springer, Berlin (2012)
5. Xu, Z.: Intuitionistic Fuzzy Aggregation and Clustering. Springer, Berlin (2013)
6. Viattchenin, D.A.: A Heuristic Approach to Possibilistic Clustering: Algorithms and Applications. Springer, Berlin (2013)
7. Viattchenin, D.A., Shyrai, S.: Intuitionistic heuristic prototype-based algorithm of possibilistic clustering. Commun. Appl. Electron. **1**(8), 30–40 (2015)
8. Shyrai, S., Viattchenin, D.A.: Clustering the intuitionistic fuzzy data, detection of an unknown number of intuitionistic fuzzy clusters in the allotment. In: Proceedings of the International Conference on Information and Digital Technologies (IDT'2015), IEEE Service Center, Piscataway, pp. 302–311 (2015)
9. Burillo, P., Bustince, H.: Intuitionistic fuzzy relations (Part I). Mathw. Soft Comput. **2**(1), 5–38 (1995)

10. Burillo, P., Bustince, H.: Intuitionistic fuzzy relations (Part II). Effect of Atanassov's operators on the properties of the intuitionistic fuzzy relations. Mathw. Soft Comput. **2**(2), 117–148 (1995)
11. Hung, W.-L., Lee, J.-S., Fuh, C.-D.: Fuzzy clustering based on intuitionistic fuzzy relations. Int. J. Uncertainty, Fuzziness Knowl.-Based Syst. **12**(4), 513–529 (2004)

Aggregation of Inconsistent Expert Opinions with Use of Horizontal Intuitionistic Membership Functions

Andrzej Piegat and Marek Landowski

Abstract Single expert opinion expressed in form of an intuitionistic membership function (IMF) has uncertainty of the second order because it consists of the membership—$\mu(x)$ and of the non-membership function $\nu(x)$. Two different, considerably inconsistent expert opinions have an increased uncertainty order. Often we do not know, which of the opinion is more or less credible. Hence, IMF representing both aggregated opinions cannot be a standard IMF. It should have an increased order of uncertainty. Possibility of appropriate modeling aggregated opinions offers theory of fuzzy sets type-2 developed mainly by J. Mendel. In this paper authors show how application of this theory in connection with horizontal version of IMFs allows for constructing of an aggregated IMF of two inconsistent intuitionistic expert opinions.

Keywords Intuitionistic fuzzy sets · Type-2 fuzzy sets · Expert opinion aggregation · Horizontal membership functions · RDM—Relative Distance Measure

1 Introduction

Expert opinions usually are only partly inconsistent. But sometimes they can be considerably inconsistent. Examples of two such opinions A and B are shown in Fig. 1. What can be the reason of such inconsistency? If e.g. the opinions concern a firm value then reason of inconsistency can lie in the difference of information about the firm possessed by expert A and B. Expert A can possess more negative and B more positive information about the firm. Therefore expert B evaluated the firm higher

A. Piegat
Faculty of Computer Science, West Pomeranian University of Technology,
Zolnierska 49, 71-210 Szczecin, Poland
e-mail: apiegat@wi.zut.edu.pl

M. Landowski (✉)
Maritime University of Szczecin, Waly Chrobrego 1-2, 70-500 Szczecin, Poland
e-mail: m.landowski@am.szczecin.pl

K.T. Atanassov et al. (eds.), *Novel Developments in Uncertainty Representation and Processing*, Advances in Intelligent Systems and Computing 401,
DOI 10.1007/978-3-319-26211-6_18

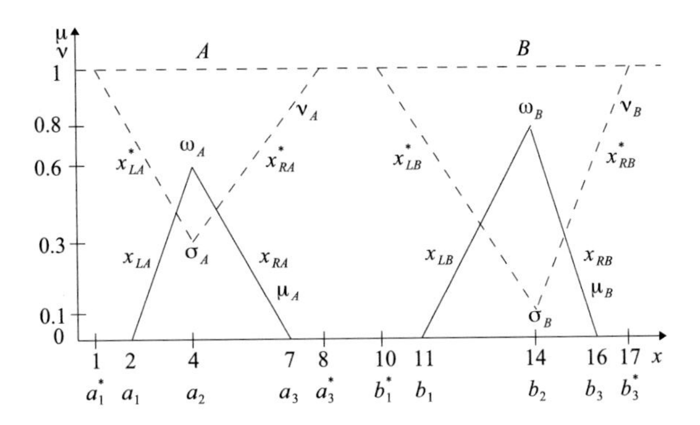

Fig. 1 Aggregated inconsistent expert opinions A and B expressed in form of intuitionistic membership $\mu_A(x)$, $\mu_B(x)$ and non-membership functions $\nu_A(x)$, $\nu_B(x)$. x_{LA}, x_{LB}, x_{RA}, x_{RB}—the minimal (*left*) and the maximal (*right*) function borders

and A lower. Because both experts had partly different information about the firm we cannot say that one evaluation is better or worse. Both evaluations are important, should be taken into account and aggregated in one general evaluation. The difficult problem of expert opinions aggregation is described e.g. in [11]. The book shows multitude of aggregation methods and their disputability. Authors of this paper have not meet methods of opinions aggregation expressed in form of IFSs. Hence they propose own method. Notion of intuitionistic fuzzy sets (IFSs) has been introduced by Atanassov [4]. In [3] he gives basic operations on IFSs such as logical sum AUB of two sets. However, such operation cannot be used in the case of inconsistent IFSs [11]. In [5] following definition of IFSs is given: "IFS theory basically defies the claim that from fact that an element x "belongs" to a given degree (say $\mu_A(x)$) to a fuzzy set A, naturally follows that x should "not belong" to A to the extent $1 - \mu_A(x)$, an assertion implicit in the concept of a fuzzy set. On the contrary, IFSs assign to each element of the universe both a degree of membership $\mu_A(x)$ and one of non-membership $\nu_A(x)$ such that $\mu_A(x) + \nu_A(x) \leq 1$, thus relaxing the enforced duality $\nu_A(x) = 1 - \mu_A(x)$ from fuzzy set theory". A method of identification of memberships (MFs) $\mu_A(x)$ and non-memberships (NMFs) $\nu_A(x)$ can be found in e.g. [2]. Standard formula of MF is of vertical character $\mu_A(x) = f(x)$. In this paper a horizontal model of IFSs based on Relative Distance Measure (RDM) will be used [12, 13]. As IFSs triangle functions described e.g. in [10] will be applied. Authors will explain the aggregation method on example of two intuitionistic expert opinions shown in Fig. 1.

Type-2 fuzzy sets introduced by L. Zadeh and developed mainly by J. Mendel and co-workers [7, 8, 10, 11] open new possibilities for aggregation of inconsistent IFSs. Inconsistent opinions can be interpreted as follows: the experts in different way evaluate position of the minimal (left) x_L and of the maximal (right) border x_R of their evaluations. Thus, the left border x_{LE} and the right border x_{RE} of the aggregated evaluation AgB is uncertain (g means operation of aggregation), Fig. 2.

A possible, left border x_{LE} of the aggregated IMF $\mu_{AgB}(x)$ in terms of interval FSs Type-2 (FST2) is called left embedded border and a possible right border x_{RE} is called embedded right border. Figure 3 shows one of possible, embedded aggregated IMFs type-1 generated by two possible left and right borders x_{LE} and x_{RE}. Figure 3 also shows denotations used in further formula derivations.

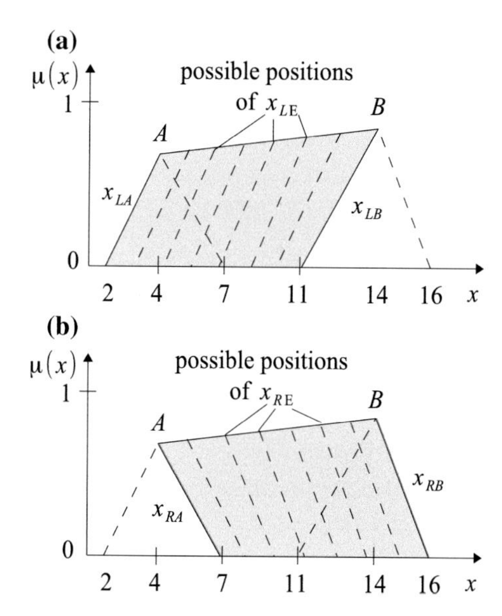

Fig. 2 Uncertainty zone (zone of possible position) of the *left border* x_{LE} and the *right border* x_{RE} of the aggregated opinion *AgB*

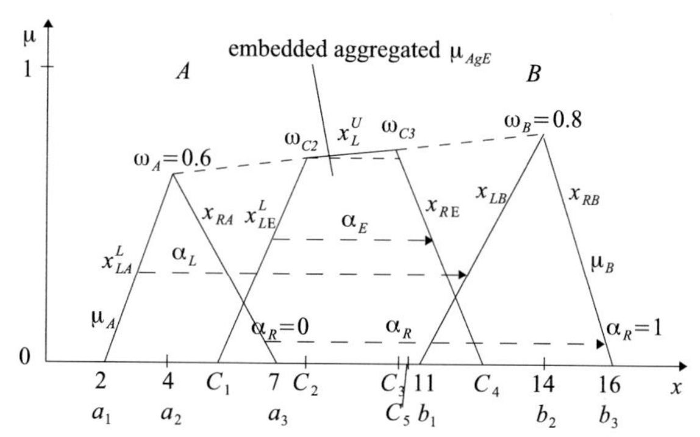

Fig. 3 Intuitionistic membership functions $\mu_A(x)$, $\mu_B(x)$ and one of possible, embedded, aggregated functions $\mu_{AgB}(x)$ of Type-1, α_L-RDM variable of the *left border* transformation, α_R-RDM variable of the *right border* transformation, α_E—inner RDM variable of the embedded IMF Type-1

2 Determining the Aggregated, Intuitionistic, Horizontal Membership Function (IMF) x_{AgB}

Formulas (1)–(4) give values x for points C_1, C_2, C_3, C_4 which characterize embedded IMFs—Type-1.

$$C_1 = a_1 + \alpha_L(b_1 - a_1) = 2 + 9\alpha_L,\ \alpha_L \in [0, 1],\ \alpha_L \le \alpha_R \tag{1}$$

$$C_2 = a_2 + \alpha_L(b_2 - a_2) = 4 + 10\alpha_L \tag{2}$$

$$C_3 = a_2 + \alpha_R(b_2 - a_2) = 4 + 10\alpha_R,\ \alpha_R \in [0, 1],\ \alpha_R \ge \alpha_L \tag{3}$$

$$C_4 = a_3 + \alpha_R(b_3 - a_3) = 7 + 9\alpha_R \tag{4}$$

Formulas (5) and (6) allow for calculation of $\omega_{C_2}, \omega_{C_3}$ which inform about two important characteristic points of embedded IMFs Type-1.

$$\omega_{C_2} = \omega_A + \alpha_L(\omega_B - \omega_A) = 0.6 + 0.2\alpha_L,\ \alpha_L \in [0,1],\ \alpha_L \leq \alpha_R \tag{5}$$

$$\omega_{C_3} = \omega_A + \alpha_R(\omega_B - \omega_A) = 0.6 + 0.2\alpha_R,\ \alpha_R \in [0,1],\ \alpha_R \geq \alpha_L \tag{6}$$

On the basis of formulas (1)–(6) for $\mu \leq \omega_{C_2}$ horizontal model (7) of the left uncertain border x_{LE}^L of the embedded, aggregated IMF is achieved.

$$\begin{aligned} &x_{LE}^L = [a_1 + \alpha_L(b_1 - a_1)] + \tfrac{[(a_2-a_1)+\alpha_L(a_1+b_2-a_2-b_1)]\mu}{\omega_A+\alpha_L(\omega_B-\omega_A)} \\ &\alpha_L \in [0,1],\ \mu \in [0,\omega_{C_2}],\ \alpha_L \leq \alpha_R \end{aligned} \tag{7}$$

For concrete parameter values of functions shown in Fig. 3 formula (7) takes form (8).

$$\begin{aligned} &x_{LE}^L = (2 + 9\alpha_L) + \tfrac{(2+\alpha_L)\mu}{0.6+0.2\alpha_L} \\ &\alpha_L \in [0,1],\ \mu \in [0,\omega_{C_2}],\ \alpha_L \leq \alpha_R \end{aligned} \tag{8}$$

The right, uncertain border x_{RE}^L of the aggregated IMF is determined by (9).

$$\begin{aligned} &x_{RE}^L = [a_3 + \alpha_R(b_3 - a_3)] - \tfrac{[(a_3-a_2)+\alpha_R(a_2+b_3-a_3-b_2)]\mu}{\omega_A+\alpha_R(\omega_B-\omega_A)} \\ &\alpha_R \in [0,1],\ \mu \in [0,\omega_{C_2}],\ \alpha_R \geq \alpha_L \end{aligned} \tag{9}$$

For numerical values of function parameters given in Fig. 3 formula (9) takes form (10).

$$\begin{aligned} &x_{RE}^L = (7 + 9\alpha_R) - \tfrac{(3-\alpha_R)\mu}{0.6+0.2\alpha_R} \\ &\alpha_R \in [0,1],\ \mu \in [0,\omega_{C_2}],\ \alpha_R \geq \alpha_L \end{aligned} \tag{10}$$

RDM variable $\alpha_E \in [0,1]$ transforms (Fig. 3) the left border x_{LE}^L in the right function border x_{RE}^L. The full, complete, horizontal model of the aggregated IMFs $x_{AgB}^{\mu} = f(\alpha_L, \alpha_R, \alpha_E)$ is determined by (11).

$$x_{AgB}^L = x_{LE} + \alpha_E(x_{RE} - x_{LE}^L),\ \alpha_E \in [0,1],\ \mu \in [0,\omega_{C_2}] \tag{11}$$

Model (11) can be used only for $\mu \in [0,\omega_{C_2}]$. For values $\mu \in [\omega_{C_2},\omega_{C_3}]$ another formula should be used what is explained in Fig. 3. In this range of membership μ the left border x_{LE}^U is different than for $\mu \in [0,\omega_{C_2}]$. For distinction the left IMF-border for the range $\mu \in [\omega_{C_2},\omega_{C_3}]$ will be denoted as x_{LE}^U. Instead, the right border x_{RE} is identical as for the range $\mu \in [0,\omega_{C_2}]$, Fig. 3, formulas (9) and (10). The left border is determined by function (12).

$$x_{LE}^{U} = a_2 + (b_2 - a_2)\frac{\mu - \omega_A}{\omega_B - \omega_A}, \quad \mu \in [\omega_{C_2}, \omega_{C_3}] \tag{12}$$

For numerical parameter values of function (12) shown in Fig. 3 x_{LE}^{U} is given by (13).

$$x_{LE}^{U} = -26 + 50\mu, \quad \mu \in [\omega_{C_2}, \omega_{C_3}] \tag{13}$$

The full upper part x_{LE}^{U} of the embedded IMF for the range $\mu \in [\omega_{C_2}, \omega_{C_3}]$ is determined by (14). RDM variable $\alpha_E \in [0, 1]$ transforms in this formula the left uncertain border x_{LE}^{U} given by (12), into the right border x_{RE} given by (9).

$$x_{AgB}^{U} = x_{LE}^{U} + \alpha_E(x_{RE} - x_{LE}^{U}), \quad \alpha_E \in [0, 1] \tag{14}$$

Summarizing: the full horizontal model of the aggregated IMF consists of two parts. The lower part x_{AgB}^{L} which concerns the range $\mu \in [0, \omega_{C_2}]$, formula (11), and the upper part x_{AgB}^{U}, formula (14). Aggregated IMF is shown in Fig. 4.

How the achieved result can be interpreted? Inconsistent IFSs A and B generate one IFS Type-2 being a family of embedded IFST1. It means that the true but precisely unknown x-value that was evaluated by the experts A and B can be contained in one of IFSsT1 imbedded in the achieved IFST2, which membership function is visualized in Fig. 4. Each possible IMFT1 can be achieved by choice of values of RDM variables $\alpha_L, \alpha_R \in [0, 1]$, $\alpha_R \geq \alpha_L$. Example the pair $\alpha_L = 0$, $\alpha_R = 0$ generates triangle, embedded IMFT1 determined by formula $x_E = (2 + 3.333\mu) + \alpha_E(5 - 8.333\mu)$ shown in Fig. 5a. The pair $\alpha_L = 0.5$ and $\alpha_R = 0.5$ generates also triangle IMFT1 determined by formula $x_E = (6.5 + 3.57\mu) + \alpha_E(5 - 7.143\mu)$ shown in Fig. 5a. The pair $\alpha_L = 1$ and $\alpha_R = 1$ generates triangle, embedded IMFT1 determined by formula $x_E = (11 + 3.75\mu) + \alpha_E(5 - 6.25\mu)$ shown in Fig. 5a. The pair $\alpha_L = 0.25$ and $\alpha_R = 0.75$ generates trapezoidal, embedded IMFT1 determined by formula $x_E^L = (4.25 + 3.46154\mu) + \alpha_E(9.5 - 6.46154\mu)$ for $\mu \in [0, 0.65]$ and $x_E^U = (-26 + 50\mu) + \alpha_E(39.75 - 53\mu)$ for $\mu \in [0.65, 0.75]$, Fig. 6. The pair $\alpha_L = 0$ and $\alpha_R = 1$ generates the outer trapezoidal, embedded IMF determined by formula $x_E^L = (2 + 3.333\mu) + \alpha_E(14 - 5.8333\mu)$ for $\mu \in [0, 0.6]$ and $x_E^U = (-26 + 50\mu) + \alpha_E(4252.5\mu)$ for $\mu \in [0.6, 0.8)$ shown in Fig. 5b.

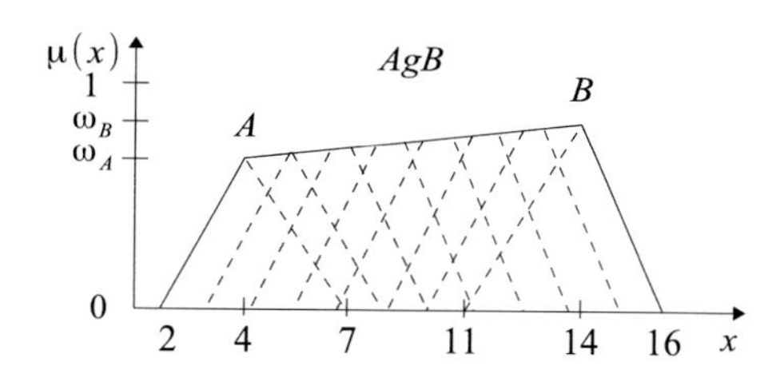

Fig. 4 Visualisation of IMF Type-2 x_{AgB}^{U} representing aggregation of IMFs A and B

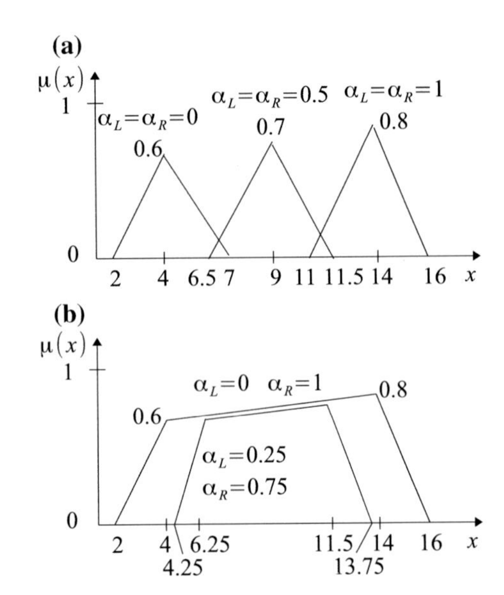

Fig. 5 Examples of IMFs T1 embedded in IMFT2 x_{AgB} generated by different pairs of RDM variables (α_L, α_R) from the horizontal, intuitionistic membership function x_{AgB} of aggregated sets A and B

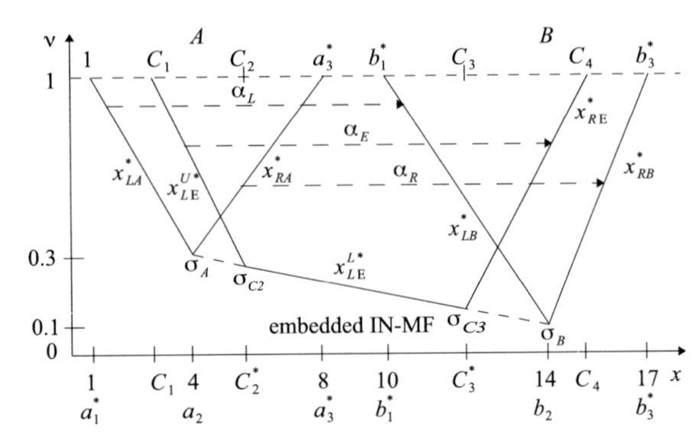

Fig. 6 IN-MFs of aggregated sets A and B, example of a trapezoidal, embedded IN-MF type-1 of AgB and denotations used in formula derivation

3 Determining of the Aggregated, Intuitionistic, Horizontal Non-membership Function (IN-MF) x^*_{AgB}

Figure 6 shows IN-MFs of type-1 ν_A and ν_B of the aggregated sets A and B and one of trapezoidal, embedded IN-MFs T1 with denotations.

Parameters shown in Fig. 6 can be calculated according to (15).

$$
\begin{aligned}
&c_1^* = a_1^* + \alpha_L(b_1^* - a_1^*) = 1 + 9\alpha_L,\ \alpha_L \in [0,1],\ \alpha_L \le \alpha_R \\
&c_2^* = a_2 + \alpha_L(b_2 - a_2) = 4 + 10\alpha_L,\ \alpha_L \in [0,1],\ \alpha_L \le \alpha_R \\
&c_3^* = a_2 + \alpha_R(b_2 - a_2) = 4 + 10\alpha_R,\ \alpha_R \in [0,1],\ \alpha_R \ge \alpha_L \\
&c_4^* = a_3^* + \alpha_R(b_3^* - a_3^*) = 8 + 9\alpha_R,\ \alpha_R \in [0,1],\ \alpha_R \ge \alpha_L \\
&\sigma_{C_2} = \sigma_A - \alpha_L(\sigma_B - \sigma_A) = 0.3 - 0.2\alpha_L,\ \alpha_L \in [0,1],\ \alpha_L \le \alpha_R \\
&\sigma_{C_3} = \sigma_A - \alpha_R(\sigma_B - \sigma_A) = 0.3 - 0.2\alpha_R,\ \alpha_R \in [0,1],\ \alpha_R \ge \alpha_L
\end{aligned}
\tag{15}
$$

RDM variable α_L with its increasing values transforms the left border x^*_{LA} of the IN-MF ν_A in the left border x^*_{LB} of function ν_B. RDM variable α_R transforms the right border x^*_{RA} of function ν_A in the right border x^*_{RB} of function ν_B. RDM variable $\alpha_E \in [0, 1]$ transforms the left, uncertain border x^*_{LE} of the embedded IN-MF of in its right border x^*_{RE}. For the range $\nu \in [\sigma_{C_2}, 1]$ the left border x^*_{LA} of function ν_A is determined by (16).

$$x^*_{LA} = a^*_1 + (a_2 - a^*_1)\left(\frac{1-\nu}{1-\sigma_A}\right) = 5.1857 - 4.2857\nu,\ \nu \in [0, 1] \tag{16}$$

The left border x^*_{LB} of function ν_B is determined by formula (17).

$$x^*_{LB} = b^*_1 + (b_2 - b^*_1)\left(\frac{1-\nu}{1-\sigma_B}\right) = 14.4444 - 4.4444\nu,\ \nu \in [\sigma_{C_2}, 1] \tag{17}$$

The left uncertain, upper border x^{U*}_{LE} of the aggregated IN-MF ν_{AgB} of sets A and B is given by (18).

$$\begin{aligned} x^{U*}_{LE} &= x_{LA} + \alpha_L(x_{LB} - x_{LA}),\ \alpha_L \in [0, 1],\ \alpha_L \le \alpha_R \\ x^{U*}_{LE} &= (5.1857 - 4.2857\nu) + \alpha_L(9.2587 - 0.1587\nu),\ \nu \in [\sigma_{C_2}, 1] \end{aligned} \tag{18}$$

The right border of IN-MF ν_A is given by formula (19).

$$x^*_{RA} = a^*_3 - (a^*_3 - a_2)\left(\frac{1-\nu}{1-\sigma_A}\right) = 2.2857 + 5.7143\nu,\ \nu \in [\sigma_{C_2}, 1] \tag{19}$$

The right border x^*_{RB} of IN-MF ν_B is determined by formula (20).

$$x^*_{RB} = b^*_3 - (b^*_3 - b_2)\left(\frac{1-\nu}{1-\sigma_B}\right) = 13.6667 + 3.3333\nu,\ \nu \in [\sigma_{C_2}, 1] \tag{20}$$

The right uncertain border of embedded IN-MF ν_{AgB} determines formula (21).

$$\begin{aligned} x^*_{RE} &= x_{RA} + \alpha_R(x_{RB} - x_{RA}),\ \alpha_R \in [0, 1],\ \alpha_R \ge \alpha_L \\ x^*_{RE} &= (2.2857 + 5.7143\nu) + \alpha_R(11.3809 - 2.3810\nu),\ \nu \in [\sigma_{C_3}, 1] \end{aligned} \tag{21}$$

RDM variable $\alpha_E \in [0, 1]$ transforms with its increasing values the left uncertain border x^{U*}_{LE} in the right upper border x^{U*}_{RE} of the aggregated function ν_{AgB}, formula (22), where x^{U*}_{LE} is determined by (18) and x^{U*}_{RE} by (21).

$$x^{U*}_{E} = x^{U*}_{LE} + \alpha_E(x^*_{RE} - x^*_{LE}),\ \alpha_E \in [0, 1],\ \nu \in [\sigma_{C_2}, 1] \tag{22}$$

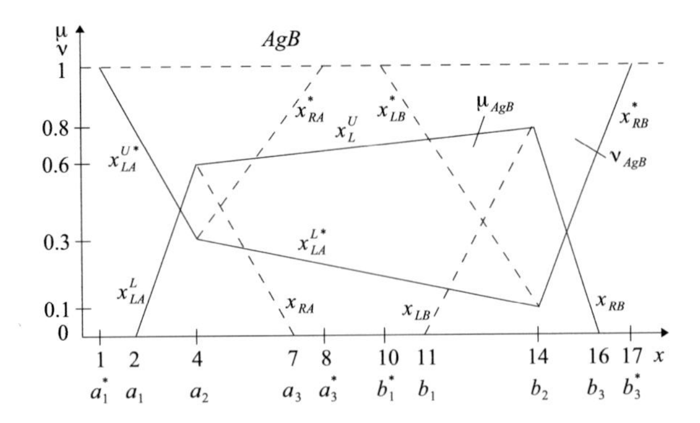

Fig. 7 Visualisation of the membership function of type 2 μ_{AgB} and of the non-membership function ν_{AgB} of two aggregated, inconsistent intuitionistic fuzzy sets *A* and *B*

For the range $\nu \in [\sigma_{C_3}, \sigma_{C_2}]$ formula for x_E^* has other form than (22). It results from the fact that the left, lower border x_{LE}^{L*} is different than for the range $\nu \in [\sigma_{C_2}, 1]$, formula (23).

$$x_{LE}^{L*} = b_2 - \left(\frac{b_2 - a_2}{\sigma_A - \sigma_B}\right)(\nu - \sigma_B) = 19 - 50\nu, \;\; \nu \in [\sigma_{C_3}, \sigma_{C_2}] \tag{23}$$

In the range $\nu \in [\sigma_{C_3}, \sigma_{C_2}]$ formula for x_{EL}^* has form (24), where x_{LE}^{L*} is determined by (23) and x_{RE}^* by (21).

$$x_E^{L*} = x_{LE}^{L*} + \alpha_E(x_{RE}^* - x_{LE}^*), \;\; \alpha_E \in [0, 1], \;\; \nu \in [\sigma_{C_3}, \sigma_{C_2}] \tag{24}$$

Formulas (22) and (24) together define the aggregated IN-MFT2 ν_{AgB}. For chosen values α_L, α_R, $\alpha_R \geq \alpha_L$, they generate an infinitive set of embedded IN-MFs of type-1 of triangle and trapezoidal, possible IN-MFs, similarly as for MFs, as shown in Figs. 4 and 5. Figure 7 shows both aggregated functions: the IMF μ_{AgB} given in its horizontal form by formulas (11) and (14) and IN-MF ν_{AgB} given by (22) and (24).

4 Conclusions

Interval-valued fuzzy sets theory and horizontal RDM membership functions allow for aggregation of uncertain intuitionistic fuzzy sets which express inconsistent expert opinions or inconsistent measurements. Interval-valued FST2-theory allows for new understanding and interpretation of inconsistent expert opinions. Each opinion delivers right and left border of a fuzzy set. Two opinions deliver two right and two left borders. It means that the aggregated borders are uncertain and that they generate a fuzzy set with uncertain borders. Hence, the aggregated IMF has uncertainty of higher order than each of the single component opinions.

References

1. Atanassov, K., Gargov, G.: Interval-valued intuitionistic fuzzy sets. Fuzzy Sets Syst. **31**(3), 343–349 (1989)
2. Atanassov, K., Szmidt, E., Kacprzyk, J.: On some ways of determining membership and non-membership functions characterizing intuitionistic fuzzy sets. In: Proceedings of Sixth International Workshop on IFSs, Banska Bistrica, Slovakia, 10 October 2010, NIFS, vol. 16, pp. 26–30 (2010)
3. Atanassov, K.: New operations defined over the intuitionistic fuzzy sets. In: Fuzzy Sets and Systems, vol. 61, pp. 137–142. North Holland, Amsterdam (1994)
4. Atanassov, K.: Intuitionistic fuzzy sets. Fuzzy Sets Syst. **20**, 87–96 (1986)
5. Cornelius, C., Atanassov, K., Kerry, E.: Intuitionistic fuzzy sets and interval-valued fuzzy sets: a critical comparison. In: Proceedings of International Conference EUSFLAT'03, pp. 159–163 (2003)
6. Mendel, J.: Advances in type-2 fuzzy sets and systems. Inf. Sci. **177**, 84–110 (2007)
7. Mendel, J., John, R.: Type-2 fuzzy sets made simple. IEEE Trans. Fuzzy Syst. **10**(2), 117–127 (2002)
8. Mendel, J., John, R., Liu, F.: Interval type-2 fuzzy logic systems made simple. IEEE Trans. Fuzzy Syst. **14**(6), 808–821 (2006)
9. Mendel, J., Rajati, M.: On computing normalized interval type-2 fuzzy sets. IEEE Trans. Fuzzy Syst. **22**(5), 1335–1340 (2014)
10. Mondal, S., Roy, T.: Non-linear arithmetic operation on generalized triangular intuitionistic fuzzy numbers. Notes Intuition. Fuzzy Sets **20**(1), ISSN 1310–4926, 9–19 (2014)
11. O'Hagan, A., et al.: Uncertain Judgements: Eliciting Experts' Probabilities. Wiley, Chichester (2006)
12. Piegat, A., Landowski, M.: Horizontal membership function and examples of its applications. Int. J. Fuzzy Syst. **17**(1), 22–30 (2015)
13. Tomaszewska, K., Piegat, A.: Application of the horizontal membership function to the uncertain displacement calculation of a composite massless rod under a tensile load. In: Proceedings of International Conference on Advanced Computer Systems, Miedzyzdroje, Poland, (also in Research Gate), pp. 1–8 (2014)

Intuitionistic Fuzzy Evaluations of the Elbow Joint Range of Motion

Simeon Ribagin, Anthony Shannon and Krassimir Atanassov

Abstract Following (Ribagin et al. 2015, In: 19th International Workshop on IFSs, [8]), in this paper it is proposed a technique to evaluate the functional capacity of the elbow joint during a complex movement using intuitionistic fuzzy and interval valued intuitionistic fuzzy sets. The membership and non-membership values are not always possible up to our satisfaction, but in deterministic (hesitation) part has more important role here, the fact that in decision making, particularly in case of orthopedic physical assessment, there is a fair chance of the existence of a non-zero hesitation part at each moment of evaluation. Based on our previous study here we will introduce intuitionistic fuzzy estimations of flexion-extension and pronation-supination movements of the elbow joint.

Keywords Elbow joint · Interval-valued intuitionistic fuzzy set · Intuitionistic fuzzy set · Range of motion

AMS Classification 03E72

S. Ribagin (✉) · K. Atanassov
Institute of Biophysics & Biomedical Engineering, Bulgarian Academy of Sciences, Acad. G. Bonchev Str., Block 105, 1113 Sofia, Bulgaria
e-mail: sim_ribagin@mail.bg

K. Atanassov
e-mail: krat@bas.bg

A. Shannon
Faculty of Engineering & IT, Univesity of Technology, Campion College, P.O. Box 3052, Toongabbie East, Sydney, NSW 2007 2146, Australia
e-mail: t.shannon@campion.edu.au; Anthony.Shannon@uts.edu.au

K.T. Atanassov et al. (eds.), *Novel Developments in Uncertainty Representation and Processing*, Advances in Intelligent Systems and Computing 401,
DOI 10.1007/978-3-319-26211-6_19

1 Introduction and Biological Motivation

The elbow is the anatomic area that joints the arm or "brachium" with the forearm or "antebrachium". The bony structures of the elbow are the distal part of the humerus and the proximal ends of the forearm bones (radius and ulna). The distal end of the humerus is formed by the pulley-shaped trochlea medially and the convex capitulum laterally. The trochlea of the humerus articulates with the deep trochlear notch found on the proximal end of the ulna. The superior surface of the head of the radius is concave for articulation with the capitulum, with the raised margin articulating with the capitulutrochlear groove. Structurally, the joint is classed as a synovial joint, and functionally as a hinge joint. Like all synovial joints the elbow joint has a capsule enclosing the joint. The elbow joint capsule wraps all three bones and both functional joints. Within the elbow complex there are three separate synovial articulations: humeroulnar joint, humeroradial joint and the proximal radioulnar joint. The humeroulnar joint is the articulation between the trochlea of the humerus and the trochlear notch of the ulna. The bones of this joint are shaped so that the axis of movement is not horizontal but instead passes downward and medially, going through an arc of movement. This position leads to carrying angle at the elbow. Typically the normal range of the carrying angle is between 10° and 15° [5]. The humeroradial joint is a uniaxial hinge joint and consist of the spheroidal capitulum of the humerus and the proximal surface of the head of the radius. The proximal radioulnar joint is the articulation between the head of the radius and the radial notch of the proximal ulna. This joint is classed as a uniaxial cone-shaped pivot joint. Stability of the elbow complex is provided mainly by the ligamentous apparatus surrounding the joint and the interlocking mechanism of the articulating surfaces. The elbow complex allows two degrees of freedom in the sagittal plane: flexion-extension and pronation-supination. In [8] we discussed the flexion-extension movements of the elbow. Now we will briefly describe the pronation-supination movements of the elbow and forearm.

The pronation-supination movements of forearm occur at the distal and proximal radioulnar joints of the elbow. The rotational movements are a pair of unique movements possible only in the forearms and hands, allowing the human body to flip the palm either face up or face down. The muscles, bones, and joints of the human forearm are specifically arranged to permit these unique and important rotations of the hands. Movement takes place around the longitudinal axis of the forearm [7], which passes obliquely from the distal aspect of ulnar head at the fovea, to the center of the radial head proximally. The normal pronation-supination arc of the intact elbow ranges from 160 to 180°. Pronation and supination should be assessed with the elbow in 90° of flexion, with the thumb-up position considered as a neutral. In the neutral position the radius and ulna lie next to each other, but in full pronation the radius has crossed over the ulna diagonally. This is possible only because of the direction of the fibers of the interosseous membrane. The radial tuberosity thereby turns towards the ulna. Active pronation is considered to be approximately 80–90° [3], so that the palm faces down. This movement is limited by stretching of the interosseous membrane and squeezing of the flexor muscles. As the radius crosses over in pronation, the distal end of the ulna moves laterally [4]. The opposite occurs during supination. In full supination the two bones lie parallel to each other and the movement is limited by the interosseous membrane, oblique cord, anterior ligament of the distal radioulnar joint and the pronator muscles. Active supination should be 90° [7], so that the palm faces up. Although the radius rotates during pronation and supination, the radial head remains otherwise in a fixed position relative to the ulna. The relative position and movement of the radius about the elbow is essential for the proper functioning of the elbow and forearm. For both pronation and supination, only about 75° of movement occurs in the forearm articulations, the remaining 15° is the result of wrist action [5].

Forearm rotation permits a rotatory displacement of objects that are grasped by the hand and proves essential for a number of manual tasks, including personal hygiene, tool use and feeding. Supination occurs in many functional activities that require the palm to be turned up, such as feeding, washing the face, opening a door etc. Pronation, in contrast, is involved with activities such as grabbing an object, drinking from a cup or cutting with a knife. Most of these activities can be accomplished with a functional range of 100° of forearm rotation, with 50° of both pronation and supination [6]. Elbow and forearm motion serves to position the hand in space. For a normal upper extremity functioning is required a freely mobile and stable elbow joint. Loss of elbow and forearm motion can impact most of the activities of daily living. The purpose of the present study is to give a possible example for evaluation of the complex elbow movements using IFD and IVIFSs.

2 Intuitionistic Fuzzy Evaluations of the Flexion-Extension and Pronation-Supination Movements of the Elbow

In intuitionistic fuzzy logic (IFL) [1, 2], if p is a variable then its truth-value is represented by the ordered pair

$$V(p) = \langle M(p), N(p) \rangle, \tag{1}$$

so that M(p), N(p), $M(p) + N(p) \in [0, 1]$, and M(p) and N(p) are respectively degrees of validity and of non-validity of p. These values can be obtained applying different formula depending on the problem considered. If we like to use Interval-Valued IF (IVIF) values (see, e.g., [1]), then (1) obtains the same form, but now

$$M(p) = [\inf M(p), \sup M(p)] \subset [0, 1],$$
$$N(p) = [\inf N(p), \sup N(p)] \subset [0, 1],$$

and

$$\sup \mathrm{M}(\mathrm{p}) + \sup \mathrm{N}(\mathrm{p}) \leq 1.$$

Now, we shall give an IF estimation of the functional capacity in the elbow joint during the pronation-supination movement together with the flexion-extension movement of the person *p*. In our previous study (see [8]) we assume that the anatomical barrier of the elbow joint is in interval [–10°, 160°] and the active range of motion is in interval [0°, −145°].

Let the movement of person *p* be between angles $\alpha(p)$ and $\beta(p)$, where $\alpha(p) < \beta(p)$. Then

$$\omega(p) = \beta(p) - \alpha(p).$$

Therefore,

$$M(p) = \frac{\beta - \alpha}{170°}.$$

$$N(p) = 1 - \frac{\beta - \alpha}{170°}.$$

Let the active range of motion during the physical examination of a person *p* be:

$$\omega(\mathrm{p}) = \beta_1(\mathrm{p}) - \alpha_2(\mathrm{p}) < 145°.$$

Therefore,

$$M(p) = \frac{\beta_1 - \alpha_2}{170°} < \frac{\mathbf{145°}}{170°}.$$

The restricted range of motion which will not be accomplished even with an additional passive force applied is:

$$\varphi(\mathrm{p}) = 170° - (\beta_2(\mathrm{p}) - \alpha_1(\mathrm{p})) > 25°$$

and

$$N(p) = 1 - \frac{\beta - \alpha}{170°} > \frac{\mathbf{35°}}{170°}.$$

So, for patient *p* we determine his/her personal IF estimation is the form $\langle M(p), N(p)\rangle$.

Now we shall apply the same approach by giving an IF estimation of the functional capacity in the elbow joint during the pronation-supination movement.

The expected maximal range of pronation-supination movements of the elbow and forearm is R: 90° −0° −90°. Let p be a patient for whom $\gamma(p) \in [-90°, 0°]$ and $\delta(p) \in [0°, -90°]$.

Therefore,

$$R(p) = \frac{90° + \gamma}{180°}$$

and

$$S(p) = \frac{90° - \delta}{180°}$$

Because,

$$0 \leq R(p) + S(p) = \frac{180° + (\gamma - \delta)}{180°} = 1 - \frac{(\gamma - \delta)}{180°} < 1 - \frac{(90° - \delta)}{180°} \geq 1,$$

then the pair $\langle R(p), S(p) \rangle$ is an intuitionistic fuzzy pair. Hence, we can define function,

$$T(p) = 1 - R(p) - S(p) = \frac{(\gamma - \delta)}{180°}$$

If we like to give an IVIF-estimation, it will have the form $\langle M(p), N(p), R(p), S(p) \rangle = \langle [\alpha_1, \alpha_2], [\beta_1, \beta_2], [\gamma], [\delta] \rangle$, where $\alpha_1, \alpha_2, \beta_1, \beta_2, \gamma, \delta$ are above described estimations for patient (p).

3 Intuitionistic Fuzzy Estimation of a Complex Movement of the Elbow

Let us have a patient (p) with concrete α, β, γ and δ parameters. Therefore, we can estimate his/her capacity of the elbow joint during a complex movement. The estimation can obtain different values and can be interpreted by different ways. Here we introduce the following five estimations:

- Strong pessimistic estimation: $\langle M(p).R(p), N(p) + S(p) - N(p).S(p) \rangle$,
- Pessimistic estimation: $\langle \min(M(p), R(p)), \max(N(p) + S(p)) \rangle$,
- Average estimation: $\langle \frac{M(p) + R(p)}{2}, N(p) + S(p) \rangle$,
- Optimistic estimation: $\langle \max(M(p), R(p)), \min(N(p) + S(p)) \rangle$,
- Strong optimistic estimation: $\langle M(p) + R(p) - M(p).R(p), N(p).S(p) \rangle$.

4 Conclusions

Somatic dysfunction of the elbow joint occurs when there is a restriction of motion occurring within the normal range of movement. A thorough and accurate assessment of a subject's elbow joint complex range of motion is essential for correct diagnosis, treatment and determination of rehabilitation potential. However, all assessment and testing in clinical practice is based on the assumption of uncertainty. By employing IF evaluations of we can express a hesitation concerning examined objects. The method proposed in this article, assigning IF values and numeric grades of flexion-extension and pronation-supination movements of the elbow joint will significantly improve the overall assessment of the elbow joint range of motion. Appling the propose intuitionistic fuzzy estimations will give the possibility of improving the overall assessment method by considering the possibility of assigning relative weight coefficients for all movement values since for some joint dysfunctions some particular joint motions are more important than others.

References

1. Atanassov, K.: Intuitionistic Fuzzy Sets: Theory and Applications. Springer, Heildelberg (1999)
2. Atanassov, K.: On Intuitionistic Fuzzy Sets Theory. Springer, Berlin (2012)
3. Chila, A.G.: Foundations of Osteopathic Medicine, pp. 650–651. Lippincott Williams & Wilkins, Philadelphia (2010)
4. Hamill, J., Knutzen, K.M.: Biomechanical Basis of Human Movement, pp. 148–149. (Lippincott Williams & Wilkins, Philadelphia 2006)
5. Magee, D.J.: Orthopedic Physical Assessment, pp. 392–393. Elsevier, Philadelphia (2013)
6. Morry, B.F., et al.: A biomechanical study of normal functional elbow motion. J. Bone Joint Surg. **63**, 872–877 (1981)
7. Ombregt, L.: A System of Orthopaedic Medicine, pp. 94–95. Elsevier, Philadelphia (2013)
8. Ribagin, S., Shannon, A., Krawczak, M., Atanassov, K.: Intuitionistic fuzzy evaluations of the elbow joint range of motion in the sagittal plane. In: 19th International Workshop on IFSs. Notes on Intuitionistic Fuzzy Sets, vol. 21, No. 2, pp. 134–139. Burgas, 4–6 June 2015, ISSN: 1310–4926

Using Phi Coefficient to Interpret Results Obtained by InterCriteria Analysis

Lyudmila Todorova, Peter Vassilev and Jivko Surchev

Abstract The authors propose an algorithm for assessment of the estimates of "correspondence" and "opposition" obtained by InterCriteria Analysis (ICA) in the form of intuitionistic fuzzy vector pairs. For this aim the modified Pearson coefficient of Karl Pearson, called φ coefficient ("mean square contingency coefficient"). The algorithm is applied on real data from neurosurgery. The statistical significance of the relations between the considered criteria is verified by data found in literature. The authors believe this approach for data exploration may prove useful in many areas.

Keywords InterCriteria analysis · Pearson phi · Correlation · Intuitionistic fuzzy vector pairs

1 Introduction and Preliminary Definitions

The term "correlation" is introduced for statistical purposes by Francis Galton [1]. Despite the fact that the correlation dependencies are objective, the variation of the variable Y in the general case is not completely determined by the variation of the factor X. The correlation relation is not occurring in all of the observed case, but only in the mass occurrence of the events. In other words not every variation of the factor leads to a predetermined change in the result. The magnitude and direction of

L. Todorova · P. Vassilev (✉)
Institute of Biophysics and Biomedical Engineering, Bulgarian Academy of Sciences, 105 Acad. G. Bonchev Str., 1113 Sofia, Bulgaria
e-mail: peter.vassilev@gmail.com

L. Todorova
e-mail: lpt@biomed.bas.bg

J. Surchev
Department of Neurosurgery, Medical University Sofia, Sofia, Bulgaria
e-mail: j_surchev@abv.bg

K.T. Atanassov et al. (eds.), *Novel Developments in Uncertainty Representation and Processing*, Advances in Intelligent Systems and Computing 401,
DOI 10.1007/978-3-319-26211-6_20

influence between two statistical variables is measured by the correlation coefficient of Karl Pearson:

$$r = \frac{\sum_{i=1}^{n} (x_i - \bar{x})(y_i - \bar{y})}{\sqrt{\sum_{i=1}^{n} (x_i - \bar{x})^2}\sqrt{\sum_{i=1}^{n} (y_i - \bar{y})^2}} \quad (1)$$

where the sets $X = x_1, \ldots, x_n$ and $Y = y_1, \ldots, y_n$, of the dependent and independent variable, respectively, have n values.

The values for the correlation coefficients calculated in accordance with formula (1) vary between –1 and 1. The values close to those two extreme values correspond to the near-linear relationship between the variables. The sign of the coefficient is positive when the two variables are changed in the same "direction" - namely, the increase/decrease of one is related to the increase/decrease of the other. A negative sign corresponds to the opposite relationship increase/decrease of one leads to decrease/increase of the other, respectively. Two scales to assess the degree of correlation are commonly applied (Tables 1 and 2): When both variables are discrete dichotomies, the correlation is calculated based on:

- The modified coefficient of Karl Pearson, called phi-factor φ, by the formula:

$$\varphi = \frac{ad - bc}{\sqrt{(a+b)(c+d)(a+c)(b+d)}} \quad (2)$$

where: a, b, c, d—are the frequencies (number of individuals) characterized in an appropriate manner with respect to the two features X, Y as shown in Table 3. The table columns are values of the independent variable, in the rows of the dependent. A restriction for the application of the Pearson correlation coefficient is that it is

Table 1 First correlation scale

$0 < R < 0.3$	Weak correlation
$0.3 < R < 0.5$	Moderate correlation
$0.5 < R < 0.7$	Significant correlation
$0.7 < R < 0.9$	High correlation
$0.9 < R < 1.0$	Very high correlation

Table 2 Second correlation scale

$0 < R < 0.2$	Weak correlation
$0.2 < R < 0.4$	Moderate correlation
$0.4 < R < 0.6$	Significant correlation
$0.6 < R < 0.8$	High correlation
$0.8 < R < 1.0$	Very high correlation

Table 3 Tabular representation of Dependent vs. Independent variable

		X		
		Independent variable		
		Yes	No	Totals
Y	Yes	a	b	$a+b$
	No	c	d	$c+d$
Dependent variable	Totals	$a+c$	b + d	n

applicable for a linear kind of dependence. The InterCriteria Analysis (ICA) [2] also estimates the relation between parameters without imposing such restriction.

Given an IM (see [3]) with index sets consisting of the names of the criteria (for rows) and objects (for columns) with real numbers as components, for any two "criteria" a vector of "agreement" of their internal comparisons may be constructed.

Let O denote the set of all objects being evaluated (in our case these correspond to the patients), and $C(O)$ denote the set of values assigned by a given criteria C to theses objects, i.e.

$$O \stackrel{\text{def}}{=} \{O_1, O_2, \ldots, O_n\}, \ C(O) \stackrel{\text{def}}{=} \{C(O_1), C(O_2), \ldots, C(O_n)\}.$$

Let $C^*(O) \stackrel{\text{def}}{=} \{\langle x, y\rangle |\, x \neq y \,\&\, \langle x, y\rangle \in C(O) \times C(O)\}$.

To construct the "vector of correspondence" of two criteria, the array of all internal comparisons which must fulfill exactly one of three relations R, $\overline{R}$ and $\tilde{R}$ is first constructed. It is assumed that for a fixed criterion C and any ordered pair $\langle x, y\rangle \in C^*(O)$ it is true:

$$\langle x, y\rangle \in R \Leftrightarrow \langle y, x\rangle \in \overline{R}, \tag{3}$$

$$\langle x, y\rangle \in \tilde{R} \Leftrightarrow \langle x, y\rangle \notin (R \cup \overline{R}), \tag{4}$$

$$R \cup \overline{R} \cup \tilde{R} = C^*(O). \tag{5}$$

For the effective calculation of the array of internal comparisons (denoted further by $V(C)$), only lexicographically ordered pairs $\langle x, y\rangle$ need to be considered, since from (3)–(5) it follows that if the relation between x and y is known, the relation between y and x is known as well. Denoting $C_{i,j} = \langle C(O_i), C(O_j)\rangle$, for a fixed criterion C the following array is obtained:

$$V(C) = \{C_{1,2}, C_{1,3}, \ldots, C_{1,n}, C_{2,3}, C_{2,4}, \ldots, C_{2,n}, C_{3,4}, \ldots, C_{3,n}, \ldots, C_{n-1,n}\}.$$

It is easy to see that it has $\frac{n(n-1)}{2}$ elements. To simplify our considerations, the vector $V(C)$ is replaced with $\hat{V}(C)$, where for the k-th component ($1 \leq k \leq \frac{n(n-1)}{2}$) it is true:

$$\hat{V}_k(C) = \begin{cases} 1 \text{ iff } V_k(C) \in R, \\ -1 \text{ iff } V_k(C) \in \overline{R}, \\ 0 \text{ otherwise.} \end{cases}$$

To each two criteria we juxtapose two vectors–of "correspondence" ("agreement") and of ("opposition") based on $\hat{V}(C)$ and $\hat{V}(C')$ as shown in the pseudocode in **Algorithm 1**.

We note that the obtained vectors V_{corr} and V_{opp} can be regarded as intuitionistic fuzzy vector pairs by analogy with intuitionistic fuzzy pairs [4] in the following sense:

$$\hat{0} \leq V_{\text{corr}} + V_{\text{opp}} \leq \hat{1},$$

Algorithm 1 Correspondence and opposition vectors between two criteria

Require: Vectors $\hat{V}(C)$ and $\hat{V}(C')$

1: **function** CORRESPONDENCE($\hat{V}(C), \hat{V}(C')$)
2: $V \leftarrow \hat{V}(C) - \hat{V}(C')$
3: $V_{\text{corr}} \leftarrow \underbrace{[0, 0, \ldots, 0]}_{\frac{n-1}{2}\text{ - times}}$
4: **for** $i \leftarrow 1$ to $\frac{n(n-1)}{2}$ **do**
5: **if** $V(i) = 0$ **then**
6: $V_{\text{corr}}(i) \leftarrow 1$
7: **end if**
8: **end for**
9: **return** V_{corr}
10: **end function**

11: **function** OPPOSITION($\hat{V}(C), \hat{V}(C')$)
12: $V \leftarrow \hat{V}(C) - \hat{V}(C')$
13: $V_{\text{opp}} \leftarrow \underbrace{[0, 0, \ldots, 0]}_{\frac{n-1}{2}\text{ - times}}$
14: **for** $i \leftarrow 1$ to $\frac{n(n-1)}{2}$ **do**
15: **if** abs($V(i)$) $= 2$ **then** ▷ abs: absolute value
16: $V_{\text{opp}}(i) \leftarrow 1$
17: **end if**
18: **end for**
19: **return** V_{opp}
20: **end function**

where

$$\hat{0} \stackrel{\text{def}}{=} \underbrace{[0, 0, \ldots, 0]}_{\frac{n-1}{2}\text{ - times}};\ \hat{1} \stackrel{\text{def}}{=} \underbrace{[1, 1, \ldots, 1]}_{\frac{n-1}{2}\text{ - times}}$$

2 Proposed Algorithm and Example

Further we propose the following algorithm

1. For the considered classes the above considered variant of ICA is used. As a result for each class binary vector for every feature is obtained.
2. The values for a, b, c, and d are determined as in Table 4
3. The Phi coefficient is calculated by formula (2)
4. Based on the obtained values the correlation between the considered features is interpreted in accordance with one of the scales presented in Tables 1 and 2.

The developed algorithm is applied to the data from neurosurgery and covers a contingent of 106 patients who underwent due to infantile hydrocephalus. The valves were implanted in the Department of Neurosurgery at the University Hospital "St. Ivan Rilski", Medical University—Sofia in the period 1984–2003. All patients included in the study were shunted in childhood (up to 18 years), and have undergone at least one revision.

Hydrocephalus is defined [5] as an abnormal accumulation of fluid in the ventricular system (in whole or in part of it), which is associated with increased intracranial pressure. This pathology is particularly common in childhood and especially—in nursing. It is worked for the surgical treatment of hydrocephalus, and after the introduction of the shunts surgeries—and in the treatment of their complications, which generally can be divided into three groups:

- infections;
- mechanical complications;
- functional complications (due to inadequate drainage of normal functioning shunt systems).

Table 4 Determining the binary values from the variant ICA

	1	0	Totals
1	a	b	$a+b$
0	c	d	$c+d$
Totals	$a+c$	$b+d$	n

Shunt infections lead to increased risk of mental retardation, development of locular hydrocephalus and even death [6–8]. Non-infection shunt complications are associated with a low but real risk of death and require surgical revision, with all related surgical risks [9–11]. Despite the continuous development of neurosurgical science and the introduction of other forms of treatment outside shunts, they retain an important place in the treatment of pediatric hydrocephalus. This is the reason to analyze their complications, to seek ways and means of reducing their number. This will contribute to reduced morbidity and will have a huge socio-economic impact due to the large financial resources used for the treatment of complications.

In this work the following relations:

- ("Criterion 1") age at implantation—("Criterion 2") frequency (number) of complications (following revisions).
- length of the interval between the implantation and the first revision—("Criterion 3") number of complications.
- age at implantation—length of the interval between the implantation and the first revision.

The contingent of patients is divided in two classes:

(a) patients who underwent only one revision for the whole observation period (53 in number);
(b) patients who underwent 2 or more revisions for the observed period (53 in number).

2.1 Results

According to the proposed algorithm in Sect. 2 for the considered classes we obtain the following:

Case I. For each class the obtained binary vectors for the features has "1" in the case of "correspondence" and "0" otherwise

Case II. For each class the obtained binary vectors for the features has "1" in the case of "opposition" and "0" otherwise

The obtained values for a, b, c, d, as well as the modified coefficient of Karl Pearson φ are given in Table 5.

For comparison the correlation between the three parameters is calculated as (see Table 6):

- Pearson Correlation (P.C.) [12]
- Kendall rank correlation coefficient (Kendall's tau coefficient) (K.T.C.) [13]
- Spearman's rank correlation (Spearman's rho) (S.R.) [14].

Table 5 Obtained by the variant ICA values for a, b, c, d and φ

		a	b	c	d	φ
Case I	Criterion 1/Criterion 2	128	1250	480	898	−0.3080
	Criterion 1/Criterion 3	606	772	688	690	−0.0596
	Criterion 2/Criterion 3	1	1377	510	868	−0.4752
Case II	Criterion 1/Criterion 2	0	1378	475	903	−0.4563
	Criterion 1/Criterion 3	645	733	606	772	0.02842
	Criterion 2/Criterion 3	0	1378	493	885	−0.4668

Table 6 Pearson Coefficient, Kendall's tau coefficient and Spearman's rho for the criteria

		Criterion 1	Criterion 2	Criterion 3
P.C.	Criterion 1	1	0.002	−0.123
	Criterion 2	0.002	1	−0.348[a]
	Criterion 3	−0.123	−0.348[a]	1
K.T.C.	Criterion 1	1.000	0.090	−0.024
	Criterion 2	0.090	1.000	−0.308[a]
	Criterion 3	−0.024	−0.308[a]	1.000
S.R.	Criterion 1	1.000	0.118	−0.044
	Criterion 2	0.118	1.000	−0.411[a]
	Criterion 3	−0.044	−0.411[a]	1.000

[a]Correlation is significant at the 0.01 level

2.2 *Discussion of the Results*

By applying the proposed in the present work algorithm we found statistically significant correlation between:

- Age of implantation and number of revisions. The correlation is negative, i.e. the smaller the age of the patient is, the more revisions he/she incurs subsequently. Also it varies from moderate to significant depending which of the two cases we consider and which of the two scales is applied. None of the other statistical methods provides a statistically significant difference.
- Age of implantation length of the interval between the implantation and the first revision. Our method, as well as the other three statistical methods did not find a correlation between this two features.
- the interval between the implantation and the first revision and the number of revisions. The correlation is negative, i.e. the smaller the length of the interval, the greater the number of subsequent revisions and in both cases is moderate or significant depending on the chosen scale.

From the results obtained over the considered data we can conclude, that since the correlation dependencies discovered by our method are stated by other sources [15–17], it is capable to capture more subtle relations.

3 Conclusion

In the present paper a new algorithm for assessment of the estimates of "correspondence" and "opposition" obtained by InterCriteria Analysis (ICA). The results seem to indicate that it is possible to use it in exploratory manner to investigate possible correlations between the "correspondence" and "opposition" levels of different features ("criteria").

Acknowledgments The first two authors are grateful for the support provided by Grant DFNI-I-02-5 "InterCriteria Analysis—A New Approach to Decision Making" of the Bulgarian National Science Fund.

References

1. Galton, F.: Co-relations and their measurement, chiefly from anthropometric data. Proc. Roy. Soc. Lond. **45**, 135–145 (1888)
2. Atanassov, K., Mavrov, D., Atanassova, V.: Intercriteria decision making: a new approach for multicriteria decision making, based on index matrices and intuitionistic fuzzy sets. Issues IFSs GNs **11**, 1–8 (2014)
3. Atanassov, K.: On index matrices, part 1: standard cases. Adv. Stud. Contemp. Math. **20**(2), 291–302 (2010)
4. Atanassov, K., Szmidt, E., Kacprzyk, J.: On intuitionistic fuzzy pairs. Notes IFS **19**(3), 1–13 (2013)
5. Di Rocco, F., Garnett, M., Roujeau, T., Puget, S., Renier, D., Zerah, M., Sainte-Rose, C.: Neurosurgery, European Manual of Medicine, Hydrocephalus, vol. 7, pp. 539–543. Springer (2010)
6. Hoffman, H.J., Soloniuk, D., Humphreys, R.P., et al.: A concerted effort to prevent shunt infection. In: Matsumoto, S., Tamaki, N. (eds.) Hydrocephalus: Pathogenesis and Treatment, pp. 510–514. Springer, New York (1991)
7. Schreffler, R.T., Schreffler, A.J., Wittler, R.R.: Treatment of cerebrospinal fluid shunt infections: a decision analysis. Pediatr. Infect. Dis. J. **21**, 632–636 (2002)
8. Walters, B.C., Hoffman, H.J., Hendrick, E.B., Humphreys, R.P.: Cerebrospinal fluid shunt infection. Influences on initial management and subsequent outcome. J. Neurosurg. **60**(5), 1014–1021 (1984)
9. Piatt Jr, J.H., Carlson, C.V.: A search for determinants of cerebrospinal fluid shunt survival: Retrospective analysis of a 14-year institutional experience. Pediatr. Neurosurg. **19**(5), 233–242 (1993)
10. Tuli, S., Drake, J., Lawless, J., Wigg, M., Lamberti-Pasculli, M.: Risk factors for repeated cerebrospinal shunt failures in pediatric patients with hydrocephalus. J. Neurosurg. **92**, 31–38 (2000)
11. Weprin, B.E., Swift, D.M.: Complications of ventricular shunts. Tech. Neurosurg. **7**(3), 224–242 (2002)
12. Pearson, K.: Notes on regression and inheritance in the case of two parents. Proc. R. Soc. Lond. **58**, 240242 (1895)
13. Kendall, M.: A new measure of rank correlation. Biometrika **30**(12), 81–89 (1938)
14. Spearman, C.: The proof and measurement of correlation between two things. Am. J. Psychol. **15**, 72101 (1904)
15. Korinth, M.C., Weinzierl, M.R., Gilsbach, M.: Experience with a new concept to lower non-infectious complications in infants with programmable shunts. Eur. J. Pediatr. Surg. **13**, 81–86 (2003)

16. Shannon, C.N., Acakpo-Satchivi, L., Kirby, R.S., Franklin, F.A., Wellons, J.C., Ventriculoperitoneal shunt failure: an institutional review of 2-year survival rates. Child's Nerv. Syst. 1–7, (2012)
17. Wu, Y., Green, N.L., Wrensch, M.R., Zhao, S., Gupta, N.: Ventriculoperitoneal shunt complications in California: 1990 to 2000. Neurosurgery **61**(3), 557–562 (2007)

Part IV
Generalized Nets and Neural Networks

Modeling Logic Gates and Circuits with Generalized Nets

Lenko Erbakanov, Todor Kostadinov, Todor Petkov, Sotir Sotirov and Veselina Bureva

Abstract In this paper, modeling of logic gates is presented for the first time. Four models of Generalized Nets (GN)—AND gate, a binary to decimal decoder, delay type flip-flop, n-bit binary counter and logical circuits are presented in the following paper. Here we also suggest using the recently proposed approach of InterCriteria Analysis, based on index matrices and intuitionistic fuzzy sets, which aim to detect possible correlations between pairs of criteria. We can perform the measurements, if we have a set of several logical circuits that can be used to obtain identical output data. The aforementioned logical circuits must be composed of different logical elements. By using several measurement points and different schematics, we can suggest the best solution for the considered type of task.

Keywords Generalized nets · Digital logic · Intercriteria analysis

1 Introduction

The theory of *Generalized Nets* is a proper tool to describe and test the digital circuits.

L. Erbakanov (✉) · T. Kostadinov · T. Petkov · S. Sotirov · V. Bureva
Intelligent Systems Laboratory, University Professor Dr. Assen Zlatarov, Burgas, Bulgaria
e-mail: erbakanov@btu.bg

T. Kostadinov
e-mail: kostadinov_todor@btu.bg

T. Petkov
e-mail: todor_petkov@btu.bg

S. Sotirov
e-mail: ssotirov@btu.bg

V. Bureva
e-mail: vbureva@btu.bg

K.T. Atanassov et al. (eds.), *Novel Developments in Uncertainty Representation and Processing*, Advances in Intelligent Systems and Computing 401,
DOI 10.1007/978-3-319-26211-6_21

A comprehensive description of the theory of *Generalized Nets* (GN) is provided in [1, 2]. Useful information concerning the research application of GNs can be found in [3].

Formally, every transition (Fig. 1) is described by a seven interconnected transitions:

$$Z = \langle L', L'', t_1, t_2, r, M, \gamma \rangle$$

where:

- L and L are finite, non-empty sets of the transition's input and output positions, respectively (for the transition in Fig. 1 these positions are $L' = \{l'_1, l'_2, \ldots, l'_m\}$, $L'' = \{l''_1, l''_2, \ldots, l''_n\}$);
- t_1 is the current time-moment of the transition's firing;
- t_2 is the current value of the duration of the transition's active state;
- r is the transition's *condition* that determine which tokens will transfer from the input to the output of the transition. The parameter r can be represented as an index matrix [4]:

$$r = \begin{array}{c|ccccc} & l''_1 & \ldots & l''_j & \ldots & l''_n \\ \hline l'_1 & & & & & \\ \vdots & & & r_{i,j} & & \\ l'_i & & & (r_{i,j} - \text{predicate}) & & \\ \vdots & & & (1 \le i \le m, 1 \le j \le n) & & \\ l'_m & & & & & \end{array},$$

- where $r_{i,j}$ is the predicate that gives the transfer condition from the i-th input point to the j-th output point. When $r_{i,j}$ has a value "*true*", then the token from the i-th input point can be transferred to the j-th output point; otherwise, this is impossible;
- M is an IM of the capacities of transition's arcs:

$$M = \begin{array}{c|ccccc} & l''_1 & \ldots & l''_j & \ldots & l''_n \\ \hline l'_1 & & & & & \\ \vdots & & & m_{i,j} & & \\ l'_i & & & (m_{i,j} \ge 0 - \text{natural number or } \infty) & & \\ \vdots & & & (1 \le i \le m, 1 \le j \le n) & & \\ l'_m & & & & & \end{array};$$

- γ is called transition type. It represents a Boolean expression; the variables it contains can be specified as the symbols used as labels for the transition's input points; when the value of the type is "*true*", the transition can become active, otherwise it cannot.

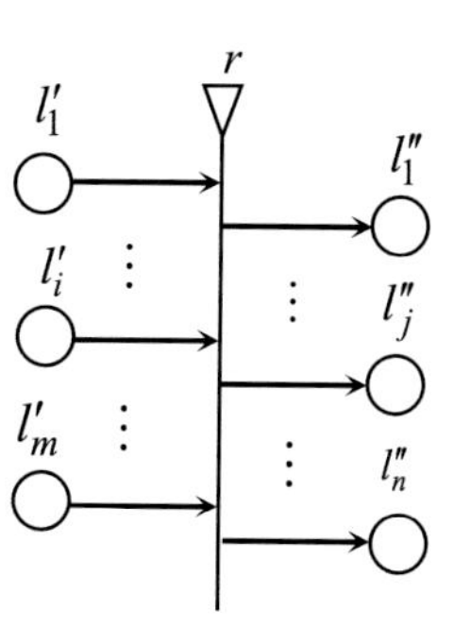

Fig. 1 GN transition in general form, featuring m inputs and n outputs

The main goal of this paper is to represent the function of several different types of logical gates by generalized networks. Then the GN models can be used to present a particular electrical circuit. The presented sample circuit uses a clock input, and a PWM (Pulse Width Modulated) output, where the duty cycle of the output PWM signal depends on the position of the switches S1–S4.

By developing several different schematics that produce equivalent output, for analysis purposes of the parameters of several points ICRA algorithm can be used in order to suggest the best logical elements for the implementation of a certain type of circuits.

2 GN Model of an AND Gate

Let's assume we have an AND gate with n inputs. We will present the gate as a single transition "Z_{AND}" that consists of $n + 1$ input and two output places (Fig. 2).

Initially at the input points X_0, X_1, … X_{n-1} the tokens α_0, α_1, … $\alpha_{n\text{-}1}$, are located respectively, with characteristics: "α_k logic state", where $k = 0 \div n - 1$. The γ token with characteristic $x_0^\gamma = 0$ is located in place C.

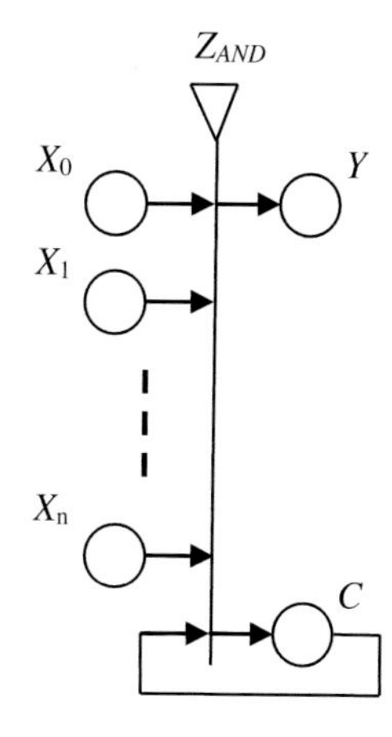

Fig. 2 GN model of n-input "AND" gate

The transition Z_{AND} is represented as follows:

$$Z_{\text{AND}} = \langle \{X_0, X_1, \ldots, X_{n-1}, C\}, \{Y, C\}, R_{AND}, \wedge(X_0, X_1, \ldots, X_{n-1}, C)\rangle$$

$$R_{AND} = \begin{array}{c|cc} & Y & C \\ \hline X_0 & false & true \\ X_1 & false & true \\ \vdots & \vdots & \vdots \\ X_{n-1} & false & true \\ C & W_{CY} & true \end{array}$$

where:

W_{CY} = "The logic function is calculated".

The tokens from places X_0, X_1, … X_{n-1} are merged in place C, where the token γ obtains the following characteristic:

$$x_{cu}^{\gamma} = x_{cu}^{\alpha_1} \wedge x_{cu}^{\alpha_2} \wedge \ldots \wedge x_{cu}^{\alpha_{n-1}},$$

where α_i ($i = 0 \div n - 1$) is the token that stays in the i-th place of the transition Z_{AND}.

The token γ from place C enters place Y without a characteristic change.

3 GN Model of a Binary to Decimal Decoder (0 of $n-1$)

This type of combinational logic circuit has multiple inputs and multiple outputs, where for every unique input bit states, only one output is activated. For example if the binary combination of the integer value k is applied to the inputs, only Y_k output would be set (Y_k = "1"), and all of the rest output states would be "0".

Assume we have a decoder with n inputs. Hence the number of outputs will be 2^n. We will present the gate as a single transition "Z_{DEC}" that consists of $n + 1$ input and $2^n + 1$ output places (Fig. 3).

Initially in the input places X_0, X_1, … X_{n-1} stay tokens α_0, α_1, … α_{n-1}, respectively, with characteristics: "α_k logic state", where $k = 0 \div n - 1$. In place C a token γ with characteristic: $x_0^{\gamma} = 0$ is positioned.

The transition Z_{DEC} is represented by the following expression:

$$Z_{DEC} = \langle \{X_0, X_1, \ldots, X_{n-1}, C\}, \{Y_0, Y_1, \ldots, Y_{2^n-1}, C\}, R_{DEC}, \wedge(X_0, X_1, \ldots, X_{n-1}, C)\rangle,$$

$$R_{DEC} = \begin{array}{c|cccccc} & Y_0 & Y_1 & \ldots & Y_{2^{n-1}} & C \\ \hline X_0 & false & false & \ldots & false & true \\ X_1 & false & false & \ldots & false & true \\ \vdots & \vdots & \vdots & \ldots & \vdots & \vdots \\ X_{n-1} & false & false & \ldots & false & true \\ C & W_{CY} & W_{CY} & \ldots & W_{CY} & true \end{array}$$

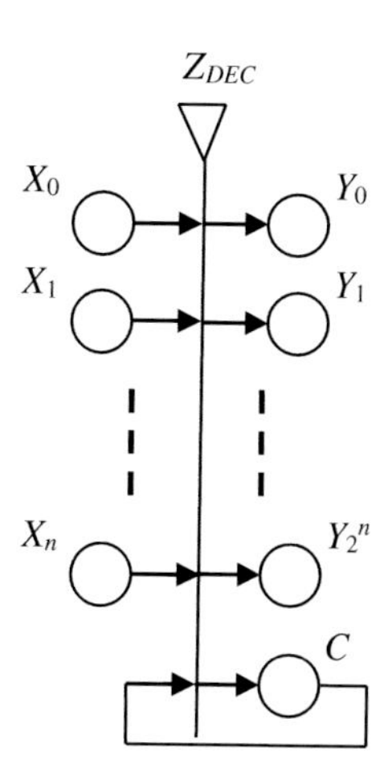

Fig. 3 GN model of n-input binary to decimal decoder

The tokens α_0, α_1, … α_{n-1} are merged in place C, where token γ obtains the following characteristic:

$$x_{cu}^{\gamma} = 2^{\sum_{i=0}^{n-1} 2^i x_{cu}^{\alpha_i}} - 1,$$

where $x_{cu}^{\alpha_i}$ is the characteristic of the α token that stays in the i-th place of the transition Z_{DEC} ($i = 0 \div n - 1$).

The token γ splits in 2^n tokens in places $Y_0, Y_1, \ldots, Y_{2^n-1}$, where the token γ_k obtains characteristic: "k-th bit of x_{cu}^{γ}" ($k = 0 \div 2^n - 1$).

4 GN Model of a Delay Type Flip-Flop (*D* Latch)

The GN model of this sequential logic circuit consists of one transition Z_D (Fig. 4). Initially, there are tokens located in places D and C, respectively:

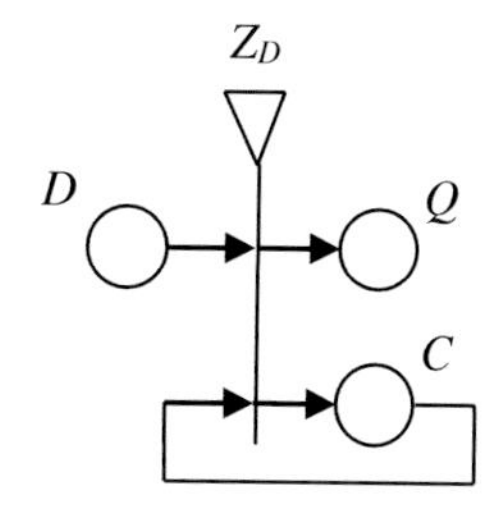

Fig. 4 GN model of D-latch

Fig. 5 GN model of n-bit counter

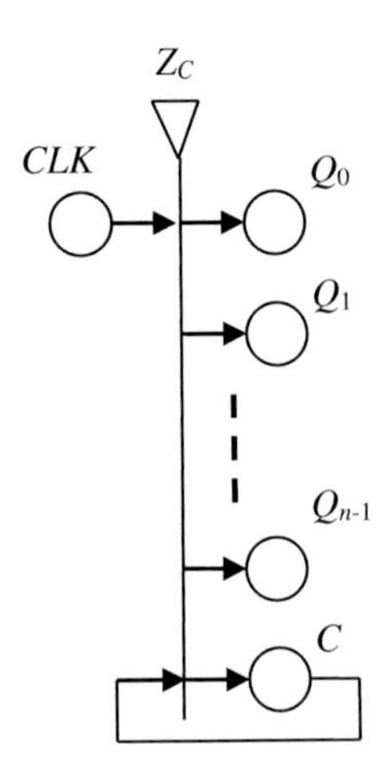

- α token with characteristic “data state”;
- β token with characteristic “clock state”.

The transition Z_D is represented by the following expression:

$$Z_D = \langle \{D, C\}, \{Q, C\}, R_D, \wedge(D, C) \rangle,$$

$$R_D = \begin{array}{c|cc} & Q & C \\ \hline D & W_{DQ} & false \\ C & false & true \end{array},$$

where

W_{DQ} = “There is a clock edge”.

The token α from place D enters place Q and does not change its characteristic.

5 GN Model of n-Bit Binary Counter

The GN model of the counter consists (Fig. 5) of one transition Z_C. Initially, there are tokens in places C and CLK, respectively:

- α token with characteristic: $x_{cu}^{\alpha} = 0$;
- β token with characteristic “clock state”.

The transition Z_C is represented by the following expression:

$$Z_C = \langle \{CLK, C\}, \{Q_0, Q_1, \ldots Q_{n-1}, C\}, R_C, \wedge(CLK, C) \rangle$$

$$R_C = \begin{array}{c|ccccc} & Q_0 & Q_1 & \dots & Q_{n-1} & C \\ \hline CLK & false & false & \dots & false & true \\ C & W_{CQ} & W_{CQ} & \dots & W_{CQ} & W_{CQ} \end{array},$$

where

W_{CQ} = “There is a clock edge”.

The token from place C splits in $n + 1$ tokens, where the token that enters place C obtains a new characteristic: $x_t^{\alpha} = x_{t-1}^{\alpha} + 1$

The token that enters place Q_i obtains characteristic: “i-th bit of x_t^{α}”, where $i = 0 \div n - 1$.

The token β enters place C and does not change its characteristic.

6 GN Model of an Example Digital Circuit

The circuit, shown on Fig. 6 has a clock input, and a PWM (Pulse Width Modulated) output, where the duty cycle of the output PWM signal depends on the position of the switches S1–S4. The duty cycle varies between 0 and 94 % (15/16).

There are four logical gates in the circuit, two 4-bit binary counters, a D-type flip-flop, and four switches. The counters have an additional reset (R) input, with an active “HIGH” level. The D-type flip-flop also possesses another output (⌐Q). The four switches define the duty cycle of the PWM output signal.

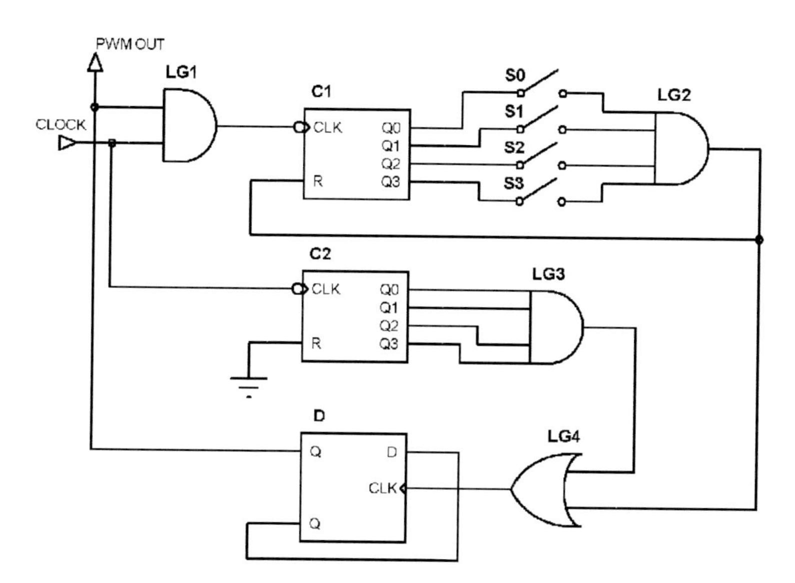

Fig. 6 Example digital circuit

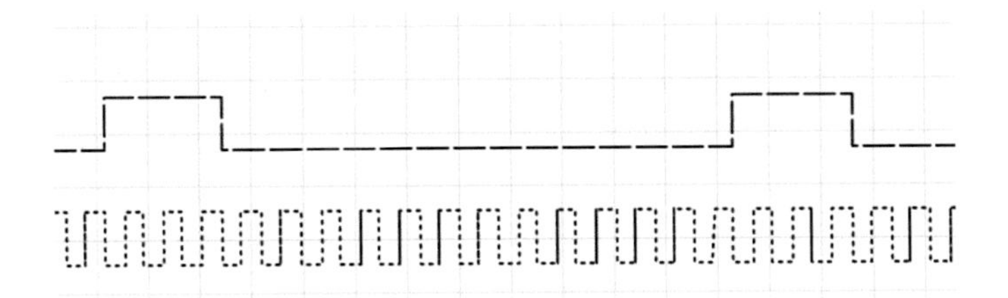

Fig. 7 Clock and output signals, when S0 and S1 are closed

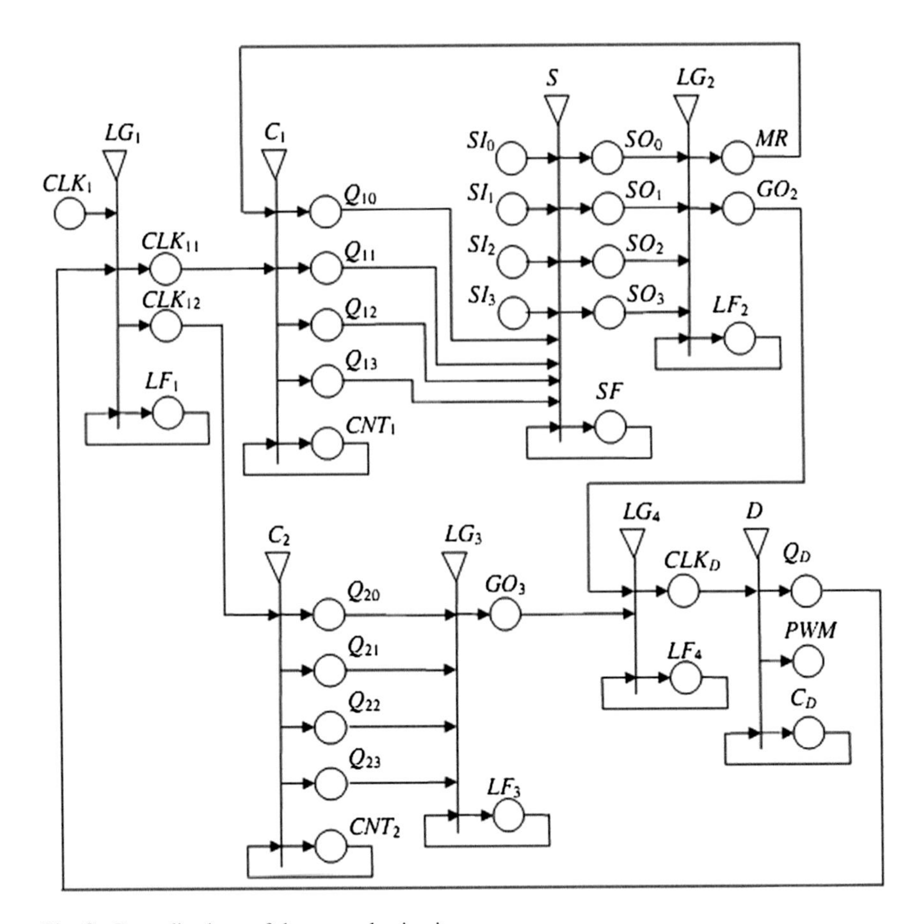

Fig. 8 Generalized net of the example circuit

The time diagram of the digital circuit is shown in Fig. 7.

The generalized net that describe the process of work of the example circuit is shown in Fig. 8. It is presented by the following set of transitions:

$$A = \{LG_1, LG_2, LG_3, LG_4, C_1, C_2, D, S\}$$

The transitions describe the following processes:

- LG_1—performs a logical operation "2-AND";
- LG_2—performs a logical operation "4-AND";
- LG_3—performs a logical operation "4-AND";
- LG_4—performs a logical operation "2-OR";
- C_1—4-bit binary counting;
- C_2—4-bit binary counting;
- D—toggle triggering;
- S—duty cycle configuration.

Initially, the following tokens are defined:

- ξ token in place CLK_1 with characteristic "clock signal";
- ω_1 token in place CNT_1 with characteristic: $x_0^{\omega_1} = 0$;
- ω_2 token in place CNT_2 with characteristic: $x_0^{\omega_2} = 0$;
- δ token in place C_D with characteristic $x_0^{\delta} = 0$;
- μ_0 token in place SI_0 with characteristic $x_0^{\mu_0} = 0$;
- μ_1 token in place SI_1 with characteristic $x_0^{\mu_1} = 0$;
- μ_2 token in place SI_2 with characteristic $x_0^{\mu_2} = 0$;
- μ_3 token in place SI_3 with characteristic $x_0^{\mu_3} = 0$;
- γ token in place SF with characteristic $x_0^{\gamma} = 0$;
- λ_1 token in place LF_1 with characteristic $x_0^{\lambda_1} = 0$;
- λ_2 token in place LF_2 with characteristic $x_0^{\lambda_2} = 0$;
- λ_3 token in place LF_3 with characteristic $x_0^{\lambda_3} = 0$;
- λ_4 token in place LF_4 with characteristic $x_0^{\lambda_4} = 0$.

The transition C_1 is represented by the following expression:

$$C_1 = \langle \{MR, CLK_{11}, CNT_1\}, \{Q_{10}, Q_{11}, Q_{12}, Q_{13}, CNT_1\}, R_{C1}, \wedge(MR, CLK_{11}, CNT_1)\rangle$$

$$R_{C_1} = \begin{array}{c|ccccc} & Q_{10} & Q_{11} & Q_{12} & Q_{13} & CNT_1 \\ \hline MR & false & false & false & false & true \\ CLK_{11} & false & false & false & false & true \\ CNT_1 & W_{C_1} & W_{C_1} & W_{C_1} & W_{C_1} & true \end{array},$$

where

W_{C_1} = "There is a clock edge".

The token ω_1 in place CNT_1 obtains characteristic: $x_t^{\omega_1} = \overline{\rho}(x_{t-1}^{\omega_1} + 1)$, where t is the current moment, $t - 1$ is the previous moment, and ρ is the token in place MR. Then the token ω_1 splits in four tokens ω_{10}, ω_{11}, ω_{12}, ω_{13}, which enter places Q_{10}, Q_{11}, Q_{12}, Q_{13} respectively.

The tokens ω_{1k} in places Q_{1k} obtain characteristics: "k-th bit of the ω_1", where $k = 0 \div 3$.

The transition C_2 is represented by the following expression:

$$C_2 = \langle \{CLK_{21}, CNT_2\}, \{Q_{20}, Q_{21}, Q_{22}, Q_{23}, CNT_2\}, R_{C2}, \wedge(CLK_{21}, CNT_2)\rangle,$$

$$R_{C_2} = \begin{array}{c|ccccc} & Q_{20} & Q_{21} & Q_{22} & Q_{23} & CNT_2 \\ \hline CLK_{21} & false & false & false & false & true \\ CNT_2 & W_{C_2} & W_{C_2} & W_{C_2} & W_{C_2} & true \end{array},$$

where

W_{C_2} = "There is a clock edge".

The token ω_2 in place CNT_2 obtains characteristic: $x_t^{\omega_2} = x_{t-1}^{\omega_2} + 1$, where t is the current time instance and $t - 1$ is the previous one. Then the token ω_2 splits in four tokens ω_{20}, ω_{21}, ω_{22}, ω_{23}, which enter places Q_{20}, Q_{21}, Q_{22}, Q_{23} respectively.

The tokens ω_{2k} in places Q_{2k} obtain the characteristic: "k-th bit of the ω_2", where $k = 0 \div 3$.

The transition S is represented by the following expression:

$$S = \langle \{SI_0, SI_1, SI_2, SI_3, Q_{10}, Q_{11}, Q_{12}, Q_{13}, SF\}, \{SO_0, SO_1, SO_2, SO_3, SF\}, R_S, \wedge(SI_0, SI_1, SI_2, SI_3, Q_{10}, Q_{11}, Q_{12}, Q_{13}, SF)\rangle,$$

$$R_S = \begin{array}{c|ccccc} & SO_0 & SO_1 & SO_2 & SO_3 & SF \\ \hline SI_0 & false & false & false & false & true \\ SI_1 & false & false & false & false & true \\ SI_2 & false & false & false & false & true \\ SI_3 & false & false & false & false & true \\ Q_{10} & false & false & false & false & true \\ Q_{11} & false & false & false & false & true \\ Q_{12} & false & false & false & false & true \\ Q_{13} & false & false & false & false & true \\ SF & W_S & W_S & W_S & W_S & true \end{array},$$

where

W_S = "There is a new input configuration".

The tokens from places SI_0, SI_1, SI_2, SI_3, Q_{10}, Q_{11}, Q_{12} and Q_{13} are merged in place SF. Then the token γ from place SF splits in four tokens γ_0, γ_1, γ_2, γ_3, located in places SO_0, SO_1, SO_2 and SO_3 respectively, where they obtain characteristics:

$$x_{cu}^{\gamma_k} = (SI_k \wedge Q_{1k}) \vee \overline{SI_k} \quad \text{where } k = 0 \div 3.$$

The transition LG_2 is represented by the following expression:

$$LG_2 = \langle \{SO_0, SO_1, SO_2, SO_3, LF_2\}, \{MR, GO_2, LF_2\}, R_{LG2}, \wedge(SO_0, SO_1, SO_2, SO_3, LF_2)\rangle,$$

$R_{LG_2} =$

	MR	GO_2	LF_2
SO_0	*false*	*false*	*true*
SO_1	*false*	*false*	*true*
SO_2	*false*	*false*	*true*
SO_3	*false*	*false*	*true*
LF_2	W_{LF_2}	W_{LF_2}	*true*

,

where

W_{LF_2} = "There is a new input configuration".

The tokens γ_0, γ_1, γ_2, γ_3 from places SO_0, SO_1, SO_2 and SO_3 are merged in place LF_2, where the token λ_2 obtains characteristic: $x_{cu}^{\lambda_2} = x_{cu}^{\gamma_0} \wedge x_{cu}^{\gamma_1} \wedge x_{cu}^{\gamma_2} \wedge x_{cu}^{\gamma_3}$.

The token λ_2 splits in two tokens ρ and λ_{20}, in places MR and GO_2 respectively, where it does not change its characteristic: $x_{cu}^{\rho} = x_{cu}^{\lambda_{20}} = x_{cu}^{\lambda_2}$.

The transition LG_3 is represented by the following expression:

$$LG_3 = \langle \{Q_{20}, Q_{21}, Q_{22}, Q_{23}, LF_3\}, \{GO_3, LF_3\}, R_{LG3}, \wedge(Q_{20}, Q_{21}, Q_{22}, Q_{23}, LF_3)\rangle,$$

$R_{LG_3} =$

	GO_3	LF_3
Q_{20}	*false*	*true*
Q_{21}	*false*	*true*
Q_{22}	*false*	*true*
Q_{23}	*false*	*true*
LF_3	W_{LF_3}	*true*

,

where

W_{LF_3} = "There is a new input configuration".

The tokens from places Q_{20}, Q_{21}, Q_{22} and Q_{23} are merged in place LF_3, where the token λ_3 obtains characteristic: $x_{cu}^{\lambda_3} = x_{cu}^{\omega_{20}} \wedge x_{cu}^{\omega_{21}} \wedge x_{cu}^{\omega_{22}} \wedge x_{cu}^{\omega_{23}}$.

The token λ_3 enters place GO_3, and does not change its characteristic.

The transition LG_4 is represented by the following expression:

$$LG_4 = \langle \{GO_2, GO_3, LF_4\},\ \{CLK_D, LF_4\}, R_{LG4}, \wedge(GO_2, GO_3, LF_4)\rangle,$$

$$R_{LG_4} = \begin{array}{c|cc} & CLK_D & LF_4 \\ \hline GO_2 & false & true \\ GO_3 & false & true \\ LF_4 & W_{LF_4} & true \end{array},$$

where

W_{LF_4} = "There is a new input configuration".

The tokens λ_{20} and λ_3 from places GO_2 and GO_3 are merged in place LF_4, where the token λ_4 obtains characteristic: $x_{cu}^{\lambda_4} = x_{cu}^{\lambda_{20}} \vee x_{cu}^{\lambda_3}$.

The token λ_4 enters place CLK_D and does not change its characteristic.

The transition D is represented by the following expression:

$$D = \langle \{CLK_D, C_D\}, \{Q_D, PWM, C_D\}, R_D, \wedge(CLK_D, C_D)\rangle,$$

$$R_D = \begin{array}{c|ccc} & Q_D & PWM & C_D \\ \hline CLK_D & false & false & true \\ C_D & W_{QD} & W_{QD} & true \end{array},$$

where

W_{QD} = "if $x_t^{\lambda_4} = 1$ and $x_{t-1}^{\lambda_4} = 0$", and the $x_t^{\lambda_4}$ and $x_{t-1}^{\lambda_4}$ are the new and the old characteristic of the token λ_4 respectively.

The tokens δ in place C_D obtains characteristic: $x_t^{\delta} = \overline{x_{t-1}^{\delta} x_{t-1}^{\lambda_4}} = 0$", where the x_t^{δ} and x_{t-1}^{δ} are the new and the old characteristic of the token δ respectively.

The token δ splits in two tokens, in places Q_D and PWM, without changing the characteristic.

The transition LG_1 is represented by the following expression:

$$LG_1 = \langle \{CLK_1, Q_D, LF_1\}, \{CLK_{11}, CLK_{12}, LF_1\}, R_{LG1}, \wedge(CLK_1, Q_D, LF_1)\rangle$$

$$R_{LG_1} = \begin{array}{c|ccc} & CLK_{11} & CLK_{12} & LF_1 \\ \hline CLK_1 & false & true & true \\ Q_D & false & false & true \\ LF_1 & W_{LF_1} & false & true \end{array},$$

where

W_{LF_1} = "There is a new input configuration".

The token ξ from place CLK_1 splits in two tokens. First one enters place CLK_{12} always and has the same characteristic as ξ. The second one merges with the token δ from place Q_D in place LF_1, where the token λ_1 obtains characteristic: $x_{cu}^{\lambda_1} = x_{cu}^{\xi} \wedge x_{cu}^{\delta}$.

The token λ_1 from place LF_1 enters place CLK_{11} always and does not change its characteristic.

7 InterCriteria Analysis Method Application

The ICRA-method [5–7] is based on two concepts: intuitionistic fuzzy sets and index matrices.

The Intuitionistic Fuzzy Sets (IFSs, see [8–11]) represent an extension of the concept of fuzzy sets, as defined by Zadeh [12], exhibiting function $\mu_A(x)$ defining the membership of an element x to the set A, evaluated in the [0; 1]-interval. The difference between fuzzy sets and intuitionistic fuzzy sets (IFSs) is in the presence of a second function $\nu_A(x)$ defining the non-membership of the element x to the set A, where $\mu_A(x) \in [0; 1]$, $\nu_A(x) \in [0; 1]$, under the condition

$$\mu_A(x) + \nu_A(x) \in [0; 1].$$

The final step of the algorithm is to determine the degrees of correlation between the criteria, depending on the user's choice of μ and ν. We call these correlations between the criteria: 'positive consonance', 'negative consonance' or 'dissonance'. For practical purposes, it carries the most information when either the positive or the negative consonance is as large as possible, while the cases of dissonance are less informative and are skipped.

If it is possible to propose different schematic solutions that satisfy the required conditions (or the requested result) by the analysis of logical circuits, certain parameters can be used as criteria. By reaching a certain number of schematics criteria—ICRA [13], a recommendation for the implementation of circuits of a certain type can be applied.

8 Conclusions

A novel approach for medeling of logical circuits is presented in the present article. Four Generalized Nets models of the logical circuits of an AND gate, binary to decimal decoder, a D-type flip flop and a n-bit binary counter have been presented. We proposed the application of the recently discovered InterCriteria Analysis approach, based on index matrices and intuitionistic fuzzy sets. This algorithm aims to discover possible correlations between the criteria pairs. As a result, if there are more than one circuits composed of different set of logical gates available that can obtain identical results, the measurements can be taken.

In the presence of several measurement points and different set of circuits, the best solution of the considered task can be suggested by the use of ICRA method.

Acknowledgements The authors are grateful for the support provided by the National Science Fund of Bulgaria under grant DFNI-I-02-5/2014 and the Project NIH-355, 2015 of University "Prof. Asen Zlatarov".

References

1. Atanassov, K.: Generalized Nets. World Scientific, Singapore (1991)
2. Atanassov, K.: On Generalized Nets Theory, Sofia, "Prof. M. Drinov", Academic Publishing House, (2007)
3. Alexieva, J., Choy, E., Koycheva, E: Review and bibliography on generalized nets theory and applications. In: Choy, E., Krawczak, M., Shannon, A., Szmidt, E. (eds.) A Survey of Generalized Nets Raffles KvB Monograph, No. 10, pp. 207–301, (2007)
4. Atanassov, K.: Index Matrices: Towards an Augmented Matrix Calculus. Springer, Cham (2014)
5. Atanassov, K., Mavrov, D., Atanassova, V.: InterCriteria decision making. A new approach for multicriteria decision making, based on index matrices and intuitionistic fuzzy sets. Issues in Intuitionistic Fuzzy Sets and Generalized Nets, Vol. 11, 2014, pp. 1–8
6. Atanassova, V., Mavrov, D., Doukovska, L., Atanassov, K.: Discussion on the threshold values in the InterCriteria decision making approach. Notes on Intuitionistic Fuzzy Sets **20**(2), 94–99 (2014)
7. Atanassova, V., Doukovska, L., Atanassov, K., Mavrov, D.: InterCriteria decision making approach to EU member states competitiveness analysis. In: Proceedings of the International Symposium on Business Modeling and Software Design—BMSD'14, pp. 289–294. Luxembourg, Grand Duchy of Luxembourg, 24–26 June 2014
8. Atanassov, K.: Intuitionistic fuzzy sets. In: Proceedings of VII ITKR's Session, Sofia, June 1983 (in Bulgarian)
9. Atanassov, K.: Intuitionistic fuzzy sets. Fuzzy Sets Syst Elsevier **20**(1), 87–96 (1986)
10. Atanassov, K.: Intuitionistic Fuzzy Sets: Theory and Applications. Physica-Verlag, Heidelberg (1999)
11. Atanassov, K.: On Intuitionistic Fuzzy Sets Theory. Springer, Berlin (2012)
12. Zadeh, L.A.: Fuzzy Sets. Inf. Control **8**, 333–353 (1965)
13. InterCriteria Research Portal, http://intercriteria.net/

Generalized Net Model of Person Recognition Using ART2 Neural Network and Viola-Jones Algorithm

Todor Petkov, Sotir Sotirov and Stanimir Surchev

Abstract In this paper we present a method for the purpose to detect a certain person in an image. We use the tools of neural networks and face recognition algorithm to achieve our goal. The type of neural network is unsupervised adaptive resonance theory 2 (ART2). It is trained by the set of person images and divided into two clusters—the first cluster represents the human who has to be found and the second one represents the other people. The algorithm which is used for face detection is Viola-Jones and the combination with neural networks helps to identify the person. The generalized net model is used to describe the recognition process.

Keywords ART2 neural network · Face recognition · Generalized nets

1 Introduction

In present years a lot of papers were devoted to the field of neural networks [2, 6] and face recognition [5, 12], here we combine them to achieve the results that could be significant in many areas. In order to achieve it, first the neural network has to be trained with a set of images of the person that we want to detect and another set with faces of other persons. The algorithm that is used for face detection is viola-jones [8, 9, 11], thus when a face is detected it has to be tested and the result of the procedure is whether the face belongs to the person whom we search for or it belongs to the other group of persons.

T. Petkov (✉) · S. Sotirov · S. Surchev
Intelligent Systems Laboratory, University Professor Dr. Assen Zlatarov Burgas, Burgas, Bulgaria
e-mail: todor_petkov@btu.bg

S. Sotirov
e-mail: ssotirov@btu.bg

S. Surchev
e-mail: stanimir_surchev@btu.bg

K.T. Atanassov et al. (eds.), *Novel Developments in Uncertainty Representation and Processing*, Advances in Intelligent Systems and Computing 401,
DOI 10.1007/978-3-319-26211-6_22

ART2 [3, 4, 6] (Fig. 1) is an unsupervised [7, 10] neural network which is designed to preform operations over continuous valued input vectors or binary input vectors that have noise. Basically the network consists of two layers composed of neurons that are fully connected with a set of weights also known as bottom-up and top-down and an orienting sub system. The first layer ($F1$) consists of three sub layers of neurons and each one of them supports a combination of normalization of the vector and suppression of noise. The second layer ($F2$) is a competitive one which means that the neuron with maximum value is going to learn its weights according to the vector. The orienting sub—system is responsible for taking decision about the current neuron with maximum value and if it responds to the criteria it is going to learn its weights, if not—the winner is going to be rejected. The learning algorithm is expressed in [6].

The viola jones is the most popular algorithm [8, 9] that is used for face recognition in many human applications. The algorithm consists basically of four procedures that are applied to the image:

1. Image features also known as Haar-like features (Fig. 2) represent the human characteristics and the pixel values in white region are subtracted from the pixel values of the black region.
2. The integral image makes feature extraction easier and the value at pixel (x, y) is the sum of the pixel above and to the left, thus instead to compute all pixels we have one value for each rectangle and the computation process speeds up.
3. All possible haar-like features are 160,000 but not all of them are relevant (Fig. 3) so in order to decrease them the adaboost algorithm is applied.
4. The final step is to apply Haar-like features in cascade way.

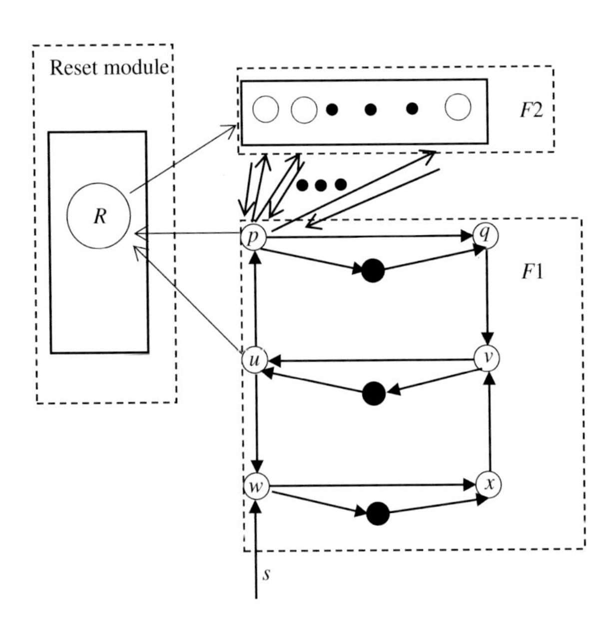

Fig. 1 The structure of ART2 neural network

Fig. 2 Haar-like features

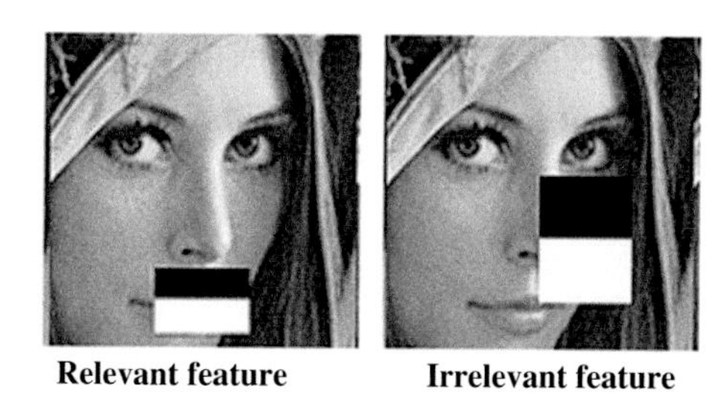

Fig. 3 An example of relevant and irrelevant features

2 Results and Discussion

In order to train the network a set of images with size of 35×40 pixels for people is used, in Table 1 are shown some of them from both sets. The first set contains images of the person who has to be found and the second set contains the images of other people. The set of images represents matrices with three pixel values of R, G, B (red, green, blue) colors. At first the red layer from the images are extracted and the matrix with pixel values is converted into a vector. When the procedure is applied over images, all vectors for training enter into the ART2 neural network which is divided into two clusters. When the network is trained the clusters have to be verified to see which cluster responds to the desired person. The results show that first cluster is for the desired person and second one is for the other people. The

Table 1 Images for training

First set			
Second set			

Table 2 Parameters for ART2 NN

Parameters	θ	ρ	α	Units
Value	0.0295	0.8157	0.6	2

The origin image

Resulted image

Fig. 4 The origin and resulted images

significant parameters of the ART2 neural network are shown in Table 2.
where

Θ—noise suppression parameter;
P—vigilance threshold;
A—learning rate;
Units—number of clusters.

When the network is verified a random image is taken then the viola-jones algorithm is applied on it in order to find faces. The red layer from each face is extracted then the matrix of pixels is normalized with size of 35 × 40 pixels. The normalized images at the final step before entering into the network have to be converted into vectors. When the identified faces are tested the ART2 neural network returns three answers—the face belongs to first cluster, the face belongs to the second one or neither of the clusters responds to the image criteria. In Fig. 4 is shown the test image—on the left side is the origin image and on the right side is the image after its proceeding through the process of person detection. It can be seen that the person is recognized and the result is that its face is depicted in black rectangle.

3 GN-Model

Initially the following tokens enter the Generalized Net (GN) [1].
In place L_1 there is one α token with characteristic "input image for testing";
In place L_5 there is one β token with characteristic "input training set";

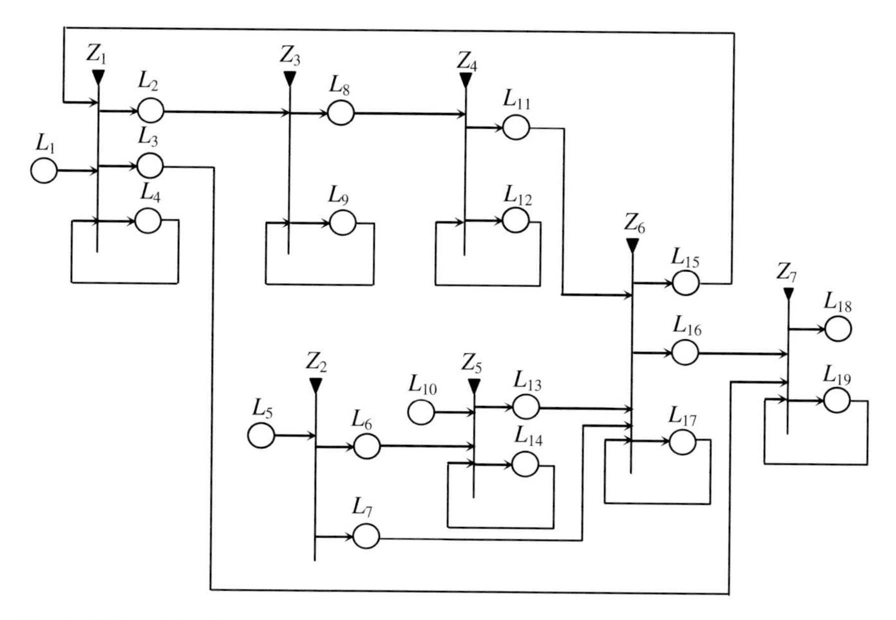

Fig. 5 GN model of the process of human recognition using ART2 neural network

In place L_{10} there is one γ token with characteristic "values for ART2 neural network".

The GN shown in Fig. 5 is introduced by the following set of transitions:

Z_1 = "Applying viola-jones algorithm and extraction of the face from the image";
Z_2 = "Division the training set";
Z_3 = "Normalization of the face and extraction of the red layer";
Z_4 = "Converting the matrix into vector";
Z_5 = "Training the neural network";
Z_6 = "Testing the neural network";
Z_7 = "Visualization of the results in the image".

GN–model consists of seven transitions with the following description:

$$Z_1 = \langle \{L_1, L_4, L_{15}\}, \{L_2, L_3, L_4\}, R_1, \vee(L_1, \wedge(L_4, L_{15}))\rangle,$$

where

$$R_1 = \begin{array}{c|ccc} & L_2 & L_3 & L_4 \\ \hline L_1 & false & false & true \\ L_4 & W_{4,2} & W_{4,3} & true \\ L_{15} & false & false & true \end{array},$$

$W_{4,2}$ = "There is activation token from place L_{15}";
$W_{4,3}$ = "The components are tested"

Token α from place L_1 splits in two tokens and enters places L_3 and L_4. The tokens do not obtain any new characteristic.

Token ζ from place L_{15} unites with token α in place L_4 and obtain characteristic

$$\delta = \text{“Extracted new face component”}$$

Token δ from place L_4 that enters place L_2 does not obtain new characteristic.

$$Z_2 = \langle \{L_5\}, \{L_6, L_7\}, R_2, \vee (L_5) \rangle,$$

where

$$R_2 = \begin{array}{c|cc} & L_6 & L_7 \\ \hline L_5 & true & true \end{array},$$

Token β from place L_5 splits in two tokens (β', β'') and enters places L_6 and L_7 accordingly.

$$Z_3 = \langle \{L_2, L_9\}, \{L_8, L_9\}, R_3, \vee (L_2, L_9)) \rangle,$$

where

$$R_3 = \begin{array}{c|cc} & L_8 & L_9 \\ \hline L_2 & false & true \\ L_9 & W_{9,8} & true \end{array},$$

$W_{9,8}$ = “The face is processed”

Token δ from place L_2 that enters place L_9 does not obtain new characteristic.
Token δ from place L_9 that enters place L_8 obtains characteristic

$$\delta' = \text{“Normalized image component”}.$$

$$Z_4 = \langle \{L_8, L_{12}\}, \{L_{11}, L_{12}\}, R_4, \vee (L_8, L_{12}) \rangle,$$

where

$$R_4 = \begin{array}{c|cc} & L_{11} & L_{12} \\ \hline L_8 & false & true \\ L_{12} & W_{12,11} & true \end{array},$$

$W_{12,11}$ = “The matrix is converted”.

Token δ' from place L_8 that enters place L_{12} does not obtain new characteristic.

Token δ' from place L_{12} that enters place L_{11} obtains characteristic

$$\delta'' = \text{"Converted vector for testing".}$$

$$Z_5 = \langle \{L_6, L_{10}, L_{14}\}, \{L_{13}, L_{14}\}, R_5, \vee(L_{10}, \wedge(L_6, L_{14}))\rangle,$$

where

$$R_5 = \begin{array}{c|cc} & L_{13} & L_{14} \\ \hline L_6 & false & true \\ L_{10} & false & true \\ L_{14} & W_{14,13} & true \end{array},$$

$W_{14,13}$ = "The network is learned".

Token γ from place L_{10} that enters place L_{14} does not obtain new characteristic.

Token β' from place L_6 unites with token γ in place L_{14} and enters place L_{13} with characteristic

$$\varepsilon = \text{"Learned ART2 neural network".}$$

$$Z_6 = \langle \{L_7, L_{11}, L_{13}, L_{17}\}, \{L_{15}, L_{16}, L_{17}\}, R_6, \vee(\wedge(L_{11}, L_{17}) \wedge (L_7, L_{13}))\rangle,$$

where

$$R_6 = \begin{array}{c|ccc} & L_{15} & L_{16} & L_{17} \\ \hline L_7 & false & false & true \\ L_{11} & false & false & true \\ L_{13} & false & false & true \\ L_{17} & W_{17,15} & W_{17,16} & true \end{array},$$

$W_{17,15}$ = "The current component is tested";

$W_{17,16}$ = "The all components are tested".

Tokens β'' and ε from places L_7 and L_{13} unites in place L_{17} with characteristic

$$\varepsilon' = \text{"Verified ART2 neural network".}$$

Tokens δ'' and ε' from places L_{11} and L_{17} unites and enters place L_{15} with characteristic

$$\zeta = \text{“Request for new face component”}.$$

Token from place L_{15} that enters place L_{16} obtains characteristic

$$\zeta' = \text{“Human information”}.$$

$$Z_7 = \langle \{L_3, L_{16}, L_{19}\}, \{L_{18}, L_{19}\}, R_7, \vee (L_{19} \wedge (L_3, L_{16})) \rangle,$$

where

$$R_7 = \begin{array}{c|cc} & L_{18} & L_{19} \\ \hline L_3 & false & true \\ L_{16} & false & true \\ L_{19} & W_{19,18} & true \end{array},$$

$W_{19,18}$ = “The coordinates are applied”.

Tokens α and ζ' from places L_3 and L_{16} unites and enters place L_{19} with characteristic

$$\eta = \text{“Applied coordinates of human information”}.$$

Token η from place L_{19} enters place L_{18} with characteristic

$$\eta' = \text{“Visualization of the results”}.$$

4 Conclusion

In this paper was described a process that combines the use of ART2 neural network and Viola-Jones algorithm. Viola-jones algorithm is useful for the purpose of finding faces in the image, the ART2 neural network was used to identify a certain person among others. It was seen that both algorithms successfully achieved their goals thus with their combination there can be resolved different problems in many areas. In this paper the ART2 neural network was trained with a set of images of a certain person and a set of images of other persons. The Viola-Jones algorithm was used in order to find a face in an image. When the face is found, several procedures were applied and then it enters into the network. It was seen that the network recognized the person successfully and the goal was achieved. The generalized net model is used in order to describe the process of recognition, which can be analyzed easily in more details.

References

1. Atanassov, K.: Generalized Nets. World Scientific, Singapore, New Jersey, London (1991)
2. Beale M., Demuth H., Hagan M.: Neural Network Design, PWS Publishing Company (1996)
3. Ben, K., van der Smagt, P.: An Introduction to Neural Networks, Chapter 6 8th edn. University of Amsterdam (1996)
4. Carpenter, G.A., Grossberg, S.: The ART of adaptive pattern recognition by a self-organizing neural network. Computer **21**(3), 77–88 (1988)
5. Chelali, F., Djeradi, A.: Face recognition system using neural network with Gabor and discrete wavelet transform parameterization. In: 6th International Conference on Soft Computing and Pattern Recognition, SoCPaR (2014)
6. Fausett, L.: Fundamentals of Neural Networks; Architecture, Algorithms and Applications (1993)
7. Kramer, O.: Dimensionality Reduction with Unsupervised Nearest Neighbors, Springer, Intelligent systems reference library, vol. 51, (2013)
8. Murphy, T., Schultz, R.: A Viola-Jones based hybrid face detection framework. In: Proceedings of SPIE—The International Society for Optical Engineering, vol. 9025, (2014)
9. Rashidan, M., Mustafah, Y.M.: Analysis of artificial neural network and Viola-Jones algorithm based moving object detection. In: Proceedings—5th International Conference on Computer and Communication Engineering: Emerging Technologies via Comp-Unication Convergence, ICCCE (2014)
10. Tian, D., Liu, Y.H., Shi, J.R.: Dynamic Clustering Algorithm Based on Adaptive Resonance Theory, Springer, 2006
11. Viola, P., Jones, J.: Robust real-time face detection. Int. J. Comput. Vision **57**(2), 137–154 (2004)
12. Wójtowicz, W., Ogiela, M.R.: Biometric watermarks based on face recognition methods for authentication of digital images. Secur. Commun. Netw. **8**(9), 1672–1687 (2015)

Optimization of the LVQ Network Architectures with a Modular Approach for Arrhythmia Classification

Jonathan Amezcua and Patricia Melin

Abstract In this paper, the optimization of LVQ neural networks with modular approach is presented for classification of arrhythmias, using particle swarm optimization. This work focuses only in the optimization of the number of modules and the number of cluster centers. Other parameters, such as the learning rate or number of epochs are static values and are not optimized. Here, the MIT-BIH arrhythmia database with 15 classes was used. Results show that using 5 modules architecture could be a good approach for classification of arrhythmias.

Keywords Classification · PSO · LVQ · Neural networks · Arrhythmias

1 Introduction

In this paper the optimization of a modular LVQ neural network architecture [1, 4] with Particle Swarm Optimization (PSO) for arrhythmia classification [22, 24] is presented. The optimization focuses on the number of modules and the number of cluster centers. Other parameters such as the epochs or learning rate are static values, obtained from a previous research on optimization of parameters [2].

LVQ is an adaptive learning method used to solve classification problems; although it uses supervised training, LVQ applies unsupervised data clustering techniques, to pre-process the dataset and obtain the centers of the clusters [13].

Particle Swarm Optimization (PSO) is used for the architecture optimization, which is a stochastic optimization technique based on the social behavior of animals. In this work PSO is applied to optimize the number of modules in the

J. Amezcua (✉) · P. Melin
Tijuana Institute of Technology, Tijuana, Mexico
e-mail: jonathan.aguiluz@yahoo.com

P. Melin
e-mail: pmelin@tectijuana.mx

K.T. Atanassov et al. (eds.), *Novel Developments in Uncertainty Representation and Processing*, Advances in Intelligent Systems and Computing 401,
DOI 10.1007/978-3-319-26211-6_23

architecture of the LVQ network [7, 8], which is variable in a range of [2, 5]. The distribution of classes in each module depends of the number of modules in the architecture.

The rest of this paper is organized as follows, in Sect. 2 some basic concepts of PSO are described, Sect. 3 presents the problem statement of this work, Sect. 4 shows the proposed model, in Sect. 5 presents the simulation results and finally in Sect. 6 the conclusions.

2 PSO Method

Particle Swarm Optimization (PSO) [2, 6, 9, 14] is a technique based on social behaviors observed in animals. This method has gained popularity as a robust technique for solving optimization problems. In PSO, individual particles of a swarm represent possible solutions to the problem.

The position of each particle is adjusted according to its velocity and the difference between its current position and the best position found by its neighbors, and the best position found so far. The position of a particle $\boldsymbol{i}$ is updated as follows:

$$v_{ij}(t+1) = v_{ij}(t) + c_1 r_{1j}(t)\left[y_{ij}(t) - x_{ij}(t)\right] + c_2 r_{2j}(t)\left[\hat{y}_j(t) - x_{ij}(t)\right] \tag{1}$$

where $\boldsymbol{v_{ij}(t)}$ is the velocity of a particle $\boldsymbol{i}$ in dimension $\boldsymbol{j}$ at time step $\boldsymbol{t}$, $\boldsymbol{x_{ij}(t)}$ is the position of a particle $\boldsymbol{i}$ in dimension $\boldsymbol{j}$ at time step $\boldsymbol{t}$; $\boldsymbol{c_1}$ and $\boldsymbol{c_2}$ are constants used to scale the contribution of the cognitive and social components and $\boldsymbol{r_{1j}(t)}$ and $\boldsymbol{r_{2j}(t)}$ are random values in the range of [0, 1] that introduce stochastic element to the algorithm.

The personal best position $\boldsymbol{y_i}$ is the best position the particle has visited, the personal best position at a time step $\boldsymbol{t + 1}$ is calculated as follows:

$$y_i(t+1) = \begin{cases} y_i(t), & f(x_i(t+1) \geq y_i(t)) \\ x_i(t+1), & f(x_i(t+1) < y_i(t)) \end{cases} \tag{2}$$

The global best position $\boldsymbol{\hat{y}(t)}$ at a time step $\boldsymbol{t}$ is defined as:

$$\hat{y}(\mathrm{t}) \epsilon \{\mathrm{y}_0(\mathrm{t}), \ldots, y_{n_s}(\mathrm{t})\} | f(\hat{y}(\mathrm{t})) = \min(\{\mathrm{y}_0(\mathrm{t}), \ldots, y_{n_s}(\mathrm{t})\}) \tag{3}$$

where $\boldsymbol{n_s}$ is the total number of particles in the swarm. The definition in Eq. (3) states that $\hat{y}$ is the best position discovered by any of the particles so far.

The algorithm for the PSO [28, 29] model is as follows, where $S.x_i$ is used to denote the position of particle i in the swarm S.

```
repeat
        for each particle i = 1,...,S.n_s do
                //Set the personal best position
                if f(S.x_i) < f(S.y_i) then
                        S.y_i = S.x_i;
                end
                //Set the global best position
                if S.y_i < S.ŷ_i then
                        S.ŷ_i = S.y_i;
                end
        end
        for each particle i = 1,...,S.n_s do
                update velocity;
                update position;
        end
until stopping condition is true;
```

3 Problem Statement

As mentioned above, PSO is a bio-inspired optimization method that has proved to be successful in the implementation of various problems, such as the optimization of membership functions of fuzzy systems, parameters optimization of neural networks, and fuzzy systems as well, and, in this case, the optimization of a modular neural network architecture for classification. In this work, such optimization consists into find the architecture for the minimum error accuracy in the classification of arrhythmias [3, 5, 21, 23].

On the other hand, classification tasks consists into assigning objects to only one of many predefined categories, a widespread problem that covers many different areas of application, such as spam detection in emails, classification of galaxies, among others [24]. In this research, modular LVQ network [12, 17, 18] was used as classification technique, Fig. 1 shows the architecture of a LVQ network [15].

3.1 Arrhythmia Dataset

Arrhythmias are expressed as changes in the normal sequences of electrical hearth impulses, these impulses may happen too fast, to slowly or erratically, and can be measured with a Holter device in ECG signals [10]. Figure 2 shows an example of one of these ECG signals [11].

The MIT-BIH [19] arrhythmia dataset was used for this research. This database consists of 15 different types of arrhythmias. This database contains 48 half-hour

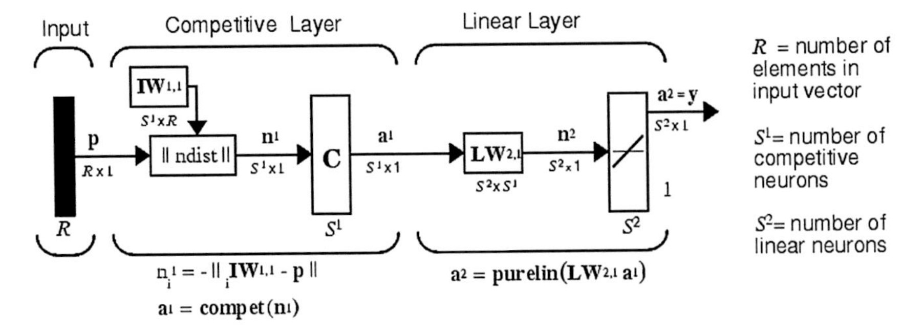

Fig. 1 Architecture of LVQ network

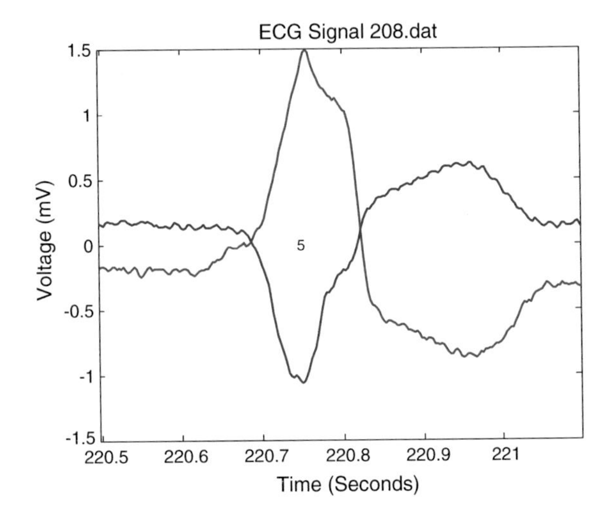

Fig. 2 Example of an ECG signal

excerpts of ECG recordings, obtained from 47 subjects studied by BIH Arrhythmia Laboratory. The recordings were digitized at 360 samples per second per channel with 11-bit resolution over a 10 mV range. The recordings were preprocessed at as follows:

- Analog outputs were filtered using a passband from 0.1 to 100 Hz. The passband-filtered signals were digitized at 360 Hz per signal.
- Taking the R point [27] as referencc, in the obtained signals was located the start and end point for each wave.
- The 38 higher voltages and 38 lower voltages of each vector were taken, resulting in vectors of 76 voltages.

In the next section we discuss the proposed optimization model for modular LVQ network architecture.

4 Proposed Model

In this work, the development of a PSO [16, 20] model for the optimization of a modular LVQ network architecture for classification is presented. In this section, we explain in depth the architecture of the PSO model and its parameters as well, and how are used for the optimization of a modular LVQ network architecture.

Within the architecture of modular LVQ network there is a parameter that defines how many cluster centers the network will be training with. Therefore we have a 2-dimension problem, the optimal number of modules in the modular architecture, and the number of cluster centers for each of the modules.

Table 1 shows the parameters, range, minimal and maximal values that PSO model works with. The values for the number of cluster centers were taken from [2], where the optimization of parameters for LVQ [25, 26] networks is presented, and it was with this range of cluster centers that the best classification accuracy was achieved.

About the number of modules in the architecture, in the work of [18] many architectures were developed to work with this same arrhythmia database, one of the best architectures with high classification accuracy was composed by 5 modules, so for this reason was it decided to develop this research working with a range of [2, 5] for the number of modules. Table 2 shows the rest of parameters for the PSO method.

So far we have that PSO model works with 2-dimension problem, the optimization of the number of modules in a modular architecture, and the number of cluster centers. In Sect. 4.1 we discuss how the database was partitioned, since it was quite difficult because of the dynamical number of modules and the information was to be evenly distributed in each module.

Table 1 Parameters of the architectures included in PSO model

Parameter	Minimal value	Maximal value
Number of modules	2	5
Number of clusters centers	15	30

Table 2 PSO parameters

Parameter	Value
Population size	15
Maximum number of iterations	15
C_1	2
C_2	$4 - C_1$
Inertia weight	Linear decreasing

4.1 Partitioning the Database

As mentioned earlier, MIT-BIH arrhythmia database consists of 15 classes. In this work was considered to have the classes the most evenly distributed in each module. The classes are labeled from 1 to 15, so to solve this particular problem, a method was developed; this method takes the total number of classes and divided it by the number of modules, depending on the architecture to evaluate. For example, if the PSO method is evaluating a five-module architecture, then this method distributes 3 classes in each module.

When PSO method is evaluating a two-module architecture, then classes are distributed 8 classes in the first module, and 7 classes in the second one. The classes are consecutively distributed, which means that in a two-module architecture, first module contains classes 1–8, and the second one, classes 9–15, and so on for each of the architectures to evaluate by the PSO method.

The rest of the LVQ network architecture, such as the learning rate (LR), and the maximum number of epochs, were left static, in *0.0994* and *200* respectively, taking into account that these parameters were already optimized in [2] achieving a good classification percentage.

5 Simulation Results

A set of 15 experiments were conducted using de PSO model described above, Table 3 shows the obtained results, where ***Time*** is expressed in HH:MM format. Experiments were performed in a Windows 7 PC x64, Quad-Core i7 processor and 16 GB of RAM.

Notice that for all experiments, the best LVQ network architecture consisted of 5 modules, which makes sense since the smaller number of data in each module is easier to train the LVQ network. The errors reached in each experiment are very similar, so basing on the obtained error the best experiment was the number 3.

6 Conclusions

In this paper, a PSO model to find an optimal modular LVQ [30] network architecture for classification of arrhythmias [1, 22] was presented. The results show that a five modules LVQ network architecture can be an optimal architecture for the classification of arrhythmias.

The main reason for these results is that, like many other methods, if a LVQ network module receives several records for training, the classification accuracy for that module tends to be lower, in this case this is for the similarity grade between certain records; there are many training records that are very similar each other but

Table 3 Simulation results

	Modules	Clusters	Epochs	Error	Time
1	5	17	97	1.66×10^{-8}	06:31
2	5	25	30	1.33×10^{-7}	07:22
3	**5**	**29**	**76**	$\mathbf{1.25 \times 10^{-8}}$	**06:45**
4	5	30	55	1.66×10^{-8}	06:29
5	5	17	51	1.38×10^{-7}	07:33
6	5	23	28	1.66×10^{-8}	07:18
7	5	17	51	1.34×10^{-7}	06:30
8	5	20	50	1.34×10^{-7}	06:57
9	5	25	61	1.44×10^{-8}	06:47
10	5	30	104	1.22×10^{-7}	07:28
11	5	24	38	1.66×10^{-8}	06:55
12	5	27	94	1.72×10^{-8}	06:32
13	5	19	37	1.33×10^{-7}	06:41
14	5	21	55	1.66×10^{-8}	07:12
15	5	24	62	1.66×10^{-8}	06:45

they belong to different classes in the network module, therefore, with less training records in each module this grade of similarity is low, and the classification percentage tends to be higher.

Regarding with the PSO model, it has proved to be a good approach for the optimization of the LVQ network model presented, for the classification of arrhythmias.

References

1. Amezcua J., Melin P.: A modular LVQ neural network with fuzzy response integration for arrhythmia classification. In: IEEE 2014 Conference on Norbert Wiener in the 21st century, Boston, June 2014
2. Amezcua J., Melin P.: Optimization of Modular Neural Networks with the LVQ Algorithm for Classification of Arrhythmias using Particle Swarm Optimization, Recent Advances on Hybrid Approaches for Designing Intelligent Systems. Studies in Computational Intelligence Book 547, pp. 307–314. Springer, 2014
3. Anuradha B., Veera-Reddy V.C.: Cardiac arrhythmia classification using fuzzy classifiers. J. Theor. Appl. Inform. Technol. 353–359 (2005)
4. Biswal, B., Biswal, M., Hasan, S., Dash, P.K.: Nonstationary power signal time series data classification using LVQ classifier. Appl. Soft Comput. Elsevier **18**, 158–166 (2014)
5. Castillo, O., Melin, P., Ramirez, E., Soria, J.: Hybrid intelligent system for cardiac arrhythmia classification with Fuzzy K-Nearest Neighbors and neural networks combined with a fuzzy system. J. Expert Syst. Appl. **39**(3), 2947–2955 (2012)
6. Cavuslu, M., Karakuzu, C., Karakaya, F.: Nueral Identification of dynamic systems on FPGA with improved PSO learning. Appl. Soft Comput. Elsevier **12**, 2707–2718 (2012)
7. Frasconi, P., Gori, M., Soda, G.: Links between LVQ and backpropagation original research article. Pattern Recogn. Lett. **18**(4), 303–310 (1997)

8. Grbovic M., Vucetic S.: Regression learning vector quantization. In: 2009 Ninth IEEE International Conference on Data Mining, Miami, December 2009
9. Hashemi, A.B., Meybodi, M.R.: A note on the learning automata based algorithms for adaptive parameter selection in PSO. Appl. Soft Comput. Elsevier **11**, 689–705 (2011)
10. Hu Y.H., Palreddy S., Tompkins W.: A patient adaptable ECG beat classifier using a mixture of experts approach. IEEE Trans. Biomed. Eng. 891–900 (1997)
11. Hu, Y.H., Tompkins, W., Urrusti, J.L., Afonso, V.X.: Applications of ANN for ECG signal detection and classification. J. Electrocardiol. **28**, 66–73 (1994)
12. Kim J., Sik-Shin H., Shin K., Lee M.: Robust algorithm for arrhythmia classification in ECG using extreme learning machine. Biomed. Eng. Online (2009)
13. Kohonen T.: Improved versions of learning vector quantization. In: International Joint Conference on Neural Networks, vol. 1, pp. 545–550, San Diego (1990)
14. Krisshna N.L., Kadetotad Deepak V., Manikantan K., Ramachandran S.: Face recognition using transform domain feature extraction and PSO-based feature selection. Appl. Soft Comput. Elsevier. **22**, 141–161 (2014)
15. Learning Vector Quantization Networks. http://www.mathworks.com/help/nnet/ug/bss4b_l-15.html. Last Accessed 24 June 2014
16. Lee, C., Leu, Y., Yang, W.: Constructing gene regulatory networks from microarray data using GA/PSO with DTW. Appl. Soft Comput. Elsevier **12**, 1115–1124 (2012)
17. Martín-Valdivia, M.T., Ureña-López, L.A., García-Vega, M.: The learning vector quantization algorithm applied to automatic text classification tasks. Neural Netw. **20**(6), 748–756 (2007)
18. Melin, P., Amezcua, J., Valdez, F., Castillo, O.: A new neural network model based on the LVQ algorithm for multi-class classification of arrhythmias. Inf. Sci. **279**, 483–497 (2014)
19. MIT-BIH Arrhythmia Database. PhysioBank, Physiologic Signal Archives for Biomedical Research. http://www.physionet.org/physiobank/database/mitdb/. Last Accessed 24 June 2014
20. Narayan, R., Chatterjee, D., Kumar-Goswami, S.: An application of PSO technique for harmonic elimination in a PWM inverter. Appl. Soft Comput. Elsevier **9**, 1315–1320 (2009)
21. Nasiri J.A., Naghibzadeh M., Yazdi H.S., Naghibzadeh B.: ECG arrhythmia classification with support vector machines and genetic algorithm. In: Third UKSim European Symposium on Computer Modeling and Simulation, 2009
22. Nouaouria, N., Boucadoum, M.: Improved global-best particle swarm optimization algorithm with mixed-attribute data classification capability. Appl. Soft Comput. Elsevier **21**, 554–567 (2014)
23. Owis M.I., Abou-Zied A.H., Youssef A.M., Kadah Y.M.: Study of features based on non-linear dynamical modeling in ECG arrhythmia detection and classification. IEEE Trans. Biomed. Eng. **49**(7) (2002)
24. Pang-Ning T., Steinbach M., Kumar V.: Introduction to Data Mining, pp. 145–148. Pearson Addison Wesley (2006)
25. Pedreira C.: Learning vector quantization with training data selection. IEEE Trans. Pattern Anal. Mach. Intell. **28**, 157–162 (2006)
26. Torrecilla, J.S., Rojo, E., Oliet, M., Domínguez, J.C., Rodríguez, F.: Self-organizing maps and learning vector quantization networks as tools to identify vegetable oils and detect adulterations of extra virgin olive oil. Comput. Aided Chem. Eng. **28**, 313–318 (2010)
27. Tsipouras M.G., Fotiadis D.I., Sideris D.: An arrhythmia classification system based on the RR-interval signal. Artif. Intell. Med. 237–250 (2005)
28. Vellasques, E., Sabourin, R., Granger, E.: Fast intelligent watermarking of heterogeneous image streams through mixture modeling of PSO populations. Appl. Soft Comput. Elsevier **13**, 3130–3148 (2013)
29. Vieira, S., Mendoca, L., Farinha, G., Souza, J.: Modified binary PSO for feature selection using SVM applied to mortality prediction of septic patients. Appl. Soft Comput. Elsevier **13**, 3494–3504 (2013)
30. Xinye Z., Jubai A., ZhiFeng Y.: A method of LVQ network to detect vehicle based on morphology. In: 2009 WRI Global Congress on Intelligent Systems, Xiamen, China, May 2009

Part V
Issues in Metaheuristic Search Algorithms and Their Applications

Imperialist Competitive Algorithm with Fuzzy Logic for Parameter Adaptation: A Parameter Variation Study

Emer Bernal, Oscar Castillo and José Soria

Abstract This paper applies the imperialist competitive algorithm (ICA) to benchmark mathematical functions with the original method to analyze and perform a study of the variation of the results obtained with the ICA algorithm as we vary the parameters manually for 4 mathematical functions. The results demonstrate the efficiency of the algorithm to optimization problems and give us the pattern for future work in dynamically adapting these parameters.

Keywords Imperialist competitive algorithm · ICA · Mathematical functions

1 Introduction

Swarm Intelligence techniques have become increasingly popular during the last two decades due to their capability to find a relatively optimal solution for complex combinatorial optimization problems. They have been applied in the fields of Engineering, Economy, Management Science, Industry, etc. Problems that benefit from the application of Swarm Intelligence techniques are generally very hard to solve optimally in the sense that there is no such exact algorithm for solving them in polynomial time. These optimization problems are also known as NP-hard problems [2].

An algorithm that is well recognized in the domain of evolutionary computation is the imperialist competitive algorithm (ICA), which was introduced by Atashpaz-Gargari and Lucas in [1]. ICA has been inspired by the concept of imperialism; where in this case powerful countries attempt to make a colony of other countries. These algorithms have recently been used in several engineering applications [4].

E. Bernal · O. Castillo (✉) · J. Soria
Tijuana Institute of Technology, Tijuana, B.C., Mexico
e-mail: ocastillo@tectijuana.mx

K.T. Atanassov et al. (eds.), *Novel Developments in Uncertainty Representation and Processing*, Advances in Intelligent Systems and Computing 401,
DOI 10.1007/978-3-319-26211-6_24

We describe the imperialist competitive algorithm in its original form. The algorithm parameters adjustment is performed manually by varying the parameters to see how the behavior of the algorithm is affected. This algorithm was applied to benchmark mathematical functions, and the results of the ICA algorithm by varying the parameters are presented in Sect. 4.

The study of the algorithm is performed in order to show the effectiveness of the imperialist competitive algorithm (ICA) when applied to optimization problems, and to take this as a basis for future works.

Some of the papers that have applied the imperialist competitive algorithm can be described as follows. In [4] the imperialist competitive algorithm combined with refined high-order weighted fuzzy time series (RHWFTS–ICA) for short term loads forecasting. In this study, a hybrid algorithm based on a refined high-order weighted fuzzy algorithm and an imperialist competitive algorithm (RHWFTS–ICA) is developed. This method is proposed to perform efficiently under short-term load forecasting (STLF) [4]. In another paper [3] an imperialist competitive algorithm with PROCLUS classifier for service time optimization in cloud computing service composition, CSSICA is proposed to make advances toward the lowest possible service time of composite service; in this approach, the PROCLUS classifier is used to divide cloud service providers into three categories based on total service time and assign a probability to each provider. An improved imperialist competitive algorithm is then employed to select more suitable service providers for the required unique services [3].

The paper is organized as follows: in Sect. 2 a description about the imperialist competitive Algorithm ICA is presented, in Sect. 3 a description of the mathematical functions is presented, in Sect. 4 the experiments results are described for we can to appreciate the ICA algorithm behavior by varying the parameters, in Sect. 5 the conclusions obtained after the study of the imperialist competitive algorithm versus mathematical functions are presented.

2 Imperialist Competitive Algorithm

In the field of evolutionary computation, the novel ICA algorithm is based on human social and political advancements [1], unlike other evolutionary algorithms, which are based on the natural behaviors of animals or physical events.

ICA starts with an initial randomly generated population, in which the individuals are known as countries. Some of the best countries are considered imperialists, whereas the other countries represent the imperialist colonies [3].

All the colonies of the initial population are divided among the mentioned imperialists based on their power. The power of an empire which is the counterpart of the fitness value in GA and is inversely proportional to its cost [1].

2.1 Forming Initial Empires (Initialization)

In order to represent an appropriate solution format, a $1 \times N_{var}$ array of variables represents a country, and a country is defined by [2]:

$$Country = [p_1, p_2, \ldots, p_{N\,var}] \tag{1}$$

where N_{var} is the number of variables to be considered of interest about a country and p_i is the value of *i-th* variable.

The variable values in the country are represented as floating point numbers. The cost of a country is found by evaluating the cost function f at the variables $(p_1, p_2, \ldots, p_n)$ then [1].

$$Cost = f(Country) = f(p_1, p_2, \ldots, p_n) \tag{2}$$

In the initialization step, we need to generate an initial population size of N_{pop}. Select N_{imp} of the most powerful countries to form empires. The remaining N_{col} population will be the colonies each of which belongs to an empire [1].

$$N_{col} = N_{pop} - N_{imp} \tag{3}$$

To form empires, the colonies are divided among the imperialist countries according to the power of the imperialists. The normalized cost of each imperialist is determined by [2].

$$C_n = \max_i \{c_i\} - C_n \tag{4}$$

where, c_n is the *n-th* imperialist's cost, and C_n is the normalized cost of *n-th* imperialist.

Therefore, the power of each imperialist is calculated based on the normalized cost [2]:

$$p_n = \frac{C_n}{\sum_{i=1}^{N_{imp}} C_i} \tag{5}$$

where p_n is the power of *n-th* imperialist. The normalized power of *n-th* imperialist is the number of colonies that are possessed by that imperialist, calculated by:

$$NC_n = Round\{p_n N_{col}\} \tag{6}$$

where, NC_n is the number of initial colonies possessed by the *n-th* imperialist; N_{col} is the total number of colonies in the initial population, and round is a function that gives the nearest integer of a fractional number.

2.2 *Moving the Colonies of an Empire Toward the Imperialist (Assimilation)*

As shown in Fig. 1 the colony moves x distance along with the d direction towards its imperialist. The moving at x distance is a random number generated by random distribution within the interval (0, βd) [2].

$$x \sim U(0, \beta d) \tag{7}$$

where β is a number greater than 1 and d is the distance between the colony and the imperialist.

As shown in Fig. 2, to search for different locations around the imperialist we add a random amount of deviation to the direction of motion, which is given by [1]:

$$\theta \sim U(-\gamma, \gamma) \tag{8}$$

Where θ is a random number with uniform distribution and γ is a parameter that adjusts the deviation from the original direction.

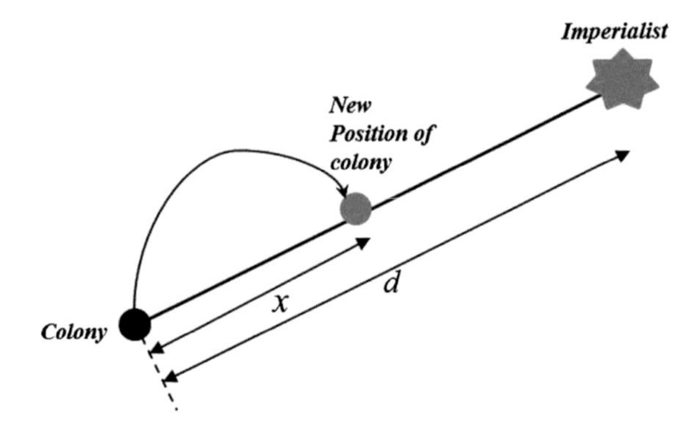

Fig. 1 Movement of the colonies toward the imperialist

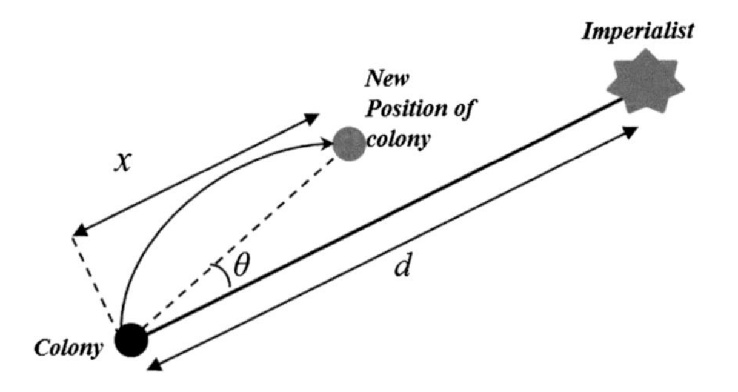

Fig. 2 Movement of the colonies toward their relevant imperialist in a randomly direction deviation

2.3 Exchanging the Positions of a Colony and an Imperialist

While moving towards the imperialist, a colony can reach a position of lower cost than the imperialistic. If so, the imperialist is moved to the position of that colony and vice versa. Then the algorithm will continue with the imperialist in a new position and the colonies begin to move toward this position [1]. In Fig. 3, the best colony of the empire is shown in darker color. This colony has a lower cost than imperialist. Figure 4 shows the whole empire after exchanging the position of the imperialist and the colony.

2.4 Total Power of an Empire

The power of an empire is calculated based on the power of its imperialist and a fraction of the power of its colonies. This fact has been modeled by defining the total Cost given by [2]:

$$TC_n = Cost(Imp) + \xi\, mean\{Cost(Col)\} \tag{9}$$

where TC_n is the total cost of *n-th* Empire and ξ is a positive number between 0 and 1.

2.5 Imperialist Competition

To start the competition, first, we find the probability of possession of each empire based on the total power. The normalized total cost is obtained by [1]:

$$NTC_n = \max_i\{TC_i\} - TC_n \tag{10}$$

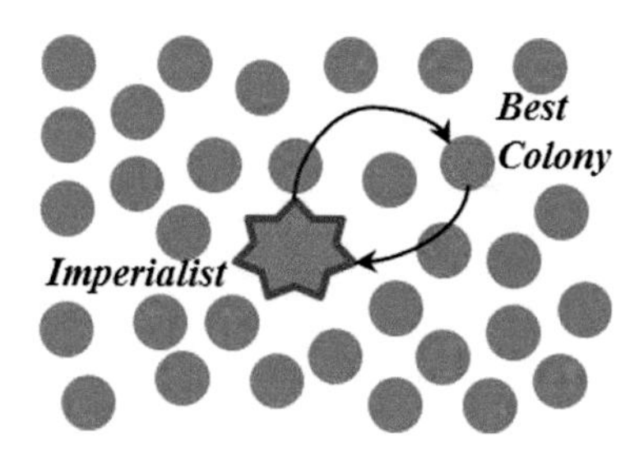

Fig. 3 Position change between the imperialist and a colony

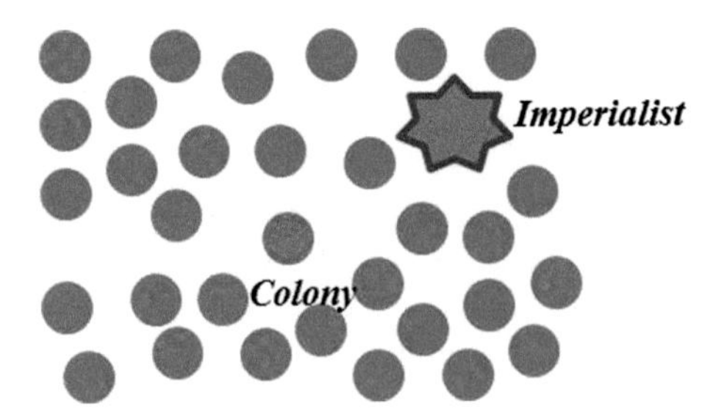

Fig. 4 Position after exchanging Empire imperialist and the colony

where NTC_n is the Normalized total cost and TC_n is the total cost of empire. Then the probability of possessing a colony is computed by:

$$p_{p_n} = \frac{NTC_n}{\sum_{i=1}^{N_{imp}} NTC_i} \tag{11}$$

where $\sum_{i=1}^{N_{imp}} p_{pi} = 1$.

2.6 Elimination of Weaker Empires

Weaker empires lose their colonies gradually to stronger empires, which in turn grow more powerful and cause the weaker empires to collapse over time. In Fig. 5, the weakest empire is eliminated by losing its last colony during the imperialist competition [2].

2.7 Convergence

Similar to other metaheuristic algorithms, ICA continues until a stopping criteria are met, such as predefined running time or a certain number of iterations. The ideal stopping criterion is when all empires have collapsed and only one (grand empire) remains (Fig. 6).

2.8 Pseudocode of ICA

The pseudocode it ICA is defined as follows:

1. Select some random points on the function and initialize the empires.
2. Move the colonies toward their relevant imperialist (assimilation).

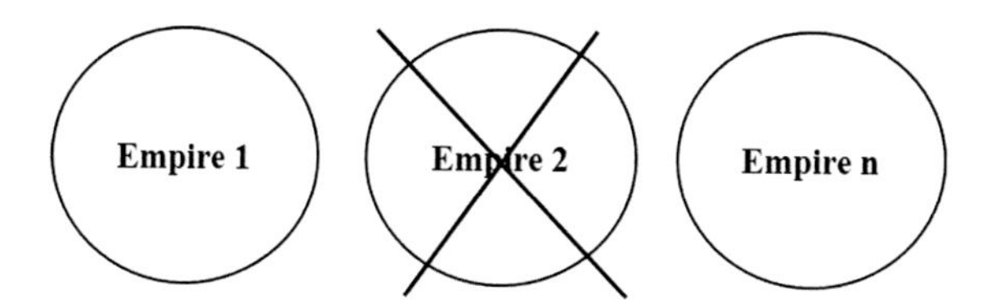

Fig. 5 Elimination of the weakest empire

3. If there is a colony in an empire which has lower cost than that of imperialist, exchange the positions of that colony and the imperialist.
4. Calculate the total cost of all empires (related to the power of both imperialist and its colonies).
5. Pick the weakest colony (colonies) from the weakest empire and give it (them) to the empire that has the most likelihood to possess it (imperialistic competition).
6. Elimínate the powerless empires.
7. If there is just one empire, stop, if not go to step 2.

3 Benchmark Mathematical Functions

In this section, the benchmark functions that are used are listed to evaluate the performance of the ICA algorithm by varying its parameters and to analyze the results obtained.

In the area of the metaheuristics for optimization the use of mathematical functions is common, and they are used in this work: consisting of an optimization algorithm based on imperialism in which the variation of its parameters will be analyzed for obtain its optimum values.

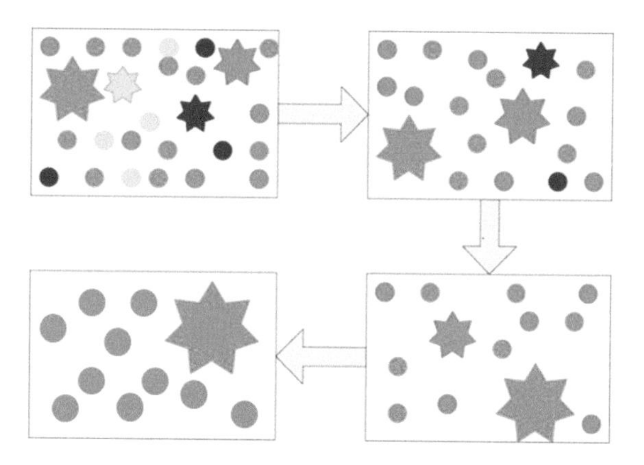

Fig. 6 Representation of convergence in ICA

The mathematical functions are shown below:

- **Sphere**

$$f(x) = \sum_{j=1}^{n_x} x_j^2 \qquad (12)$$

Witch $x_j \in [-100, 100]$ and $f^*(x) = 0.0$

- **Quartic**

$$f(x) = \sum_{j=1}^{n_x} x_j^4 + random(0, 1) \qquad (13)$$

Witch $x_i \in [-1.28, 1.28]$ and $f^*(x) = 0.0$ + noise

- **Rosenbrock**

$$f(x) = \sum_{j=1}^{n_x/2} [100(x_{2j} - x_{2j}^2)^2 + (1 - x_{2j-1})^2] \qquad (14)$$

Witch $x_j \in [-2.048, 2.048]$ and $f^*(x) = 0.0$

- **Rastrigin**

$$f(x) = \sum_{j=1}^{n_x} x_j^2 - 10\cos(2\pi x_j) + 10 \qquad (15)$$

With $x_j \in [-5.12, 5.12]$ and $f^*(x) = 0.0$

4 Simulation Results

In this section the imperialist competitive algorithm (ICA) is implemented with 4 benchmark mathematical functions with 30 variables by varying their parameters and the results obtained by the ICA algorithm are shown in separate tables by parameter.

The parameters used in the imperialist competitive algorithm are:

- Number of variables: 30
- Number of countries: 200
- Number of imperialist: 10
- Number of decades: 3000

Table 1 shows that after executing the ICA Algorithm 10 times, by varying the revolution parameter, we can find the best, average and worst results for the different mathematical functions benchmark.

Table 1 Results by varying the revolution parameter

Function		0.2	0.3	0.4	0.5	0.6	0.7	0.8	0.9
Sphere	Best	7.47E-10	3.24E-09	5.61E-10	2.52E-09	6.23E-09	2.74E-09	5.07E-10	4.34E-09
	Worst	6.69E-08	1.57E-07	1.64E-07	1.51E-07	1.39E-07	7.40E-08	1.11E-07	2.25E-07
	Mean	2.33E-08	2.72E-08	2.54E-08	4.63E-08	3.77E-08	2.65E-08	2.52E-08	3.89E-08
Quartic	Best	2.34E-13	1.91E-13	8.45E-12	9.30E-13	3.58E-14	5.81E-12	3.87E-13	1.87E-13
	Worst	3.22E-09	1.99E-09	2.87E-09	3.10E-08	6.04E-10	3.84E-10	2.88E-09	2.14E-08
	Mean	3.68E-10	4.03E-10	9.12E-10	3.26E-09	1.42E-10	9.76E-11	5.83E-10	2.79E-09
Rosenbrock	Best	2.51	11.74	19.10	1.39E-01	2.51	15.05	6.25	25.54
	Worst	81.26	83.71	80.05	84.89	88.59	85.24	94.01	98.58
	Mean	58.40	56.13	65.16	47.70	47.52	52.81	59.32	65.56
Rastrigin	Best	2.59	1.60E-03	1.12E-02	4.50E-03	3.30E-03	1.00E-03	9.42E-04	2.20E-02
	Worst	72.00	73.59	74.15	74.55	74.27	72.00	73.29	72.78
	Mean	29.84	44.29	37.41	40.96	38.11	26.66	37.38	30.73

Table 2 Results by varying the β (Beta) parameter

Function		2	1.8	1.6	1.5	1.4	1.3	1.2	1
Sphere	Best	8.12E-09	1.45E-28	5.03E-61	4.60E-83	2.03E-86	3.60E-25	6.16E-19	7.02E-18
	Worst	1.02E-08	3.97E-25	1.32E-58	1.46E-72	6.51E-73	2.08E-17	1.86E-15	1.54E-17
	Mean	6.83E-09	1.20E-25	1.29E-56	1.46E-73	1.30E-73	1.01E-17	3.81E-16	3.30E-17
Quartic	Best	2.34E-13	4.86E-41	8.15E-90	7.76E-132	8.03E-156	5.67E-37	9.75E-27	1.99E-23
	Worst	3.22E-09	8.48E-34	5.58E-85	1.13E-117	4.48E-143	1.79E-27	4.56E-22	2.40E-20
	Mean	3.68E-10	1.70E-34	1.12E-85	1.13E-118	4.48E-144	6.49E-28	9.16E-23	4.92E-21
Rosenbrock	Best	2.52	2.50	3.10E-02	7.67E-04	4.88E-04	6.68	12.36	19.06
	Worst	83.13	73.80	13.32	3.99	9.37	74.84	28.35	73.05
	Mean	59.48	24.15	5.28	1.95	2.11	28.14	22.29	32.60
Rastrigin	Best	1.60E-03	0.00	3.98	6.96	47.76	90.54	110.44	100.50
	Worst	73.59	36.00	80.95	74.80	120.46	120.39	163.54	138.30
	Mean	44.29	18.00	42.91	44.35	90.15	97.73	139.94	121.84

Table 3 Results by varying the γ (Gama) parameter

Function		1	0.9	0.7	0.6	0.5	0.4	0.2	0.1
Sphere	Best	6.47E-09	4.01E-09	2.95E-09	5.51E-09	5.51E-09	2.05E-09	2.06E-09	6.15E-09
	Worst	1.40E-07	2.68E-08	1.63E-08	1.21E-07	1.21E-07	7.63E-08	5.21E-08	5.13E-08
	Mean	2.96E-08	1.70E-08	8.91E-09	3.71E-08	3.71E-08	2.37E-08	1.39E-08	2.06E-08
Quartic	Best	1.46E-13	1.10E-13	3.06E-14	1.19E-12	2.34E-13	2.10E-13	1.86E-12	2.06E-14
	Worst	1.33E-10	1.49E-11	5.42E-11	1.61E-09	3.22E-09	5.64E-10	5.28E-11	2.84E-08
	Mean	3.10E-11	9.46E-12	1.37E-11	3.51E-10	3.68E-10	1.17E-10	2.48E-11	5.78E-09
Rosenbrock	Best	8.45	36.00	36.01	1.46E-01	6.50E-03	6.50E-03	3.95	25.02
	Worst	38.05	72.00	72.00	72.00	72.00	36.00	72.00	72.00
	Mean	28.35	54.07	54.00	46.82	29.23	27.70	33.19	55.41
Rastrigin	Best	2.73	2.71	36.00	9.99E-01	9.70E-03	4.79E-04	6.20E-04	3.95
	Worst	72.00	72.00	72.00	72.00	72.00	72.00	72.00	36.00
	Mean	40.58	27.74	46.82	46.90	36.28	25.44	32.38	31.70

Table 4 Results by varying the ξ (Eta) parameter

Function		0.02	0.1	0.2	0.3	0.5	0.7	0.8	0.9
Sphere	Best	1.40E-09	3.49E-09	7.79E-09	4.54E-09	2.00E-09	4.90E-09	2.73E-09	3.53E-09
	Worst	4.71E-08	3.00E-08	3.82E-08	1.74E-08	1.68E-08	1.66E-08	1.98E-08	4.49E-08
	Mean	1.80E-08	1.32E-08	2.26E-08	1.01E-08	8.46E-09	1.12E-08	9.18E-09	1.67E-08
Quartic	Best	2.34E-13	6.15E-13	9.42E-14	7.89E-14	2.25E-13	1.51E-13	2.65E-13	2.06E-13
	Worst	3.22E-09	1.08E-11	2.37E-12	2.54E-11	3.69E-11	1.12E-08	2.97E-11	3.94E-09
	Mean	3.68E-10	5.72E-12	1.35E-12	3.13E-12	6.35E-12	1.58E-09	5.68E-12	5.34E-10
Rosenbrock	Best	6.50E-03	1.17E-01	2.85E-04	3.60E+01	4.30E-03	4.30E-04	36.00	2.81
	Worst	72.00	72.00	108.00	72.00	108.00	72.00	72.00	71.82
	Mean	29.23	46.82	47.11	47.35	36.88	34.22	47.06	47.06
Rastrigin	Best	8.45	8.92E-02	1.17E-01	1.07	2.40E-03	18.12	5.20E-03	1.70E-03
	Worst	38.05	72.00	72.00	71.82	72.00	72.00	71.82	72.00
	Mean	28.17	36.06	39.79	47.13	26.01	40.78	36.93	32.40

Table 2 shows that after executing the ICA Algorithm 10 times, by varying the β (Beta) parameter, we can find the best, average and worst results for the different mathematical functions benchmark.

Table 3 shows that after executing the ICA Algorithm 10 times, by varying the γ (Gama) parameter, we can find the best, average and worst results for the different mathematical functions benchmark.

Table 4 shows that after executing the ICA Algorithm 10 times, by varying the ξ (Eta) parameter, we can find the best, average and worst results for the different mathematical functions benchmark

5 Conclusions

By analyzing the ICA algorithm by varying the parameters, we noticed that the parameter that affected most the operation is the β parameter in the range of 1.4–1.6 good results for the sphere and quartic functions.

For the Rosenbrock and Rastrigin functions the results are not good for 30 variables, but with fewer variables the algorithm performs better.

The remaining parameters though apparently do not affect significantly the operation of the algorithm these can be in the range of 0.2–0.5 in the revolution, from 0.4 to 0.6 for and small values of ξ the algorithm behaves better.

Acknowledgment We would like to express our gratitude to the CONACYT and Tijuana Institute of Technology for the facilities and resources granted for the development of this research.

References

1. Atashpaz-Gargari, E., Lucas, Y.C.: Imperialist competitive algorithm: an algorithm for optimization inspired by imperialistic competition. Evol. Computat. 4661–4667 (2007)
2. Hosseini, S.M., Al Khaled, Y.A.: A survey on the imperialist competitive algorithm metaheuristic: implementation in engineering domain and directions for future research. Appl. Soft Comput. J. **24**, 1078–1094 (2014)
3. Jula, A., Othman, Z., Sundararajan, Y.E.: Imperialist competitive algorithm with PROCLUS classifier for service time optimization in cloud computing service composition. Expert Syst. Appl. **42**(1), 135–145 (2014)
4. Rasul, E., Javedani Sadaei, H., Abdullah, A.H., Gani, Y.A.: Imperialist competitive algorithm combined with refined high-order weighted fuzzy time series (RHWFTS–ICA) for short term load forecasting. Energy Convers. Manag. **76**, 1104–1116 (2013)

Fuzzy Logic for Improving Interactive Evolutionary Computation Techniques for Ad Text Optimization

Quetzali Madera, Mario Garcia and Oscar Castillo

Abstract The description of a product or an ad's text can be rewritten in many ways if other text fragments similar in meaning substitute different words or phrases. A good selection of words or phrases, composing an ad, is very important for the creation of an advertisement text, as the meaning of the text depends on this and it affects in a positive or a negative way the interest of the possible consumers towards the advertised product. In this paper we present a method for the optimization of advertisement texts through the use of interactive evolutionary computing techniques. The EvoSpace platform is used to perform the evolution of a text, resulting in an optimized text, which should have a better impact on its readers in terms of persuasion.

1 Introduction

Text content is plays a very important role in e-commerce applications, as this is one of the most common ways of giving information about a commercial product to the consumers [1]. When the author of an advertisement text is an expert ad writer, the product should have a better chance of receiving a positive response from the consumers. The combination of words or phrases (blocks of text) that the experts decide to use when writing the text of an ad is important, because this particular combination could be the one that persuades the consumer into buying the product. If an inexperienced writer decides to write an advertisement text, it would be very difficult for him to choose a correct combination of the blocks of text that is

Q. Madera (✉) · M. Garcia · O. Castillo
Tijuana Institute of Technology, Calzada Tecnologico S/N, Tijuana, Mexico
e-mail: quetzalimadera@tectijuana.edu.mx

M. Garcia
e-mail: mariosky@tectijuana.mx

O. Castillo
e-mail: ocastillo@tectijuana.mx

K.T. Atanassov et al. (eds.), *Novel Developments in Uncertainty Representation and Processing*, Advances in Intelligent Systems and Computing 401,
DOI 10.1007/978-3-319-26211-6_25

successful into persuading the majority of the consumers. In this work, we propose that a writer of any level of experience can create an ad with different interchangeable blocks of text carrying the same meaning, and a third party could optimize it.

Evolutionary algorithms are commonly used to solve optimization problems [2] and that's why we decided to use this kind of techniques to optimize the advertisement texts. We believe that if a group of people can evaluate, in terms of persuasion, different combinations of the same ad, after many generations of evolution we can find an optimal ad, which will have a better impact on the majority of the consumers.

2 Basic Concepts

In this section we present some basic concepts for understanding this work, and the essential parts that compose this project. One of the most important components is EvoSpace, which turned to be a crucial tool for the implementation of the genetic algorithm used in this work.

2.1 Evolutionary Algorithms

Evolutionary algorithms are a subfield of artificial intelligence. They are mainly used in optimization problems where the search space is very large and aren't lineal. These algorithms search for solutions based on the theory of the Darwinian evolution.

The methods of this kind generate a set of individuals that represent possible solutions. These solutions are usually generated randomly at the beginning of the evolution process. After each generation, the best solutions share part of their information to create other possible, better, solutions. All of the individuals compete to be the more fit solutions; the better solutions are conserved, while the worse are destroyed, according to a fitness function that evaluates their performance [3].

2.2 Genetic Algorithms

Genetic algorithms (see Fig. 1) are inspired in biological evolution; they evolve a population of individuals by performing genetic recombination and mutation. A selection of the best solutions is made by the use of certain criterion and a fitness function, and based on their performances, the more fit individuals survive and the less fit are discarded. Optimization based on genetic algorithms is a search method based on probability [4].

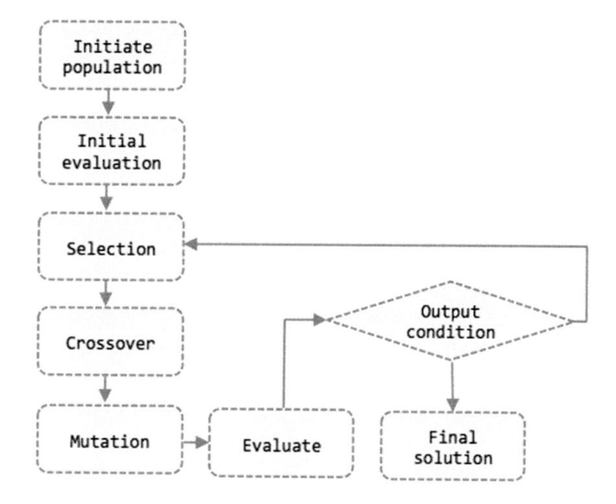

Fig. 1 Genetic algorithm diagram

This is an elitist algorithm as it always conserves the best individual of the population unchanged. As the number of generations or iterations increases, the probability of finding the optimum solution tends to increase.

2.3 *Interactive Evolutionary Computation*

Interactive evolutionary computing is a variation of evolutionary computing where the fitness of an individual is determined through the subjective evaluation performed by a human being. In traditional evolutionary computing, a human being requires a computational process to solve a problem. To do this, a person gives a problem's description as input to a solution model, and this model returns a result that has to be interpreted by a human being. But in interactive evolutionary computing the roles are inverted: there's an algorithm that asks a human being or a group of human beings to solve a problem, and then it gathers this information to interpret it later [5].

2.4 *Article Spinning*

Article spinning is a method used to create multiple versions of a text article without creating versions considered as plagiarism, due to the uniqueness achieved of the generated content. Duplicated content is not accepted by several search engines like Google, Yahoo and Bing, so this method is used to generate many different versions of a single article that have a higher probability of being considered as unique content by these search engines. Words and phrases are randomly changed by other text blocks that have the same meaning, resulting in another version of the article with the same meaning, but different text content [6].

2.5 EvoSpace

EvoSpace is a cloud's space or habitat where evolutionary algorithms can be stored, developed, tested and be put into production. EvoSpace is very versatile, as the population is independent to the evolutionary model being used, and this allows us to make modifications to the evolutionary algorithm at any time. The client processes, called EvoWorkers, interact dynamically and asynchronously, and they can be displayed in remote clients like in the platform storing the server [7].

2.6 Online Advertisement

Online advertisement is performed based on the content of a website. For the creation of this type of advertisement, which has the objective of giving information about a product on Internet, it must contain different media elements such as text, links, images, videos, animations, etc. There are companies like Google that have created systems for the creation of online advertisement campaigns, like AdSense and AdWords [8]. AdSense positions ads in websites related to the textual content being displayed in the web pages. Users owning these ads pay a certain amount of money for each click on their ads.

3 Related Work

In this section we will explain the platform EvoSpace-Interactive, which is the platform that we used to build the graphical interfaced used by the users to choose what ads they considered were more persuasive.

3.1 EvoSpace-Interactive

EvoSpace-interactive was initially tested by implementing an interactive evolutionary computation program called Shapes. This software evolved images formed by equilateral triangles that could have one of twelve possible colors. Currently, EvoSpace has made modifications to the images displayed by changing the shape to more attractive animations that last for a short period of time. For this work, we modified EvoSpace to be capable of displaying text ads instead of images [9].

4 Problem Description

If an inexperienced article writer decided to create many different versions of a text ad, where he changed some words and phrases to other text blocks of similar meaning, and he showed them to some friends and asked them to vote for the version they think is the best one, he would be performing a small optimization to his text. This is because his text isn't based only on his opinion, but on the opinion of all of his friends and his own, and now the winning text could be considered more attractive to a higher percentage of readers.

Based on this example, we believe that a more effective optimization of the text for an ad should take into consideration thousands of possible versions, and dozens of people should give their opinion on what blocks of text are better to increase the persuasion of the text.

5 Methodology

5.1 *Article Format*

The text (see Fig. 2) is contains different sections enclosed by curly braces, which contain different blocks of text (the blocks of text can be of any size, from a single word to whole sentences and paragraphs). These blocks of text are separated by bars. The text blocks that are outside of the curly braces (represented by grey text in Fig. 2) won't have any modifications when the texts evolve.

```
{Diseñamos | Creamos | Fabricamos | Construimos | Desarrollamos} {este auto |
este carro | esta pieza de arte | este impresionante transporte | este auto
único | este carro único}, {para ser | para convertirse en | para que sea |
con el fin de ser} la mejor forma {de viajar | de transportarte | para ir de
un lugar a otro | de moverte | de trasladarte} {con tu familia | con tus
amigos | a donde quieras | románticamente | cómodamente | de forma divertida |
silenciosamente | de forma segura}. Por eso está {equipado | preparado |
construido | fabricado | diseñado | creado | desarrollado} con {barras
laterales | protecciones a los costados | puertas protegidas | protecciones
laterales} contra {impacto | golpes | accidentes | choques}. Puede incluir
{frenos ABS | un sistema antibloqueo de ruedas | frenos antibloqueo | frenos
reforzados | frenos inteligentes | frenos antiderrapantes} y bolsas de aire
{frontales | al frente | delanteras | grandes | suaves | seguras}. {Y por si
fuera poco, es maniobrable, estable y eficiente | Cuenta con dirección
hidráulica en todas sus versiones | Tiene un motor de 4 cilindros de 1.6L |
Tiene transmisión manual de 5 velocidades | Tiene transmisión automática de 4
velocidades | Tiene un espacio interior y cajuela amplia | Por eso este
transporte es tu mejor opción | Compra seguro, compra inteligentemente}.
```

Fig. 2 Text format

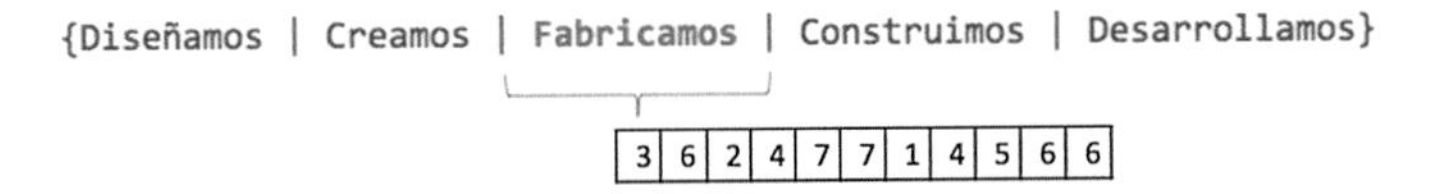

Fig. 3 Representation of the vector

A vector generates each version of the text, where each number in it represents the corresponding option. In this example (see Fig. 3), if we want to know what number is the word "Fabricamos", we have to look in the options to find out it's the third option. So, in the first position of the vector we'll have the number 3. The rest of the vector represents the rest of the options that should be printed in the generated text.

5.2 *Implementation of EvoSpace-Interactive*

EvoSpace was modified to be able to evolve advertisement texts. The tex, using the format explained before, must be analyzed by the program to determine how many text segments will be changing, and how many and what are their options. This way a vector can be generated, which would represent the chromosome of an individual. EvoSpace creates 100 random individuals when it initializes its population.

Changes were also made to the graphical user interface (see Fig. 4). The amount of likes an individual currently has was removed so it wouldn't act as a bias during the decision process of the user. By default, in EvoSpace users can create a

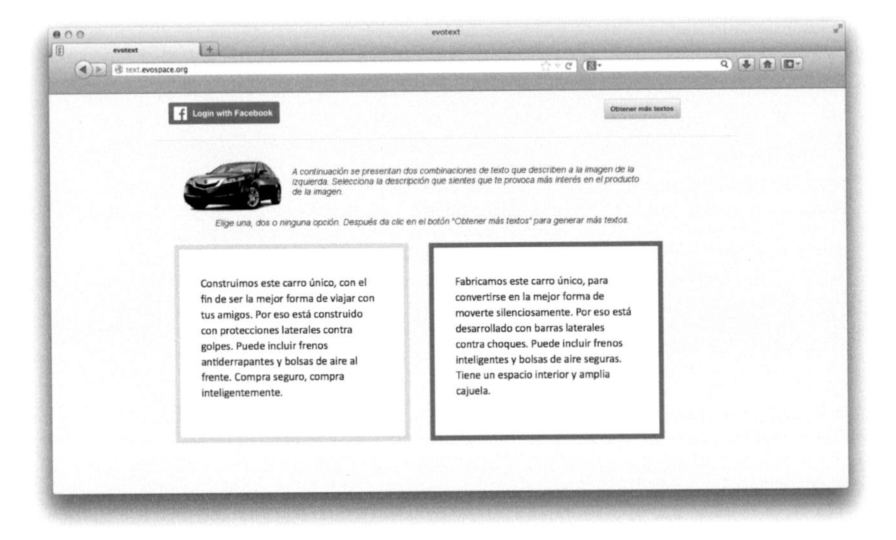

Fig. 4 User Interface

collection of their favorite individuals, and this feature was also removed as we consider this isn't necessary for our experiments. The general design was modified so the texts stand out from the rest of the elements of the interface.

5.3 System Configuration

The population is initialized with 100 randomly generated individuals. For the evaluation of the individuals, the users are presented with two texts. A user can choose what text (or texts) he considers would be more persuasive. If no text is considered as a good choice, the user can also decide not to choose any of the options. When a selection is made, the selected texts are sent to EvoSpace and stored in its database, and the user is immediately presented with two more options, so he can continue with the selection process.

6 Results

The system started operation at http://text.evospace.org since December 4th 2013, and a total of 75 users have participated in the selection process, generating more than 180 samples.

The best chromosome was extracted from the database and it was compared against the worst generated chromosome. This best chromosome was also compared against an actual ad that advertises a Chevrolet car created by an expert. This ad, similar in size to the evolved text, was taken from Chevrolet's website [10].

A total of 30 people were surveyed, showing them two paper sheets containing two texts (see Fig. 5). On the first paper sheet, the text representing our best individual and the text created by the expert were shown. The second paper sheet showed the texts generated by our best and worst individuals. The texts were shown in different positions to discard the possibility of a person choosing an option due to its position (for example, if a person likes more a text because it's on the right side).

In our first case, 60 % of the people chose the text generated by our genetic algorithm, and 40 % chose the text generated by the expert. In the second case, 63.3 % of the people surveyed chose the text represented by the best individual, while 36.7 % chose the text represented by the worst individual.

The fuzzy rules that can control the evolutionary process are:

1. Si (Tiempo es Bajo) y (CTR es Bajo) y (Precio es Bajo) entonces (Precio es Bajo)
2. Si (Tiempo es Bajo) y (CTR es Bajo) y (Precio es Medio) entonces (Precio es Bajo)
3. Si (Tiempo es Bajo) y (CTR es Bajo) y (Precio es Alto) entonces (Precio es Bajo)
4. Si (Tiempo es Bajo) y (CTR es Medio) y (Precio es Bajo) entonces (Precio es Bajo)

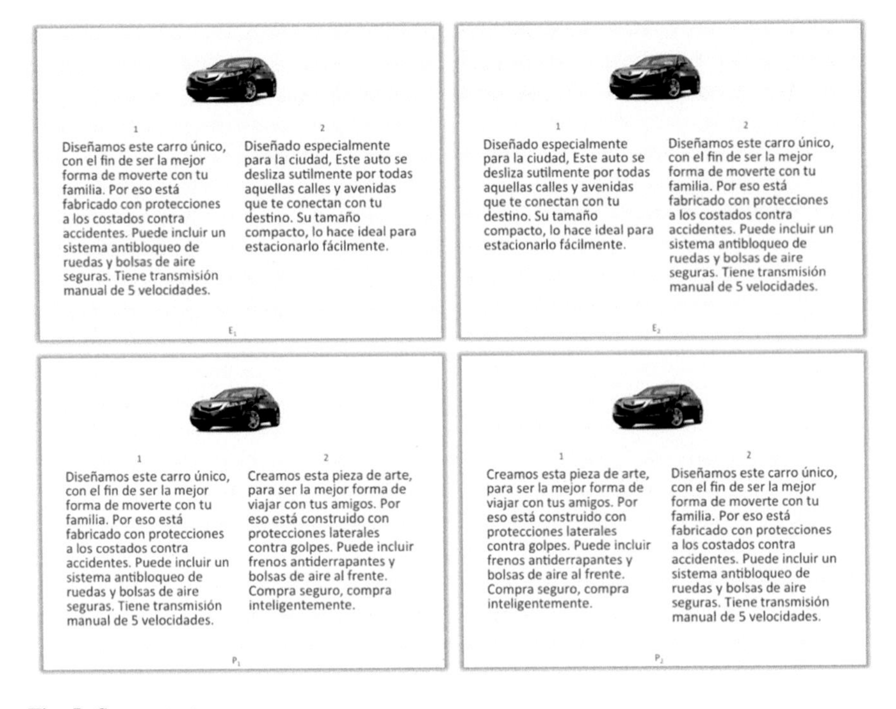

Fig. 5 Survey texts

5. Si (Tiempo es Bajo) y (CTR es Medio) y (Precio es Medio) entonces (Precio es Bajo)
6. Si (Tiempo es Bajo) y (CTR es Medio) y (Precio es Alto) entonces (Precio es Bajo)
7. Si (Tiempo es Bajo) y (CTR es Alto) y (Precio es Bajo) entonces (Precio es Bajo)
8. Si (Tiempo es Bajo) y (CTR es Alto) y (Precio es Medio) entonces (Precio es Bajo)
9. Si (Tiempo es Bajo) y (CTR es Alto) y (Precio es Alto) entonces (Precio es Bajo)
10. Si (Tiempo es Medio) y (CTR es Bajo) y (Precio es Bajo) entonces (Precio es Medio)
11. Si (Tiempo es Medio) y (CTR es Bajo) y (Precio es Medio) entonces (Precio es Alto)
12. Si (Tiempo es Medio) y (CTR es Bajo) y (Precio es Alto) entonces (Precio es *)
13. Si (Tiempo es Medio) y (CTR es Medio) y (Precio es Bajo) entonces (Precio es Medio)
14. Si (Tiempo es Medio) y (CTR es Medio) y (Precio es Medio) entonces (Precio es Alto)
15. Si (Tiempo es Medio) y (CTR es Medio) y (Precio es Alto) entonces (Precio es *)
16. Si (Tiempo es Medio) y (CTR es Alto) y (Precio es Bajo) entonces (Precio es *)
17. Si (Tiempo es Medio) y (CTR es Alto) y (Precio es Medio) entonces (Precio es *)
18. Si (Tiempo es Medio) y (CTR es Alto) y (Precio es Alto) entonces (Precio es *)
19. Si (Tiempo es Alto) y (CTR es Bajo) y (Precio es Bajo) entonces (Precio es *)

20. Si (Tiempo es Alto) y (CTR es Bajo) y (Precio es Medio) entonces (Precio es *)
21. Si (Tiempo es Alto) y (CTR es Bajo) y (Precio es Alto) entonces (Precio es *)
22. Si (Tiempo es Alto) y (CTR es Medio) y (Precio es Bajo) entonces (Precio es *)
23. Si (Tiempo es Alto) y (CTR es Medio) y (Precio es Medio) entonces (Precio es *)
24. Si (Tiempo es Alto) y (CTR es Medio) y (Precio es Alto) entonces (Precio es *)
25. Si (Tiempo es Alto) y (CTR es Alto) y (Precio es Bajo) entonces (Precio es *)
26. Si (Tiempo es Alto) y (CTR es Alto) y (Precio es Medio) entonces (Precio es *)
27. Si (Tiempo es Alto) y (CTR es Alto) y (Precio es Alto) entonces (Precio es *)

Simulation results using these fuzzy rules are encouraging, but we are still in the process of making more experiments.

7 Conclusions

The evolution of advertisement texts written by an inexperienced person in the field of marketing, through the use of interactive evolutionary techniques and fuzzy logic, is a viable alternative for the creation of texts with a higher probability of persuading consumers into buying the advertised product.

8 Future Work

We are currently working on the implementation of a clustering algorithm for grouping [11] users according to their profiles, and perform interactive evolution to generate optimal advertisement texts for each cluster of people.

References

1. McQuarrie, Edward F., David, M.: Visual rhetoric in advertising: Text-interpretive, experimental, and reader-response analyses. J. Consum. Res. **26**(1), 37–54 (1999)
2. De Jong, K.A.: Evolutionary computation–a unified approach. MIT Press, Cambridge (2006)
3. Bäck, T.: Evolutionary Algorithms In Theory And Practice: Evolution Strategies, Evolutionary Programming, Genetic Algorithms, vol. 996. Oxford University Press, Oxford (1996)
4. Whitley, D.: A genetic algorithm tutorial, Statistics and Computing, p. 65. Kluwer Academic Publishers 4.2, Norwell (1994)
5. Takagi, H.: Interactive evolutionary computation: fusion of the capabilities of EC optimization and human evaluation. Proc. IEEE **89**(9), 1275 (2001)
6. Malcolm, J.A., Peter L.: An approach to detecting article spinning. In: Proceedings of the Third International Conference on Plagiarism (2008)
7. García, M., et al.: EvoSpace: a distributed evolutionary platform based on the tuple space model. In: Applications of Evolutionary Computation, pp. 499–508. Springer, Berlin (2013)
8. Davis, H.: Google Advertising Tools: Cashing in with AdSense, AdWords, and the Google APIs. O'reilly, Cambridge (2006)

9. Fernández, F., et al.: EvoSpace-interactivo: una herramienta para el arte y diseño interactivo y colaborativo, pp. 220–228. Algoritmos Evolutivos y Bioinspirados, IX Congreso Español de Metaheurísticas (2013)
10. Chevrolet.: Chevrolet Mexico Ad description (2013). http://www.chevrolet.com.mx/spark-2014.html.Accessed 12 Dec 2013
11. Hartigan, J.: Clustering Algorithms. Wiley, New York (1975)

InterCriteria Analysis of Generation Gap Influence on Genetic Algorithms Performance

Olympia Roeva and Peter Vassilev

Abstract In this investigation InterCriteria Analysis (ICA) is applied to examine the influences of one of the genetic algorithms parameters—the generation gap (*ggap*). The investigation is carried out during the model parameter identification of *E. coli* MC4110 cultivation process. The apparatuses of index matrices and intuitionistic fuzzy sets, which are the core of ICA, are used to establish the relations between *ggap* and GAs outcomes (computational time and decision accuracy), on one hand, and cultivation process model parameters on the other hand. The obtained results after ICA application are analyzed in terms of convergence time and model accuracy and some conclusions about derived interactions are reported.

Keywords Intercriteria analysis · Genetic algorithms · Generation gap · Parameter identification · *E. coli*

1 Introduction

Microorganisms have been a subject of particular attention as a biotechnological instrument. Numerous useful bacteria, yeasts and fungi are widely available in nature. Bacteria *Escherichia coli* is now the most important model organism in biology [9, 31]. *E. coli* has come to prominence not only in academic and commercial genetic engineering, pharmaceutical production, and experimental microbial evolution, but also in the biotechnology industry [9, 18].

To provide new ways of analyzing and understanding microorganisms, modelling approaches are applied. In this paper, mathematical modelling of an *E. coli* fed-batch cultivation process is considered. To describe the bacteria growth kinetic, the Monod

O. Roeva · P. Vassilev (✉)
Institute of Biophysics and Biomedical Engineering,
Bulgarian Academy of Sciences, 105 Acad. G. Bonchev Str., 1113 Sofia, Bulgaria
e-mail: peter.vassilev@gmail.com

O. Roeva
e-mail: olympia@biomed.bas.bg

K.T. Atanassov et al. (eds.), *Novel Developments in Uncertainty Representation and Processing*, Advances in Intelligent Systems and Computing 401,
DOI 10.1007/978-3-319-26211-6_26

model is used [25]. In order to obtain a model with a high degree of accuracy, adequate estimates of the Monod model should be done. It is known that in parameter identification when the Monod kinetic is fitted to a set of experimental data, the substrate affinity constant k_S values vary with the maximum specific growth rate μ_{max} [21]. The question of whether and to what extent the observed changes in Monod model parameters are a result of this high correlation between them is still unanswered [19]. To bring some more clarity to the existing correlations and dependences between Monod model parameters, InterCriteria Analysis (ICA) [1] is applied. Any increase in the knowledge of *E. coli* expands the range of biological phenomena for which its strains can be used as models to study [9].

ICA implements the apparatuses of index matrices (IM) [3, 4] and intuitionistic fuzzy sets (IFS) [5] in order to compare some criteria or objects estimated by them. Up to now there has been one published application of ICA examining the influences of GA parameters—number of individuals (*nind*) and number of generations (*ngen*)—during the model identification of *E. coli* and *S. cerevisiae* cultivation processes [28]. In the current research genetic algorithms (GAs) [16] are considered as the most successfully performing meta-heuristics. Their effectiveness has been already demonstrated for model parameter identification of cultivation processes [8, 24, 27, 29]. To identify the *E. coli* model parameters, real-coded GAs are applied. The optimization variables are represented as floating point numbers to avoid time-consuming encoding and decoding in each step, large memory requirement and loss of precision due to quantization error [17, 24, 30]. Since the GAs parameters values have a great impact on performance and efficacy of the algorithm [15, 26], GAs require setting of the algorithm parameters values. Usually, in the parameters tuning a compromise between solution quality (J) and search time (J) should be done. Similar to research [28], the investigation here includes an analysis of the GA parameter—generation gap (*ggap*).

The parameter *ggap* is considered as a design factor with 51 levels. The analysis of the mean GA results is conducted to find the influence of the design factor and its optimal value for a fed-batch cultivation processes of bacteria *E. coli*.

The paper is organized as follows: the formulation of the identification problem is given in Sect. 2. In Sect. 3 the background of ICA is presented. Numerical results are discussed in Sect. 4, and conclusion remarks are given in Sect. 5.

2 Model Parameter Identification Problem

2.1 E. Coli Fed-Batch Cultivation Model

The model is presented by a system of non-linear differential equations [7, 27]:

$$\frac{dX}{dt} = \mu_{max}\frac{S}{k_S + S}X - \frac{F}{V}X \tag{1}$$

$$\frac{dS}{dt} = -\frac{1}{Y_{S/X}}\mu_{max}\frac{S}{k_S + S}X + \frac{F}{V}(S_o - S) \tag{2}$$

$$\frac{dV}{dt} = F, \tag{3}$$

where X is biomass concentration, [g/l]; S—substrate concentration, [g/l]; F—feeding rate, [l/h]; V—bioreactor volume, [l]; S_o—substrate concentration in the feeding solution, [g/l]; μ_{max}—the maximum values of the specific growth rate, [1/h]; k_S—saturation constant (substrate affinity constant), [g/l]; $Y_{S/X}$—yield coefficient, [-].

Using Monod kinetics [25], the growth rate relates to the concentration of a single growth-controlling substrate ($\mu = f(S)$) via two kinetic parameters: μ_{max} and k_S. The growth and substrate utilization linearly relates the stoichiometric parameter $Y_{X/S}$ (a measure for the conversion efficiency of a growth substrate into cell material) to the specific rates of biomass growth and substrate consumption [19].

The parameter vector that should be identified for the model Eqs. (1)–(3) is $u = [\mu_{max}\ k_S\ Y_{S/X}]$.

Experimental data for biomass and glucose concentration of an *E. coli* MC4110 fed-batch process are used for the purposes of model parameters identification. The detailed description of the process conditions and experimental data can be found in [27].

For the considered problem, the objective function is defined as:

$$J = \|Z\|^2 \rightarrow \min, \tag{4}$$

where $\|\|$ denotes the ℓ^2-vector norm, $Z = Z_{\text{mod}} - Z_{\text{exp}}$, $Z_{\text{mod}} \stackrel{\text{def}}{=} [X_{\text{mod}}\ S_{\text{mod}}]$ are model predictions for biomass and substrate, and $Z_{\text{exp}} \stackrel{\text{def}}{=} [X_{\text{exp}}\ S_{\text{exp}}]$ are known experimental data for both process variables.

2.2 Genetic Algorithm

Model parameter identification of cultivation process models has become a research field of particularly great interest. Since the considered problem has been known to be NP-complete, using meta-heuristic techniques can solve this problem more efficiently than exact or traditional methods [20]. Some of the most successfully performing meta-heuristics are the GAs [16].

Real-coded GAs can be regarded as GAs that operate on the actual candidate solutions (phenotype). For real-coded GAs, no genotype-to-phenotype mapping is needed. Compared to binary-coded GAs, real-coded GAs have several distinct advantages, such as: amended computation complexity, improved computation efficiency and much higher solution precision than that of binary-coded GAs [12, 20, 22, 23].

Table 1 Main GA operators and parameters

Operators and parameters	Type/value
Encoding	Real-coded
Crossover operator	Extended intermediate recombination
Mutation operator	Real-value mutation
Selection operator	Roulette well selection
ggap	0.5 : 0.01 : 1
ngen	100
nind	100
xovr	0.8
mutr	1/*nind*

GAs workability is guided mainly by different operators and parameters that can be implemented specifically in different problems. The first attempt to evaluate empirically GAs performance with overlapping populations has been done by De Jong [14]. De Jong has introduced the generation gap that defines the percentage of the population that is replaced with each GAs generation. It is found that at low values of *ggap* the algorithm has a severe loss of alleles, which results in poor search performance [11, 14].

In this research *ggap* is used as a design factor of GA performance in model identification of an *E. coli* MC4110 fed-batch cultivation process. The parameter *ggap* takes 51 different values (levels)—from 0.5 to 1, with step 0.01. The rest of the GA operators and parameters, summarized in Table 1, are tuned based on several pre-tests.

3 InterCriteria Analysis Approach

Here the idea proposed in [1] is expanded. Following [1, 5], an Intuitionistic Fuzzy Pair (IFP) [2] with the degrees of "agreement" and "disagreement" between two criteria applied on different objects is obtained. An IFP is an ordered pair of real non-negative numbers $\langle a, b\rangle$ such that: $a + b \leq 1$.

Let an IM (see [3]) whose index sets consist of the names of the criteria (for rows) and objects (for columns) be given. The elements of this IM are further supposed to be real numbers. An IM with index sets consisting of the names of the criteria (for rows and for columns) with elements IFPs corresponding to the "agreement" and "disagreement" of the respective criteria will be obtained. Two things are further assumed: (i) all criteria provide an evaluation for all objects and all these evaluations are available; (ii) all the evaluations of a given criteria can be compared amongst themselves.

The set of all objects being evaluated is denoted by O, and the set of values assigned by a given criteria C to the objects by $C(O)$, i.e.

$$O \stackrel{\text{def}}{=} \{O_1, O_2, \dots, O_n\}, \; C(O) \stackrel{\text{def}}{=} \{C(O_1), C(O_2), \dots, C(O_n)\}.$$

Let $C^*(O) \stackrel{\text{def}}{=} \{\langle x, y\rangle | \; x \neq y \;\&\; \langle x, y\rangle \in C(O) \times C(O)\}$.

In order to compare two criteria, the vector of all internal comparisons of each criteria which fulfill exactly one of three relations R, $\overline{R}$ and $\tilde{R}$ must be constructed. It is required that for a fixed criterion C and any ordered pair $\langle x, y\rangle \in C^*(O)$ it is true:

$$\langle x, y\rangle \in R \Leftrightarrow \langle y, x\rangle \in \overline{R}, \tag{5}$$

$$\langle x, y\rangle \in \tilde{R} \Leftrightarrow \langle x, y\rangle \notin (R \cup \overline{R}), \tag{6}$$

$$R \cup \overline{R} \cup \tilde{R} = C^*(O). \tag{7}$$

From the above it is seen that only a subset of $C(O) \times C(O)$ has to be considered for the effective calculation of the vector of internal comparisons (denoted further by $V(C)$), since from (5)–(7) it follows that if the relation between x and y is known, the relation between y and x is known as well. Thus, only lexicographically ordered pairs $\langle x, y\rangle$ are considered. Let, for brevity, $C_{i,j} = \langle C(O_i), C(O_j)\rangle$. Then, for a fixed criterion C the following vector is constructed:

$$V(C) = \{C_{1,2}, C_{1,3}, \dots, C_{1,n}, C_{2,3}, C_{2,4}, \dots, C_{2,n}, C_{3,4}, \dots, C_{3,n}, \dots, C_{n-1,n}\}.$$

It can be easily seen that it has exactly $\frac{n(n-1)}{2}$ elements. Further, to simplify our considerations, the vector $V(C)$ is replaced with $\hat{V}(C)$, where for each $1 \leq k \leq \frac{n(n-1)}{2}$ for the k-th component it is true:

$$\hat{V}_k(C) = \begin{cases} 1 \text{ iff } V_k(C) \in R, \\ -1 \text{ iff } V_k(C) \in \overline{R}, \\ 0 \text{ otherwise.} \end{cases}$$

Then, when comparing two criteria, the degree of "agreement" between the two is the number of matching components (divided by the length of the vector for normalization purposes). The degree of "disagreement" is the number of components of opposing signs in the two vectors (again normalized by the length).

The above described algorithm for calculating the degrees of "agreement" (μ) and degrees of "disagreement" (ν) between two criteria C and C' is realized in Matlab environment. A pseudo-code of the **Algorithm 1** used in this study is presented below.

Algorithm 1 Calculating "agreement" and "disagreement" between two criteria

```
Require: Vectors V̂(C) and V̂(C')
 1: function DEGREE OF AGREEMENT(V̂(C), V̂(C'))
 2:     V ← V̂(C) − V̂(C')
 3:     μ_{C,C'} ← 0
 4:     for i ← 1 to n(n−1)/2 do
 5:         if V_i = 0 then
 6:             μ_{C,C'} ← μ_{C,C'} + 1
 7:         end if
 8:     end for
 9:     μ_{C,C'} ← 2/(n(n−1)) μ_{C,C'}
10:     return μ_{C,C'}
11: end function

12: function DEGREE OF DISAGREEMENT(V̂(C), V̂(C'))
13:     V ← V̂(C) − V̂(C')
14:     ν_{C,C'} ← 0
15:     for i ← 1 to n(n−1)/2 do
16:         if abs(V_i) = 2 then                    ▷ abs: absolute value
17:             ν_{C,C'} ← ν_{C,C'} + 1
18:         end if
19:     end for
20:     ν_{C,C'} ← 2/(n(n−1)) ν_{C,C'}
21:     return ν_{C,C'}
22: end function
```

It is obvious that for $\mu_{C,C'}$, $\nu_{C,C'}$, we have $\mu_{C,C'} = \mu_{C',C}, \nu_{C,C'} = \nu_{C',C}$. Also, $\langle \mu_{C,C'}, \nu_{C,C'} \rangle$ is an IFP. In most of the obtained pairs $\langle \mu_{C,C'}, \nu_{C,C'} \rangle$, the sum $\mu_{C,C'} + \nu_{C,C'}$ is equal to 1. However, there may be some pairs, for which this sum is less than 1. The following difference is considered as a degree of "uncertainty":

$$\pi_{C,C'} = 1 - \mu_{C,C'} - \nu_{C,C'}. \tag{8}$$

4 Numerical Results and Discussion

A series of identification procedures of model parameters vector u using real-coded GAs has been performed. For each value of $ggap$ (51 design levels) thirty independent runs of GA have been fulfilled. Average values of the obtained model parameters estimates and resulting values of objective function (J) and total computational time (T) have been calculated. To perform ICA, an IM $A_{(ggap)}$ (Eq. 9) has been constructed, where $T, J, ggap, \mu_{max}, k_S$ and $Y_{S/X}$ are considered as criteria $C_1, C_2, \ldots, C_6$, respectively, for 51 designed levels, i.e. as objects $O_1, O_2, \ldots, O_{51}$, in accordance with ICA theory. The full IM $A_{(ggap)}$ is available at http://intercriteria.net/studies/gengap/e-coli/.

$$A_{(ggap)} = \begin{array}{c|cccccc} & C_1 & C_2 & C_3 & C_4 & C_5 & C_6 \\ \hline O_1 & 4.846773 & 34.272917 & 0.50 & 0.519918 & 0.017989 & 0.494683 \\ O_2 & 4.581270 & 35.021875 & 0.51 & 0.502164 & 0.014447 & 0.494507 \\ \dots & \dots & \dots & \dots & \dots & \dots & \dots \\ O_{29} & 4.462116 & 51.234375 & 0.78 & 0.484761 & 0.011293 & 0.494725 \\ \dots & \dots & \dots & \dots & \dots & \dots & \dots \\ O_{50} & 4.569453 & 63.318750 & 0.99 & 0.496026 & 0.013495 & 0.494479 \\ O_{51} & 4.880437 & 63.164063 & 1.00 & 0.499184 & 0.014268 & 0.494324 \end{array} \quad (9)$$

As the number of objects influences the proper ICA estimations, ICA is performed for 6 different groups of objects. In this way the influence of the object number on the ICA results has be investigated. The groups involve 6, 11, 21, 31, 41 and 51 objects, respectively. The first group G_1 has 6 objects, namely, for the following values of $ggap : G_1 = \{0.5,\ 0.6,\ 0.7,\ 0.8,\ 0.9,\ 1\}$, i.e. in the input IM only the objects $O_1, O_{11}, O_{21}, O_{31}, O_{41}$ and O_{51} are included for the corresponding six criteria. The second group G_2 consists of 11 objects, which are constructed by adding some new values of $ggap$ to the first group, namely, $G_2 = \{G_1 \cup \{x + 0.05 | x \in G_1 \setminus \{1\}\}\}$. In the same manner the other groups may be represented as $G_i = \{G_{i-1} \cup G'_i,\ i = 3, 4, 5, 6\}$, where

$$G'_i = \{x + \frac{2i-4}{100} | x \in G_1 \setminus \{1\}\} \cup \{x + \frac{2i - 4 + (-1)^{\max(0,2i-9)} * 5}{100} | x \in G_1 \setminus \{1\}\}.$$

Based on the defined six IMs, following the presented ICA algorithm (**Algorithm 1**), ninety IFPs $\langle \mu, \nu \rangle$ (for every two pairs of the considered criteria) are obtained. The results are summarized in Table 2. In all cases the degree of "uncertainty" $\pi = 0$ is observed.

Observing the obtained values of the degrees of "agreement" (μ) and the degrees of "disagreement" (ν), the criteria relations and dependences are discussed according to the scale proposed in [6] (see Table 3). In Fig. 1 a graphical representation of the ICA results is shown.

The results show that there is a clear dependence of the object numbers on the μ- and ν-values obtained by ICA. For example, μ-values of the criteria $C_2 \leftrightarrow C_4$ and $C_3 \leftrightarrow C_4$ start from $\mu = 0.07$ (NC) in case of 6 objects and finish with $\mu = 0.38$ (D) in case of 51 objects. Also, μ-values of the criteria $C_2 \leftrightarrow C_5$ and $C_3 \leftrightarrow C_5$ start from $\mu = 0.2$ (WNC) in case of 6 objects and finish with $\mu = 0.39$ (D) in case of 51 objects. This means that there should be as more as possible object numbers to perform ICA and to define the exact dependence between any two considered criteria. In the case of an insufficient number of objects the ICA could lead to some wrong conclusions. Thus, it could be considered that there is NC between the following four pair combinations: T and μ_{max} ($C_2 \leftrightarrow C_4$), T and k_S ($C_2 \leftrightarrow C_5$), $ggap$ and μ_{max} ($C_3 \leftrightarrow C_4$), $ggap$ and k_S ($C_3 \leftrightarrow C_5$). Actually, these criteria are in dissonance.

Another example is the correlation between the criteria $C_1 \leftrightarrow C_4$ and $C_1 \leftrightarrow C_5$. If the results from ICA based on 6 objects are considered, the conclusion is that these criteria are in WD. Actually, they are in PC—the model parameters identification results show that estimates of μ_{max} higher than 0.5 lead to higher values of

Table 2 Results from the ICA

Criteria correlation	Number of objects					
	6	11	21	31	41	51
	$\langle \mu, \nu \rangle$	$\langle \mu, \nu \rangle$	$\langle \mu, \nu \rangle$	$\langle \mu, \nu \rangle$	$\langle \mu, \nu \rangle$	$\langle \mu, \nu \rangle$
$C_2 \leftrightarrow C_1$	$\langle 0.33, 0.67 \rangle$	$\langle 0.42, 0.58 \rangle$	$\langle 0.53, 0.47 \rangle$	$\langle 0.51, 0.49 \rangle$	$\langle 0.47, 0.53 \rangle$	$\langle 0.47, 0.53 \rangle$
$C_3 \leftrightarrow C_1$	$\langle 0.33, 0.67 \rangle$	$\langle 0.42, 0.58 \rangle$	$\langle 0.53, 0.47 \rangle$	$\langle 0.51, 0.49 \rangle$	$\langle 0.47, 0.53 \rangle$	$\langle 0.47, 0.53 \rangle$
$C_4 \leftrightarrow C_1$	$\langle 0.73, 0.27 \rangle$	$\langle 0.85, 0.15 \rangle$	$\langle 0.89, 0.11 \rangle$	$\langle 0.88, 0.12 \rangle$	$\langle 0.88, 0.12 \rangle$	$\langle 0.89, 0.11 \rangle$
$C_5 \leftrightarrow C_1$	$\langle 0.73, 0.27 \rangle$	$\langle 0.87, 0.13 \rangle$	$\langle 0.88, 0.12 \rangle$	$\langle 0.88, 0.12 \rangle$	$\langle 0.89, 0.11 \rangle$	$\langle 0.90, 0.10 \rangle$
$C_6 \leftrightarrow C_1$	$\langle 0.60, 0.40 \rangle$	$\langle 0.45, 0.55 \rangle$	$\langle 0.42, 0.58 \rangle$	$\langle 0.43, 0.57 \rangle$	$\langle 0.39, 0.61 \rangle$	$\langle 0.38, 0.62 \rangle$
$C_3 \leftrightarrow C_2$	$\langle 1.00, 0.00 \rangle$	$\langle 1.00, 0.00 \rangle$	$\langle 1.00, 0.00 \rangle$	$\langle 1.00, 0.00 \rangle$	$\langle 0.99, 0.01 \rangle$	$\langle 1.00, 0.00 \rangle$
$C_4 \leftrightarrow C_2$	$\langle 0.07, 0.93 \rangle$	$\langle 0.27, 0.73 \rangle$	$\langle 0.41, 0.59 \rangle$	$\langle 0.40, 0.60 \rangle$	$\langle 0.37, 0.63 \rangle$	$\langle 0.38, 0.62 \rangle$
$C_5 \leftrightarrow C_2$	$\langle 0.20, 0.80 \rangle$	$\langle 0.33, 0.67 \rangle$	$\langle 0.43, 0.57 \rangle$	$\langle 0.41, 0.59 \rangle$	$\langle 0.38, 0.62 \rangle$	$\langle 0.39, 0.61 \rangle$
$C_6 \leftrightarrow C_2$	$\langle 0.73, 0.27 \rangle$	$\langle 0.78, 0.22 \rangle$	$\langle 0.72, 0.28 \rangle$	$\langle 0.71, 0.29 \rangle$	$\langle 0.69, 0.31 \rangle$	$\langle 0.68, 0.32 \rangle$
$C_4 \leftrightarrow C_3$	$\langle 0.07, 0.93 \rangle$	$\langle 0.27, 0.73 \rangle$	$\langle 0.41, 0.59 \rangle$	$\langle 0.40, 0.60 \rangle$	$\langle 0.37, 0.63 \rangle$	$\langle 0.38, 0.62 \rangle$
$C_5 \leftrightarrow C_3$	$\langle 0.20, 0.80 \rangle$	$\langle 0.33, 0.67 \rangle$	$\langle 0.43, 0.57 \rangle$	$\langle 0.41, 0.59 \rangle$	$\langle 0.37, 0.63 \rangle$	$\langle 0.39, 0.61 \rangle$
$C_6 \leftrightarrow C_3$	$\langle 0.73, 0.27 \rangle$	$\langle 0.78, 0.22 \rangle$	$\langle 0.72, 0.28 \rangle$	$\langle 0.72, 0.28 \rangle$	$\langle 0.70, 0.30 \rangle$	$\langle 0.68, 0.32 \rangle$
$C_5 \leftrightarrow C_4$	$\langle 0.87, 0.13 \rangle$	$\langle 0.95, 0.05 \rangle$	$\langle 0.98, 0.02 \rangle$	$\langle 0.98, 0.02 \rangle$	$\langle 0.99, 0.01 \rangle$	$\langle 0.99, 0.01 \rangle$
$C_6 \leftrightarrow C_4$	$\langle 0.33, 0.67 \rangle$	$\langle 0.31, 0.69 \rangle$	$\langle 0.31, 0.69 \rangle$	$\langle 0.31, 0.69 \rangle$	$\langle 0.29, 0.71 \rangle$	$\langle 0.29, 0.71 \rangle$
$C_6 \leftrightarrow C_5$	$\langle 0.33, 0.67 \rangle$	$\langle 0.33, 0.67 \rangle$	$\langle 0.31, 0.69 \rangle$	$\langle 0.31, 0.69 \rangle$	$\langle 0.29, 0.71 \rangle$	$\langle 0.29, 0.71 \rangle$

Table 3 Consonance and dissonance scale [6]

Value of $\mu_{C,C'}$	Meaning
[0–0.5]	Strong negative consonance (SND)
(0.5–0.15]	Negative consonance (ND)
(0.15–0.25]	Weak negative consonance (WND)
(0.25–0.33]	Weak dissonance (WD)
(0.33–0.43]	Dissonance (D)
(0.43–0.57]	Strong dissonance (SD)
(0.57–0.67]	Dissonance (D)
(0.67–0.75]	Weak dissonance (WD)
(0.75–0.85]	Weak positive consonance (WPC)
(0.85–0.95]	Positive consonance (PC)
(0.95–1.00]	Strong positive consonance (SPC)

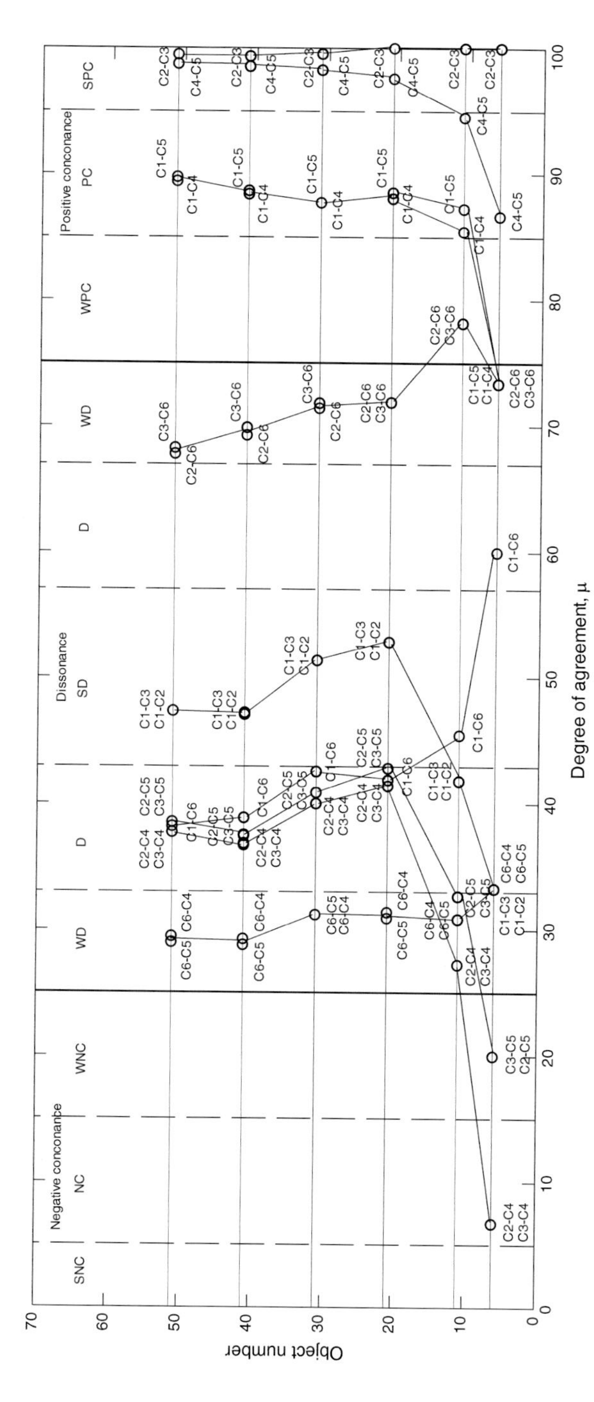

Fig. 1 Degree of "agreement" (μ) variation on the consonance and dissonance scale

the objective criteria. The estimates of 0.49 lead to higher model accuracy. The situation is the same with the model parameter k_S. As could be seen in Table 2, the parameters μ_{max} and k_S (criteria C_4 and C_5) are in SPC with $\mu = 0.99$. This means that the two parameters are not completely independent, in fact they "draw" each other during the fitting procedure as also discussed in [21]. Changing μ_{max} in such a procedure will also immediately lead to a small adjustment of k_S, and they will not be varied independently, unlike what one would expect theoretically. The μ_{max}/k_S ratio referred to as specific affinity, bridges the kinetics of enzymatic substrate uptake and microbial growth [10, 19]. Any combination of the two parameters that results in the same μ_{max}/k_S ratio will fit equally well in the parameter estimation procedure. This is a practical problem of the parameter identifiability for growth models containing non-linearities of the type Michaelis-Menten [10, 13].

The obtained high correlation for criteria $C_2 \leftrightarrow C_3$ is obvious. With the increase of the value of *ggap* floating point operations exponentially increase, as well as, consequently, the resulting computation time T. The observed dependence between *ggap* and T for all 51 values of *ggap* and for every 30 algorithm runs is presented in Fig. 2. In Fig. 3 the influence of *ggap* on the value of the objective function J is presented. As the ICA results show ($\mu = 0.47$, i.e. SD), there is no clear relation between *ggap* and J ($C_1 \leftrightarrow C_3$). Some dependences or relations between *ggap* and J can be found at ICA when the rest of GA parameters are varied at the same precision (i.e. such fine variations of the values of *nind*, *ngen*, etc.).

According to the obtained results, the criteria pairs $C_1 \leftrightarrow C_2$, $C_1 \leftrightarrow C_3$, $C_4 \leftrightarrow C_6$ and $C_5 \leftrightarrow C_6$ are in WD or SD, i.e. it can be concluded that between them there are no dependences. For example, for the pair $C_1 \leftrightarrow C_2$ such behavior is explained by

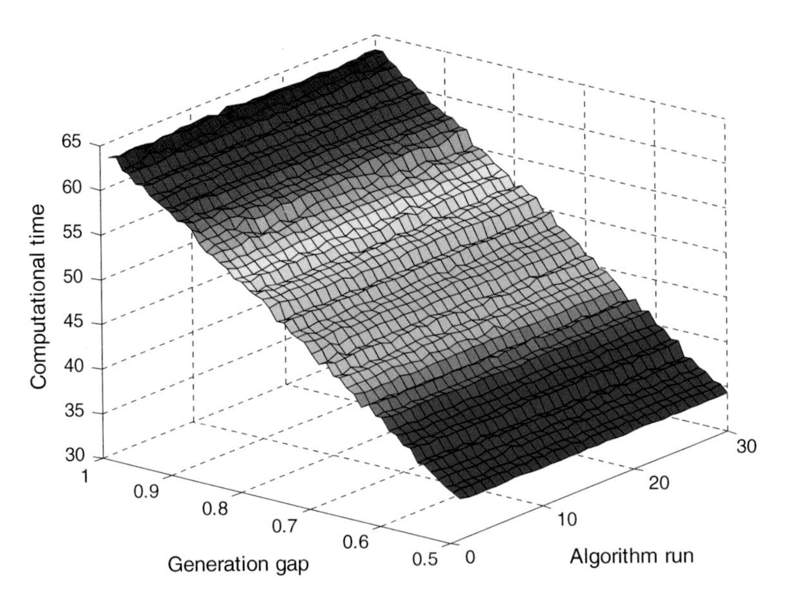

Fig. 2 Influence of *ggap* on the computational time T

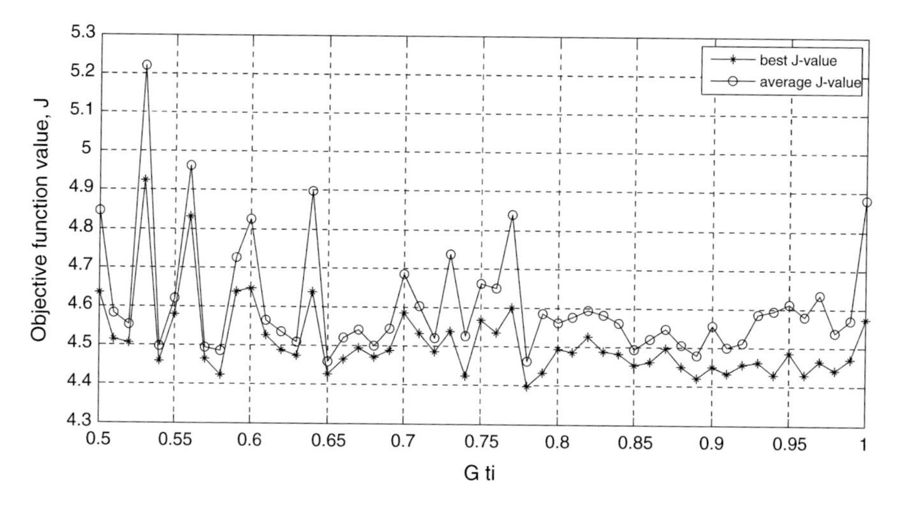

Fig. 3 Influence of *ggap* on the objective function value J

the stochastic nature of the GAs. If the algorithm is ran for a short running time, it is very likely that only bad solutions will be achieved. At the same time, long running time cannot guarantee achieving good solutions.

The physical meaning of the model parameters [7, 19, 21] explains the obtained dissonance considering the pairs $C_4 \leftrightarrow C_6$ and $C_5 \leftrightarrow C_6$, as well as the SPC considering pair $C_4 \leftrightarrow C_5$.

If the results discussed here (Table 2) are compared to those presented in [28], it can be seen that there are large discrepancies between them. This is due to the reported very high values of the "uncertainty" π. For example, for the criteria pair $C_4 \leftrightarrow C_5$ the π-value is $\pi = 0.8$ [28]. Therefore, such results cannot be regarded as credible. Although in the results in [28] higher values of π are observed and a small number of objects are considered, there are a few close matches. Similar estimates of μ-values for the pairs $C_1 \leftrightarrow C_6$, $C_2 \leftrightarrow C_4$, $C_2 \leftrightarrow C_5$ and $C_4 \leftrightarrow C_6$ have been obtained. This comparison shows once more that the number of objects in the ICA is essential for the accuracy of the obtained results.

5 Conclusion

The aim of this paper is to explore the idea of applying the recently proposed InterCriteria Analysis for establishing certain relations, considering parameters identification of an *E. coli* fed-batch cultivation model by use of Genetic algorithms. The investigation is particularly focused on the relations between the model parameters μ_{max}, k_S and $Y_{X/S}$, on one hand, and the GA parameter *ggap*, the convergence time and the model accuracy, on the other hand.

Microbial growth kinetics, i.e., the relationship between the specific growth rate (μ) of a microbial population and the substrate concentration (S), is an indispensable tool in all fields of microbiology, and a key step in cultivation processes control and optimization. Therefore, the ICA of the existing correlations between the studied kinetic and stoichiometric parameters is an important investigation.

The dependences or independences obtained from ICA can be explaned both by the physical meaning of the considered model parameters and by the stochastic nature of the used meta-heuristic techniques—GAs. Moreover, the derived additional knowledge will be useful in further identification procedures of cultivation process models to obtain more accurate estimations and achieve better GAs performance.

Acknowledgments The work is supported by the Bulgarian National Scientific Fund under the grant DFNI-I-02-5 "InterCriteria Analysis—A New Approach to Decision Making".

References

1. Atanassov, K., Mavrov, D., Atanassova, V.: Intercriteria decision making: a new approach for multicriteria decision making, based on index matrices and intuitionistic fuzzy sets. Issues IFSs GNs **11**, 1–8 (2014)
2. Atanassov, K., Szmidt, E., Kacprzyk, J.: On intuitionistic fuzzy pairs. Notes IFS **19**(3), 1–13 (2013)
3. Atanassov, K.: On index matrices, part 1: standard cases. Adv. Stud. Contemp. Math. **20**(2), 291–302 (2010)
4. Atanassov, K.: On index matrices, part 2: intuitionistic fuzzy case. Proc. Jangjeon Math. Soc. **13**(2), 121–126 (2010)
5. Atanassov, K.: On Intuitionistic Fuzzy Sets Theory. Springer, Berlin (2012)
6. Atanassov, K., Atanassova, V., Gluhchev, G.: InterCriteria analysis: ideas and problems. Notes Intuitionistic Fuzzy Sets **21**(1), 81–88 (2015)
7. Bastin, G., Dochain, D.: On-line Estimation and Adaptive Control of Bioreactors. Elsevier Scientific Publications (1991)
8. Benjamin, K.K., Ammanuel, A.N., David, A., Benjamin, Y.K.: Genetic algorithm using for a batch fermentation process identification. J. Appl. Sci. **8**(12), 2272–2278 (2008)
9. Blount, Z.D.: The Unexhausted Potential of E. coli. eLife 4, 2015, doi:10.7554/eLife.05826
10. Button, D.K.: Differences between the kinetics of nutrient uptake by micro-organisms, growth and enzyme kinetics. Trends Biochem. Sci. **8**, 121–124 (1983)
11. Cantu-Paz, E.: Selection intensity in genetic algorithms with generation gaps. In: Proceedings of the Genetic and Evolutionary Computation Conference, pp. 911–918. Morgan Kaufmann, Las Vegas, Nevada, USA (2000)
12. Chen, Z.-Q., Wang, R.-L.: Two efficient real-coded genetic algorithms for real parameter optimization. Int. J. Innovative Comput. Inf. Control **7**(8), 4871–4883 (2011)
13. Contois, D.E.: Kinetics of bacterial growth: relationship between population density and specific growth rate of continuous culture. J. Gen. Microbiol **21**, 40–50 (1959)
14. De Jong, K.A.: An Analysis of the Behavior of a Class of Genetic Adaptive Systems. Doctoral Dissertation, University of Michigan, Ann Arbor, University Microfilms No. 76–9381 (1975)
15. Eiben, A.E., Hinterding, R., Michalewicz, Z.: Parameter control in evolutionary algorithms. IEEE Trans. Evol. Comput. **3**(2), 124141 (1999)
16. Goldberg, D.E.: Genetic Algorithms in Search. Optimization and Machine Learning. Addison Wesley Longman, London (1989)

17. Goldberg, D.E.: Real-coded genetic algorithms, virtual alphabets, and blocking. Complex. Syst. **5**, 139–167 (1991)
18. Huang, C., Lin, H., Yang, X.: Industrial production of recombinant therapeutics in escherichia coli and its recent advancements. J. Ind. Microbiol. Biotechnol. **39**, 383–399 (2012)
19. Kovarova-Kovar, K., Egli, T.: Growth kinetics of suspended microbial cells: from single-substrate-controlled growth to mixed-substrate kinetics. Microbiol. Mol. Biol. Rev. **62**(3), 646–666 (1998)
20. Radu, V.: Application. In: Radu, Vasile (ed.) Stochastic Modeling of Thermal Fatigue Crack Growth. ACM, vol. 1, pp. 63–70. Springer, Heidelberg (2015)
21. Lobry, J.R., Flandrois, J.P., Carret, G., Pave, A.: Monod's bacterial growth model revised. Bull. Math. Biol. **54**, 117–122 (1992)
22. Luo, Y.-Z., Tang, G.-J., Wang, Z.G., Li, H.Y.: Optimization of perturbed and constrained fuel-optimal iimpulsive rendezvous using a hybrid approach. Eng. Optim. **38**(8), 959–973 (2006)
23. Magalhaes-Mendes, J.: A comparative study of crossover operators for genetic algorithms to solve the job shop scheduling problem. WSEAS Trans. Comput. **12**(4), 164–173 (2013)
24. Mohideen, A.K., Saravanakumar, G., Valarmathi, K., Devaraj, D., Radhakrishnan, T.K.: Real-coded genetic algorithm for system identification and tuning of a modified model reference adaptive controller for a hybrid tank system. Appl. Math. Model. **37**, 3829–3847 (2013)
25. Monod, J.: Recherches sur la Croissance des Cultures Bacteriennes. Hermann et Cie, Paris (1942)
26. Nowotniak, R., Kucharski, J.: GPU-based tuning of quantum-inspired genetic algorithm for a combinatorial optimization problem. In: Proceedings of the XIV International Conference System Modeling and Control (2011)
27. Pencheva, T., Roeva, O., Hristozov, I.: Functional State Approach to Fermentation Processes Modelling. Prof. Marin Drinov Academic Publishing House, Sofia (2006)
28. Pencheva, T., Angelova, M., Atanassova, V., Roeva, O.: InterCriteria analysis of genetic algorithm parameters in parameter identification. Notes Intuitionistic Fuzzy Sets **21**(2), 99–110 (2015)
29. Roeva, O., Fidanova, S., Paprzycki, M.: Population size influence on the genetic and ant algorithms performance in case of cultivation process modeling. Recent Adv. Comput. Optim. Stud. Comput. Intell. **580**, 107–120 (2015)
30. Valarmathia, K., Devaraja, D., Radhakrishnanb, T.K.: Real-coded genetic algorithm for system identification and controller tuning. Appl. Math. Model. **33**(8), 3392–3401 (2009)
31. Zimmer, C.: Microcosm: E. coli and the New Science of Life. Pantheon Books, New York (2008)

Proposed CAEva Simulation Method for Evacuation of People from a Buildings on Fire

Jacek M. Czerniak, Łukasz Apiecionek, Hubert Zarzycki and Dawid Ewald

Abstract This paper presents practical applications of the cellular automata theory for building fire simulation using the CAEva method. Thanks to the tests carried out using appropriately configured program, realistic results of simulated evacuation of people from the building have been achieved. The paper includes the references to actual fire disasters and provides numbers of their resulting casualties. Using such a kind of predication in civil engineering should increase the fire safety of buildings. Simulations described in this paper seem to be very useful, particularly in case of building renovation or temporary unavailability of escape routes. Using them, it is possible to visualize potential hazards and to avoid increased risk in case of fire. Inappropriate operation of buildings, including insouciant planning of renovations are among frequent reasons of tragic accidents cited by fire brigade information services. Similar problems are encountered by inspectors who assess spontaneous fire accidents or arsons during mas events, where wrong safety procedures or inappropriate attempts to cut costs resulted in tragedy. Thanks to the proposed solutions it shall be easier to envisage consequences of problematic decisions causing temporary or permanent unavailability of escape routes. This is exactly the problem analyzed by this paper. It does not take into account, by the rule, the influence of CO_2 and other gases on evacuation difficulty. The described method has been analyzed using descriptions of real life fires, the participants of which were neither asleep nor asphyxiated with carbon monoxide, while the escape was hindered by fire, room

J.M. Czerniak (✉) · Ł. Apiecionek · D. Ewald
Institute of Technology, Casimir the Great University in Bydgoszcz,
ul. Chodkiewicza 30, 85-064 Bydgoszcz, Poland
e-mail: jczerniak@ukw.edu.pl

Ł. Apiecionek
e-mail: lapiecionek@ukw.edu.pl

D. Ewald
e-mail: dawidewald@ukw.edu.pl

H. Zarzycki
Wroclaw School of Applied Informatics "Horyzont", ul. Wejherowska 28,
54-239 Wroclaw, Poland
e-mail: hzarzycki@horyzont.eu

K.T. Atanassov et al. (eds.), *Novel Developments in Uncertainty Representation and Processing*, Advances in Intelligent Systems and Computing 401,
DOI 10.1007/978-3-319-26211-6_27

layout as well as stress and number of the event participants. The results achieved for such conditions are approximate to the actual (reallife) outcomes, which proved the method to be correct.

1 Introduction

Cellular automata are classified to one of IT branches, namely to artificial intelligence. They include a network of cells, each of which is characterized by some specific state and a set of rules. Change of a current state of a given cell is the outcome of the above mentioned properties and interrelations with the neighboring cells. The theory of cellular automata was first introduced by an American scientist of Hungarian origin, John von Neumann. He showed, among other things, that even simple machines are characterized by reproduction ability; that feature was previously regarded as a fundamental feature observed in living organisms [7–9, 16, 18]. For many years cellular automata had been subject to theoretical studies only. With the development of computers and software, optimizing methods based on that attitude have been more and more frequently studied and implemented in practice. Thanks to their versatility, cellular automata are applied in many real life fields, such as: biology, physics, mathematics and in different fields of IT, such as cryptography or computer graphic.

1.1 Application of Cellular Automata

Cellular automata have been applied in practice. One of the examples of such applications is the simulation of the traffic, where specifically defined cellular automaton controls the street traffic. The traffic is controlled basically at the specific segment of a given traffic intensity [1, 11]. This applies for example to traffic intensity control in highways of the Ruhr in Germany. The monitoring centers designed especially for that purpose collect data from selected sections of highways. Then the data is analyzed and used to prepare shorttime simulations of the traffic intensity using cellular automata [2, 3]. Web sites of that project include statistical information about performed studies on behavior of drivers who were prewarned about possible traffic problems [1, 3, 9] that might occur over several following hours [6, 12]. Demographic simulations for a given region are among other examples of cellular automata applications. The aim of such simulations is to generate the structure showing the size of population at a given area in a way to create a map of forecasted population density [4]. Simulations of this type can be based on the wellknown “Game of Life” [14]. This is possible, because following some modifications of the algorithm, it can count life occurring in observed cells. Implementations of other automata include image processing, generation of textures, simulation of waves, wind as well as the program CAEva (Cellular Automata simulation of Evacuation) developed also for

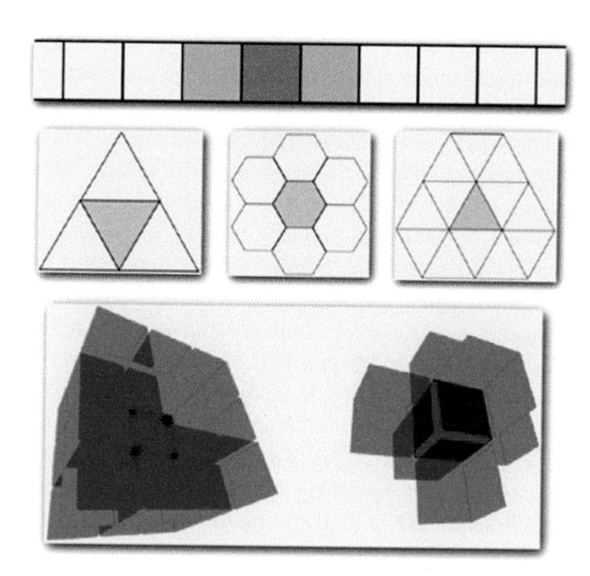

Fig. 1 Types of grids: 1D, 2D, 3D [*source* [5]]

the purposes of this study. The aim of the proposed algorithm is to simulate escape of people from the building on fire with a given number of exits and fire sources [5, 13, 15, 19].

1.2 The Grid of Cellular Automata

A grid or a discrete space, where cellular automata evolution takes place is totally built of identical cells. All the cells must be surrounded by the same number of neighbors and must be characterized by the same number of states. There are three structural factors which significantly influence the grid form and, as a consequence, the behavior of the entire cellular automaton [2, 9, 16]

- the size of a space depends on the magnitude of the studied problem, the examples of which are shown in Fig. 1 (grid 1D, 2D, 3D);
- regularity condition, which requires complete filling of the grid with identical cells;
- the number of neighbors (dependent on both the above factors).

2 Forecasting the Fire Hazard

Fire is an element against which man is often helpless, especially when the fire breaks out inside rooms. Thus the designs of residential, commercial or other public buildings must meet complicated firesafety requirements. Width of corridors, number of escape exits and permissible number of persons which can stay in a room at the same time significantly influences safety of people inside. Obviously, it is not sufficient to include escape exists in the plan, but the door must be unlocked too. Casualties in many fires were caused by locked emergency exits [10, 17, 19]. Recently, there were many disasters caused by fire in buildings, e.g. the fire of the hotel in Kamień Pomorski (PL) with 23 casualties (where emergency exit turned out to be locked) or the fire of hypermarket in Nowy Targ (PL), caused probably by welding works inside. Another example is the fire of a block of flats in Koluszki (PL), were one person died. When designing buildings, architects meet requirements of binding firesafety standards, but this is often insufficient to avoid a tragedy. Additional simulation studied could help to solve that problem. Moreover, even the best architectural design cannot prevent fire if a building is incorrectly operated. It is not rare to encounter renovation plans ignoring the fact that escape routes or some exits from a building would be temporarily unavailable during the renovation. From the statistical point of view, fire hazard increases during renovation works. The program CAEva is an implementation of the CAEva method, the pseudocode of which is shown below. It has been developed in order to test escape of people from a building as a result of fire hazard. It allows comparison of different simulation results and development of appropriate conclusions. The program has been implemented in the C++Builder environment, which is an objectoriented programming tool in Windows environment and is available free of charge at the AIRlab web site. Using the program it is possible to draw a board of any size including the plan of a singlestorey building, to locate people inside and to indicate the place of fire. The board consists of the grid of cells. Each cell can assume only one of the following states: fire, wall, person, person on fire or an empty cell. Figure 2 shows the diagram of states for a single cell in the fire simulation automaton.

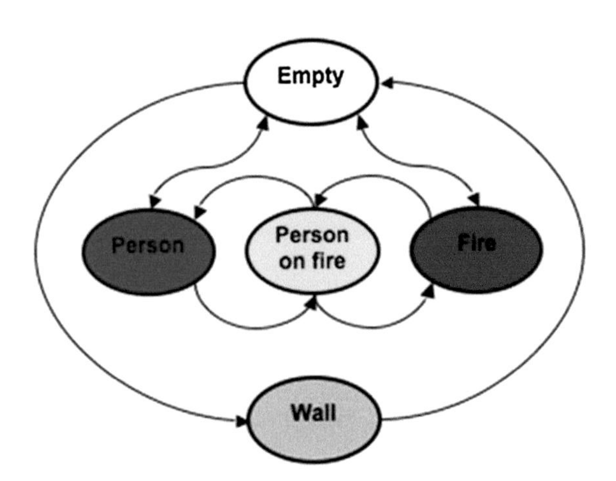

Fig. 2 Diagram of cell states

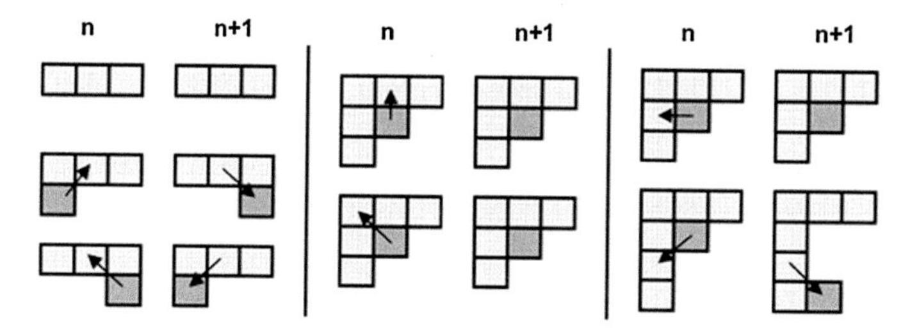

Fig. 3 Boundary conditions (rebound from the grid edges)

2.1 Boundary Conditions

The discrete space, where different evolution of cellular automata take place includes d-dimensional, theoretically infinite grid. As a grid of that type cannot be implemented into a computer application, it is represented in the form of a finite table. Thus it is necessary to set boundary conditions at the grid borders, i.e. at the table limits. The set of basic conditions is shown in Fig. 3. Those conditions are analogous for the rotation by 90°, so they were skipped as trivial ones.

The following rules were used for the simulation of the cell motion in the wall direction:

- straight motion—unchanged state of a cell,
- diagonal motion—state of a cell changes into empty one, angle of incidence equals the angle of rebound and, as a consequence, the state of a cell in the mirror image shall change into the state of the cell that initiated the motion,

motion conditions:

- motion is possible if a target cell is in the empty state. Otherwise the cell shall not change the state,
- the attempt of the motion of the cell in "person" state to the cell in "fire" state increases the number of burns of the initiating cell.

A special case is an attempt of the motion from the corner of the board. Rebound in three initiating directions does not change the state of a cell and the attempt of the motion in the remaining five directions may cause such change. It should be also noted that motion rules and conditions apply to the cells in the "person" state as well as in the "fire" state. The fields to which motion cannot be made are cells in "wall" state. Rebound conditions occur at the edge of the cellular automata grid, which constitutes a barrier from which moving virtual objects rebound (in visual sense). Those conditions are used to simulate encased empirical spaces. Figure 4 shows seven consecutive phases of cell generations visualizing rebound of objects from the grid edges.

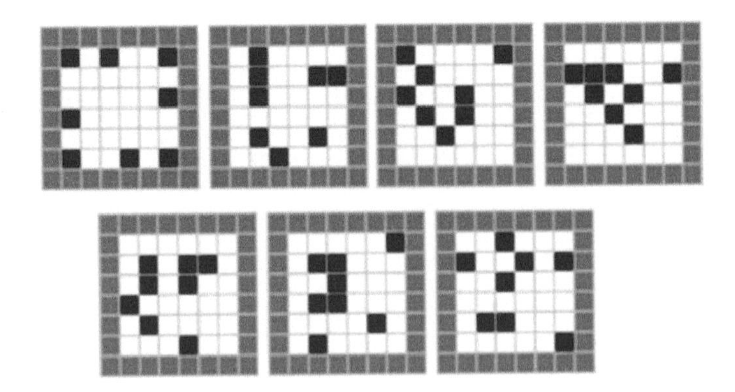

Fig. 4 Sample seven steps of the evolution

2.2 Transfer Function

Evolution of cellular automata takes place in discrete time determining consecutive processing cycles. Each discrete moment $t = \{0, 1, 2, n\}$ is used for updating the state of individual cells, thus each automaton is a dynamic object over time. In every iteration, the transfer function can process (calculate) all the cells in the grid one by one according to specific rules. Each processed cell receives its new state based on calculation of its current state and states of the neighboring cells. Transfer rules and the state space as well as defined neighborhood are inherent elements of the cellular automata evolution process.

Once executed, the program displays main screen ready to draw the building plan and to arrange individual elements inside. Once the board is drawn and all the components are arranged there, one can start configuration of fire and people parameters and setting of the group effect.

Fire parameters:

- fire goes out alone, if the number of neighbors is less than 1,
- fire goes out from overpopulation, if the number of neighbors is more than 3,
- new fire is generated when the number of neighbors is at least 3,
- but not more than 4.

Parameters of people:

- probability that a person goes towards the exit 50,
- number of burns resulting in death 5,
- group effect On/Off.

The Fig. 5 shows the result of the program after creating 50 generations of evolution. There are points in the screen simulating people escaping towards the exit and the propagating fire. All the events are recorded in the table of statistics. They include: number of people remaining within the board, saved from and died in fire or by crushing. That data shall be used to draw conclusions from experiments.

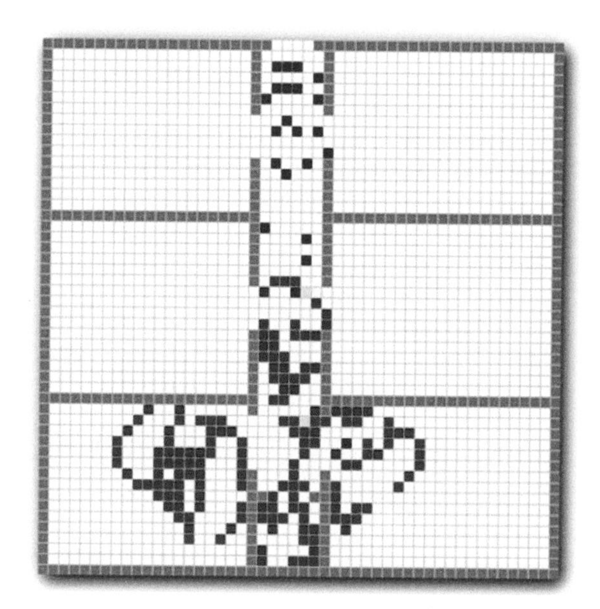

Fig. 5 The result of the program after creating 50 generations of evolution

The test including simulation of a building on fire was based on certain rules and relations. Setting of the following parameters, selection of versions and inherent rules altogether make up an environment influencing the mortality rate of people during fire of a building for the proposed simulations:

- layouts of the building floors, including the number and location of doors,
- distribution of defined number of people inside the building at specified places,
- setting the fire parameters:
  - fire goes out alone, if there are less than one neighbor,
  - fire goes out because of overpopulation, if there are more than 3 neighbors,
  - new fire is generated when there are at least 3 neighbors, but not more than 4.
- setting of the parameters for people (live cells):
  - number of burns resulting in death is by default set to 5,
- location of the fire source on the board:
  - specifying the probability, with which people go towards the exit (four options): 25, 50, 75, 100 %,
  - specifying whether people go towards the exit in groups (two options): with or without a group effect.

3 The Experiment

The authors assumed in the experiment that three boards are always used with the same arrangement of walls as well as location of fire and people, but with different number of doors (1 to 3). The board pl_1drzwi_ogień1 shown in Fig. 6 is provided with only one exit from the building and includes seven rooms, where 70 blue points simulating people were located and the fire was set using red points (lower left corner of the board). The remaining two boards pl_2drzwi_ogień1 and pl_3drzwi_ogień1 differ from each other in the number of doors. In the board pl_2drzwi_ogień1 there are two doors and in the board pl_3drzwi_ogień1 there are three doors (Fig. 6).

The results of performed experiments using CAEva program as regards the behavior of people at the moment of fire outbreak in building are presented in Table 1. Results of the experiments have been classified considering the group effect and probability with which people go towards the emergency exit. 8 tests were performed for each board and obtained numerical results of the tests concerned people, who:

- died in the fire,
- were crushed in the crowd,
- were saved from the fire.

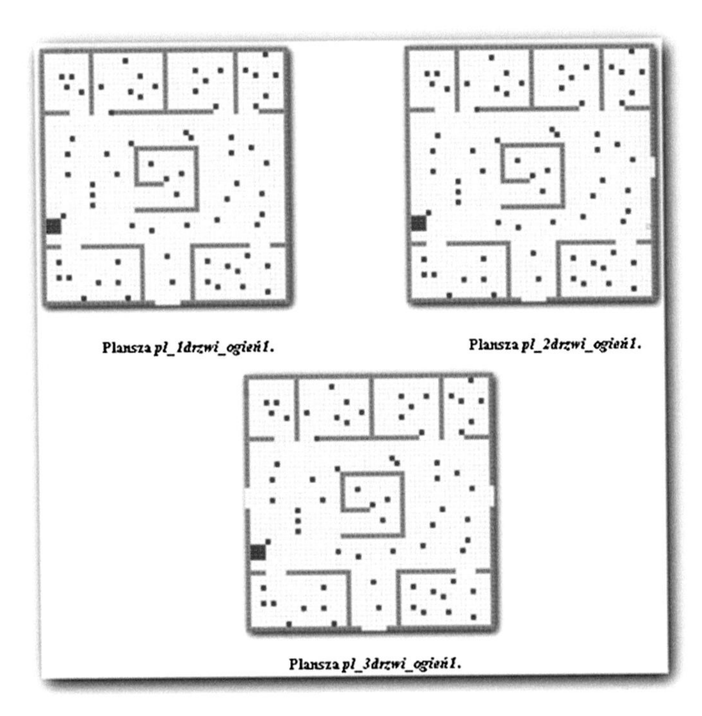

Fig. 6 Sample boards of the building

Table 1 Mortality rate of people as a result of fire

Building plans	Group effect					
	Yes			No		
	Probability that people go towards the exit			Probability that people go towards the exit		
	25 %	50 %	75 %	25 %	50 %	75 %
	Number of people die/crushed/saved from the fire					
bd_1door_fire1	64/0/6	47/0/23	9/3/58	65/0/5	38/0/32	11/0/59
bd_2doors_fire1	38/0/32	24/0/46	7/0/63	40/0/30	21/0/49	5/0/65
bd_3doors_fire1	43/1/26	24/0/46	9/0/61	39/1/30	18/0/52	11/0/59
bd_1door_fire2	57/0/13	24/2/44	3/0/67	58/0/12	16/0/54	2/0/68
bd_2doors_fire2	40/0/30	3/0/67	1/0/69	31/0/39	10/0/60	0/0/70
bd_3doors_fire2	28/0/42	1/0/69	0/0/70	23/0/47	5/0/65	0/0/70
bd_1door_fire3	59/0/11	25/2/43	1/2/67	56/0/14	16/1/53	1/0/69
bd_2doors_fire3	50/0/20	12/1/57	7/0/63	53/0/17	13/0/57	7/0/63
bd_3doors_fire3	42/0/28	3/0/66	0/0/70	41/0/29	5/0/65	0/0/70

The mortality rate depends on the place of the fire outbreak. If the fire blocks any room, then people staying there are not able to escape and to reach the exit even if they go towards the exit with 100 % probability. The group effect used in the program does not necessarily help in escape of people from a building. It can cause crowd as people are looking for other people to form groups and thus crushes can occur. When a person does not have any direction when he/she could move he/she is crushed. Figures 7 and 8 show the number of saved people who went towards the exit with the probability of 50 and 75 %.

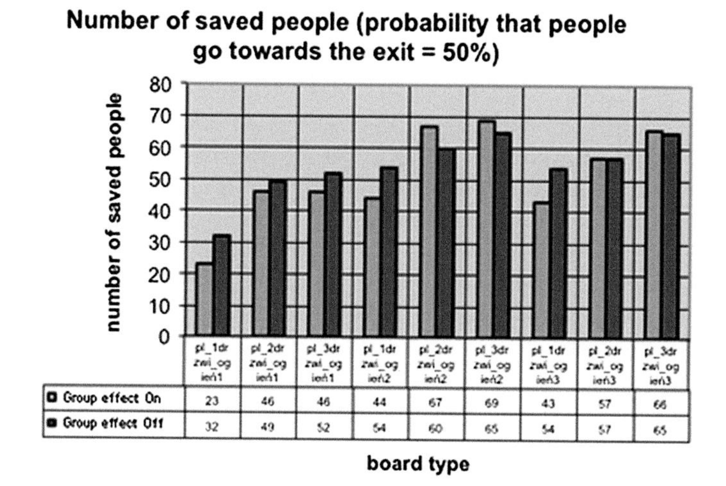

Fig. 7 Probability that people go towards the exit = 50 %

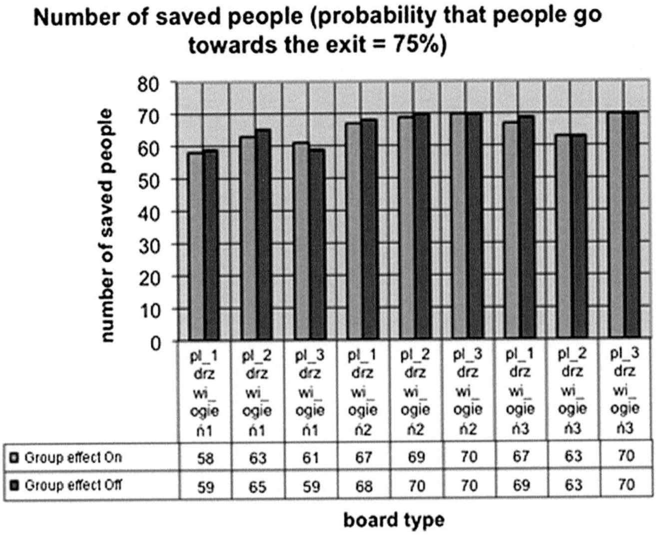

Number of saved people (probability that people go towards the exit = 75%)

board type	pl_1 drz wi_ogień1	pl_2 drz wi_ogień1	pl_3 drz wi_ogień1	pl_1 drz wi_ogień2	pl_2 drz wi_ogień2	pl_3 drz wi_ogień2	pl_1 drz wi_ogień3	pl_2 drz wi_ogień3	pl_3 drz wi_ogień3
Group effect On	58	63	61	67	69	70	67	63	70
Group effect Off	59	65	59	68	70	70	69	63	70

Fig. 8 Probability that people go towards the exit = 75 %

4 Conclusions

It is extremely difficult to simulate real fire inside a building. Behavior of people during fire can be stochastic and unpredictable. Authors of this study managed to present simulation of the escape of people from a building by means of cellular automata, the implementation of which was used in the study.

Following appropriate configuration of the program using the probability with which a person goes towards exit, setting the fire parameters and selecting proper option for the group effect one can draw the following conclusions:

Number of people saved from fire thanks to the group effect is comparable to the result without the group effect. The results differ depending on the type of the board and location of fire, but they are essentially very similar to each other.

When group effect is used in the program, the number of people who die as a result of crushing is larger than when no group effect is used. This happens when a person is not able to move in any direction. This is due to the fact that simulated individuals gathering in groups create areas of high density which results in death as a consequence of crushing.

In the implementation used for the experiments, people who go towards the exit way with 100 % probability are most likely to survive. To be realistic about the observations of people escaping from a building on fire, about their stress and the constantly increasing fire intensity, more probable value of probability with which they escape shall optimally be below 75 %. Hindrances that affect the decision making process during evacuation include, among others, limited visibility smoke resulting from combustion of flammable materials, high temperature and toxic gases. It is obvious here that no person would be able to pass the shortest way 100 % probability under such conditions during evacuation.

Number of doors on the board of the program is of high importance for the escape route. The more emergency exits the higher chance for people inside a building to escape and save.

Simulations of the presented experiments confirm the thesis that insouciant or unlawful blocking of escape routes inside buildings may result in tragic consequences at each stage of the building operation. Personnel responsible for fire safety and structural safety inspections may apply such tools to justify their decisions that sometimes could seem too strict. To make the simulation even more realistic, it is worth considering the option of automatic change of the parameter of the program related to the probability with which a person goes towards an exit. It is commonly known fact that the analysis of underlying causes and conditions of disasters show that the probability of survival decreases with the passing of time. Thus future experiments should take into account this fact.

References

1. Anonymous: expansion for the motorway network (2008, 2009). http://www.autobahn.nrw.de/
2. Burzyński, M., Cudny, W., Kosiński, W.: Traffic flow simulation cellular automata with fuzzy rules approach. In: Rutkowski L., Kacprzyk J., P.V.H. (ed.) Advances in Soft Computing, Proceedings of the Sixth International Conference on Neural Network and Soft Computing. pp. 808–813. Zakopane, Poland (June 11–15 2003)
3. Chowdhury, D., Santen, L., Schadschneider, A.: Statistical physics of vehicular traffic and some related systems. Phys. Rep. **329** (2000)
4. Czerniak, J.M., Apiecionek, Ł., Zarzycki, H.: Application of ordered fuzzy numbers in a new ofnant algorithm based on ant colony optimization. In: Beyond Databases, Architectures, and Structures, pp. 259–270. Springer International Publishing (2014)
5. D'Orazio, M., Spalazzi, L., Quagliarini, E., Bernardini, G.: Multi-agent simulation model for evacuation of care homes and hospitals for elderly and people with disabilities in motion. In: Ambient Assisted Living, pp. 197–204. Springer (2014)
6. Gardner, M.: Mathematical games. Sci. Am. **223**(4), 120–123 (1970)
7. Lee, U., Gerla, M.: A survey of urban vehicular sensing platforms. Comput. Netw. **54** (2010)
8. Lo, S., Hsu, C.: Cellular automata simulation for mixed manual and automated control traffic. Math. Comput. Model. **51** (2010)
9. Maerivoet, S., De Moor, B.: Cellular automata models of road traffic. Phys. Rep. **419** (2005)
10. Mikołajewska, E., Mikołajewski, D.: Neuroprostheses for increasing disabled patients' mobility and control. Adv. Clin. Exp. Med. **21**(2), 263–272 (2012)
11. Nagel, K., Schreckenberg, M.: A cellular automaton model for freeway traffic. J. Physique **2**, 2221–2229 (1992)
12. Płaczek, B.: Performance evaluation of road traffic control using a fuzzy cellular model. Lecture Notes in Artificial Intelligence, vol. 6679 (2011)
13. Ronchi, E., Nilsson, D.: Fire evacuation in high-rise buildings: a review of human behaviour and modelling research. Fire Sci. Rev. **2**(1), 1–21 (2013)
14. Schadschneider, A.: Statistical physics of traffic flow. Phys. A **285**(101) (2000)
15. Tam, L., Lao, S., Choi, H.: Numerical simulation of atrium fire using two cfd tools. In: Computational Methods in Engineering & Science: Proceedings of Enhancement and Promotion of Computational Methods in Engineering and Science X", 21–23 Aug 2006, Sanya, China. Springer Science & Business Media (2007)

16. Terrier, V.: Two-dimensional cellular automata and their neighborhoods. Theoret. Comput. Sci. **312** (2004)
17. Tofilo, P., Cisek, M., Lacki, K.: The study on the effects of the counter-flow on the evacuation of people from tall buildings. In: Pedestrian and Evacuation Dynamics 2012, pp. 509–520. Springer (2014)
18. Wolfram, S.A.: A New Kind of Science. Wolfram Media, pp. 64–157 (2002)
19. Yang, G.S., Peng, L.M., Zhang, J.H., An, Y.l.: Simulation of people's evacuation in tunnel fire. J. Central S. Univ. Technol. **13**(3), 307–312 (2006)

The CutMAG as a New Hybrid Method for Multi-edge Grinder Design Optimisation

Jacek M. Czerniak, Marek Macko and Dawid Ewald

Abstract This article is a part of the series dedicated to AI Methods Inspired by Nature and their implementation in the mechatronic systems. The CutMAG algorithm uses hybrid approach to optimisation, i.e. a combination of classic genetic algorithms (GA) with morphologic optimisation (M) thus creating innovative approach to optimisation of cutting disk design (Cut) for the multi-edge grinder. The input data include population of individuals. Each individual is represented by a set of cutting disks. Whereas the fitness function was assumed as a combination of several postulates of the mechanical design foundations. The method includes mechanical, design and energy aspects. Each individual constitutes a complete solution of the disk set whereas the population represents the entire class of solutions. The fitness function of an individual is calculated as the average fitness of each disk supplemented by information describing the relationship between both adjacent disks. The method for calculation of function values was selected so as to ensure its maximisation in the process of evolution. Although promising results of the genetic algorithms operation were achieved, one can consider further improvement of the method efficiency. The authors used morphological operations in order to better adopt the method to the task.

1 Introduction

Grinding is one of the key issues in polymer plastic processing and recycling as well as in the food, wood and chemical industries and the grinding method determines grinding product quality and energy consumption during the process. Many

J.M. Czerniak (✉) · D. Ewald
Casimir the Great University in Bydgoszcz, Institute of Technology,
ul. Chodkiewicza 30, 85-064 Bydgoszcz, Poland
e-mail: jczerniak@ukw.edu.pl

M. Macko
Institute of Mechanics and Applied Computer Science,
Casimir the Great University in Bydgoszcz, Bydgoszcz, Poland
e-mail: mackomar@ukw.edu.pl

K.T. Atanassov et al. (eds.), *Novel Developments in Uncertainty Representation and Processing*, Advances in Intelligent Systems and Computing 401,
DOI 10.1007/978-3-319-26211-6_28

researchers have been trying to improve the design of grinders and to increase the process efficiency [1, 4, 6, 8, 10, 11, 17, 20, 22, 24, 25]. The efficiency is influenced by innovation of the design and technology of working components as well as by the type of the material, shape of grinded parts and expected average size (geometric form) of the ground particles. It is possible to identify, implement and use the most advantageous conditions thanks to thorough analysis of energy estimators for selected design solutions and proposed universal system based on correct rules of knowledge codification. The purpose of plastic grinding is, on one hand, to reduce average size of ground parts, and on the other hand, to achieve material with specific properties influencing its further processing, i.e. mainly the process of introducing plastic into plasticizing systems of moulding machines and extruding presses. Appropriate form, size and roughness of particles results first of all from the design of the grinding assembly. Apart from the size, also qualitative characteristics of the granulated product and its form are important. The ground product should have regular, usually cylindrical (but also cubical or cubicoid etc.) shape. Results obtained from the numerical analysis are more and more used to find efficient design solutions over recent years [18, 23]. It seems to be very popular now to create 3D CAD models which are basis for preparation of detailed documentation as well as kinematical and strength analysis using finite elements method (FEM) and artificial intelligence (AI) [7, 26]. Based on the analysis of the available Polish literature on the subject (such as Brożek M., Ciesielska D., Chwiej M., Drzymała Z., Flizikowski J., Hawrylak H., Heim A., Konieczka R., Mielczarek E., Opielak M., Otwinowski H., Rokach I., Sidor J., Sikora R., Siwiec A., Sokołowski M., Tumidajski T., Zawada J. and others] as well as foreign literature [such as Armarego E.J.A., Bauer W., Berger B.S., Csöke B., Fang Q., Grellmann W., Justin A. Gantt, Melkote S.N., Mellor S.H., Pahl M.H., Pasikatan M.C., Pwu H.Y., Richman M.W., Salman A.D., Shamoto E., Schubert W., Wanibe Y., Weichert R., Yan J., Strenkowski S.J., Schubert G., Peukert W.) one can conclude that this subject has not been systematically studied so far as regards grinding models in view of further process ability of polymer materials, with AI implementations. Existing, but distributed data bases concerning design solutions of the grinders and their working characteristics for specific group of plastics as well as descriptions of the ground product properties are also only the basis for systematizing the methods of identification and assessment of the grinding quality. Also the professional literature on ant colony optimization and other selected metaheuristic methods using AI available to the authors (Dorigo M., Stützle T., Bersini H., diCaro G., Corne D., Gloger F., Siarry P., Fogel D., Michalewicz Z. et al.) [20, 22, 24, 25] do not include proven studies on application of those methods to grinder design optimization. Nevertheless, there are known engineering applications using genetic algorithms [5, 9, 13, 15, 16, 19, 21]. That potential gap seems to be a very interesting area of innovative research conducted at the meeting of two disciplines: mechanical engineering and machine operation, and artificial intelligence.

2 The Design of the Multi-edge Grinder

Thanks to proper geometry of holes in their disks, drums or strips as well as due to appropriate relationship of the movement between neighbouring edges, multiedge grinders are able to grind plastic using the neighbouring edges [1, 11, 12, 14, 17]. Similarly as for knife grinders, the main operation of multiedge grinders is cutting. Cutting takes place as a result of mating between the rotating knife (disk or drum) and the fixed one (installed in the grinder housing) (Fig. 1). During the cutting process, a plastic part is supported on the tool rake surface of the fixed knife. The design problem here consists in appropriate shape of the working space of the machine (grinder) to achieve required form of the ground product with minimum energy consumption and maximum grinding quality. Grinding process is characterised by the parameter of the unit energy consumption as the measure of energy needed to grind 1 kg of plastic and it is associated with the process efficiency issue.

To identify and assess properties of grinding products based on selected multiedge grinding technologies in recycling, the following design solutions (from the allowable set) were assumed: which maximize (efficiency, output, degree of fineness) or minimize (power demand, unit energy consumption, energy dissipation, torque, angular, linear and rotational speed) values of selected operational characteristics [12, 17]. Like for other complex technological processes, the course of grinding depends on many factors that can be classified into system and design related factors (associated with grinding assembly and its equipment) and process related factors. The first group of factors includes: applied grinding system (of periodical, continuous or periodically-cyclic type), number of grinding machines and their grouping, the system of connections between grinders and plasticizing equipment as well as characteristics of applied equipment (sort, type, peripheral speed, grinding components and their design properties etc.) [2, 3]. Integrated system allows to determine permissible range of variables (design properties) of the grinder in laboratory circumstances thanks to used computer aided experiment, design and operation of grinders [1, 11, 14]. To allow objective assessment of the efficiency, authors used the measuring system that features recording of momentary values of torque and momentary values of rotational speed of the drive shaft. The test station was designed to allow replacement of the working assembly (knife-, disk- or beater system) as well as change of orientation (from horizontal to vertical). The solution presented here

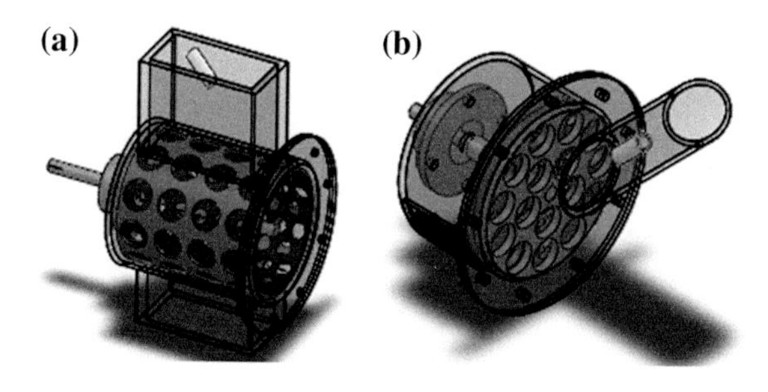

Fig. 1 The overview of design solutions used in multiedge grinders—of drum type (**a**) and disk type (**b**)

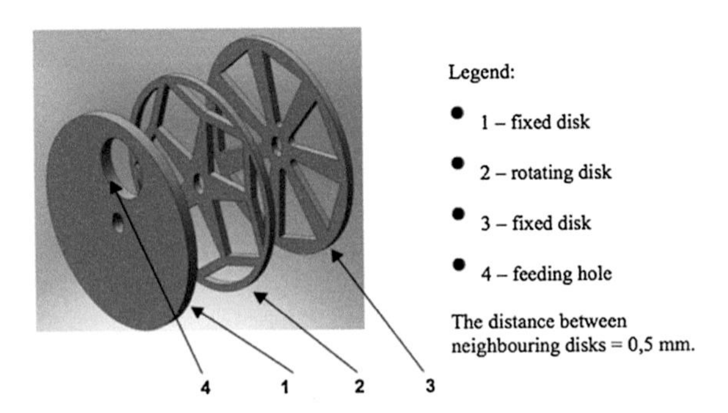

Fig. 2 The layout of the laboratory grinder disks

uses one of the variants with three disks installed in the grinder chamber. Disk no. 2 is mounted on the rotating shaft whereas disks 1 and 3 are fixed (Fig. 2).

Based on geometric analysis performed using CAD applications, disks with the following hole geometry and layout were proposed.

3 The Research Methodology

The following assumptions were made in the research methodology used for multi-edge grinders:

1. The multi-edge assembly is a set of disks with holes made over coaxial perimeters. The disks are made of steel plates and are brought into rotation by the system of bushes and couplings permanently fixed with them.
2. The disks move relative to each other with constant angular speed (gradient) and angular speed of each disk is different but constant.
3. According to available literature and own research of the authors, it was noted that grinding using the multi-edge grinder provides potential possibility to improve operational characteristics of the process compared to other disintegration methods thanks to introduction of design modification of the working assembly. The solutions concern the area of expertise and description of operational characteristics of grinding process from an independent range of design properties (C_k):

   - number of disks (l_t),
   - number of disk holes (l_{otw}),
   - number of hole rows in the first and in the remaining disks (l_{rz}),
   - the gap between adjacent disks (s_{i-k}),
   - hole diameters and geometry (d_{otw}),
   - disk holes pitch diameters (D_{ro}),
   - angles of cutting edges (β_{ij}),
   - grinding speed (v_{ri}).

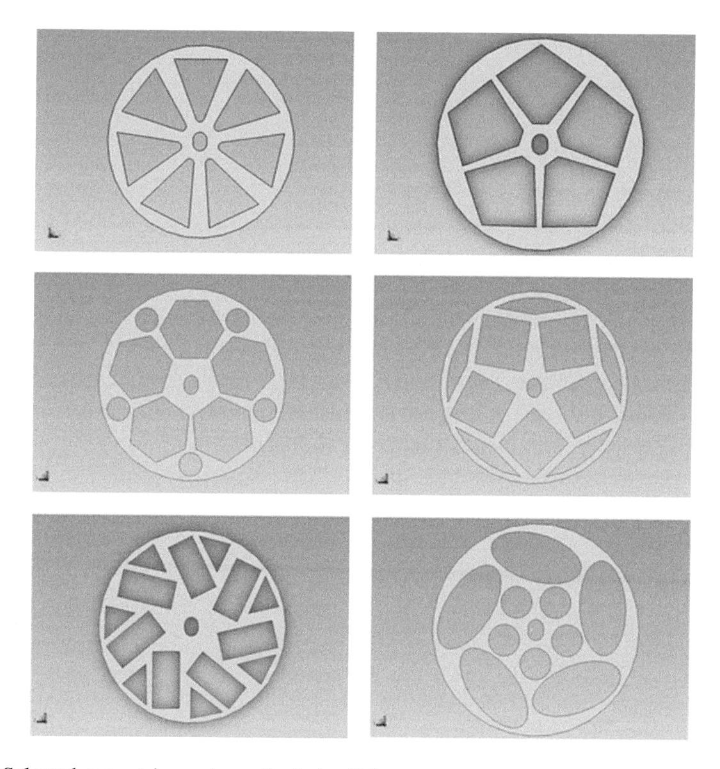

Fig. 3 Selected geometric versions of grinder disks

3.1 Implementation of CutMAG Algorithm

The algorithm uses hybrid approach to optimisation, i.e. a combination of classic genetic algorithms (GA) with morphologic optimisation (M) thus creating innovative approach to optimisation of cutting disk design (Cut) for the multi-edge grinder. The input data include population of individuals. Each individual is represented by a set of cutting disks. The fitness function of an individual is calculated as the fitness average of each disk supplemented by information describing the relationship between both disks. The method for calculating function values was selected to ensure its maximisation in the process of evolution (Fig. 3).

$$F(t) = F_{T1}(t) + F_{T2}(t) + F_{T1T2}(t)$$
$$F_{Tx} = F_{PW}(t) + F_{MIN}(t) + F_{MINO}(t) \qquad (1)$$

where,

F(t)—resultant fitness function,
F_{T1}(t), F_{T2}(t)—fitness functions for a particular disk,
F_{T1T2}(t)—fitness functions for a set of disks,

F_{PW}(t)—the function of free area (holes),
F_{MIN}(t)—function of the minimum thickness of the material,
F_{MINO}(t)—the function of the minimum hole.

The modified roulette wheel method was used as a way of selection. This will allow for further improvement of the algorithm efficiency in the future by using more optimum methods. Crossing and mutation operators are described in the paragraphs below. The algorithm is terminated when the population includes the individual with the fitness function value not lower than the assumed one. If there is no such an individual and the algorithm stops at a relative extreme, the algorithm will be automatically terminated when 10,000 populations are exceeded. It is also worth mentioning that the proposed approach meets J.H. Hollan's postulate and represents classic algorithm without memory, characterized by overwriting (replacement) of the parent population with the population of children.

Base structures of the algorithm The cutting disk design optimisation algorithm is a genetic algorithm. Compared to a classic genetic algorithm, this algorithm differs in the block of "Morphologic optimisation". Morphological transformations mentioned there are slightly similar to filters. However, an image element is not always modified, but only if a specified condition is satisfied. Such a conditional modification applicable only to those image points the surrounding of which complies with the structural element makes it possible to plan transformations very precisely.

Classic morphological transformations include dilatation, erosion and their combination, i.e. opening and closing. Those transformations turned out to be very useful in solving the cutting disk problem, as some disks received numerous small holes as a result of crossing and mutation operations. Although such holes can be machined using modern laser machine tools, but usefulness of that seems to be doubtful.

The distribution of results obtained from the selection of individuals should correspond to actual distribution occurring in the reality. So overmuch determinism is not desirable here. Although the selection process is random, it is conducted so that individuals with the highest value of the fitness function were most likely to be selected for reproduction. This method has already been mentioned in the introduction to this paragraph, but as it constitutes the base for modification used in the CutMAG algorithm, it seems to be advisable to provide more details. Consecutive phases of that selection method are as follows: calculating the fitness function $eval(v_i)$ for each chromosome v_i, where $i \in [1, max\_pop]$ calculating total fitness of the population

$$F = \sum_{i=1}^{max\_pop} eval(v_i), \tag{2}$$

calculating the probability of selection p_i for each chromosome v_i, where $i \in [1, max\_pop]$ calculating total fitness of the population

$$\forall_{i \in [1, max\_pop]}, p_i = \frac{eval(v_i)}{F} \tag{3}$$

calculating the cumulative distribution function q_i for each chromosome v_i, where $i \in [1, max\_pop]$

$$q_i = \sum_{j=1}^{i} p_i \tag{4}$$

generating a random real number r from the range [0, 1] if $r < q_1$, then the chromosome vi should be selected; otherwise select the chromosome v_i, where $i \in [2, max\_pop]$, for which $q_{i-1} < r \leq qi$. The figure below shows the visualisation of the roulette wheel method. As it may be seen, the name of the method seems to stem from the scheme described in the literature. Individuals are searched for in the roulette wheel calibrated proportionally to the fitness factors achieved by individual chromosomes.

Below we presented modified form of the classic roulette wheel method. Calculating the fitness function strength(v_i) (maximal values of strength were disks or drums will not be damaged) for each individual v_i, where $i \in [1, max\_pop]$, calculating the sum of all fitness function values

$$F = \sum_{i=1}^{max\_pop} strength(v_i), \tag{5}$$

improvement of the pseudocode generator properties by rescaling F = F*100, sorting individuals in the ascending order with regard to fitness function

$$\exists_{Z(\Psi)} \left\{ \begin{array}{c} Z(\Psi) = \Phi, where\,(\Psi = (v_i, .., v_k, .., v_j)) \wedge \\ \left[\begin{array}{c} (\Phi = \Psi) \wedge \\ (strength(v_1) \leq \cdot\cdot \leq strength(v_i)) \end{array} \right] \end{array} \right\} \tag{6}$$

random selection of an integral number from the range [0, F]

$$\exists_{r \in C}[(0 \leq r) \wedge (r \leq F)] \tag{7}$$

selection of the first individual that fulfils the relationship below

$$\exists_{i \in [1, max\_pop]} \sum_{i=1}^{max\_pop} [(S = S + strength(v_i)), \\ ((r \leq S * 100) \Rightarrow (i = max\_pop))] \tag{8}$$

Selection of chromosomes designed for reproduction is done according to modified method of the roulette wheel. It is also worth mentioning that the situation resembles the X-axis with a scale. Starting from zero in the direction of the sum, the system checks whether the previously randomly selected number is located within the range (Fig. 4). If it exceeds the range, the searching is continued. Otherwise it

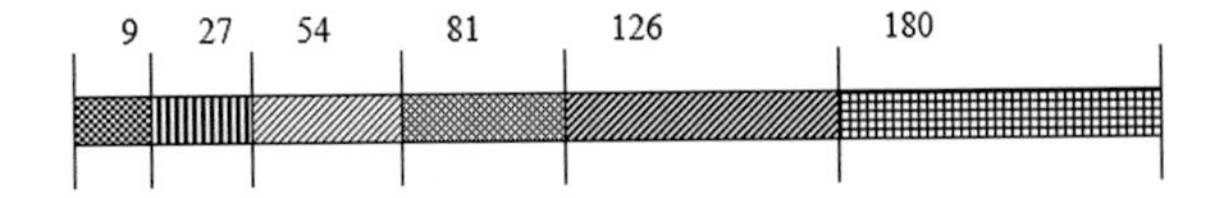

Fig. 4 Visualisation of the modified roulette wheel method

Table 1 Sample data set for selection (for the roulette wheel)

	a	b	c	d	e	f
Quantity	18	45	9	27	54	27
[%]	10	25	5	15	30	15

indicates that an individual for procreation has been selected. Visualisation of the roulette wheel method presented in the previous paragraph was based on the data set included in the table below. To give better picture of it, the visualisation of the modified roulette wheel method was based on the same data. This visualisation is shown in the last figure of the paragraph (Table 1).

Crossing and mutation operators The core of the method consists in application of crossing and mutation adequate to the problem. Figure 5a shows result of crossing of two disks. The layout of the disks is presented and described in section one. Disks are crossed at corresponding positions. Random selection of the disk out of the pair has not been implemented and its impact on the convergence of the algorithm has not been analysed, but this can be the subject of future research. The disk presented in the next figure is generated as a result of the crossing. In order to optimise the geometry of cutting holes, one can use implemented morphologic operations. The

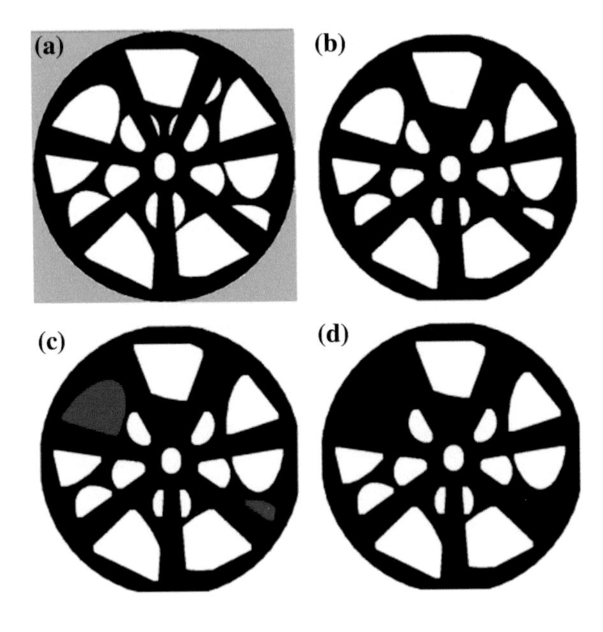

Fig. 5 Consecutive stages of the grinder disk modification

figure below shows four stages of processing for the same disk. Point Fig. 5a shows a "raw" result of the crossing. It includes small holes, which, although technologically feasible, do not make any difference for the structure. Point Fig. 5b presents the disk after morphological optimisation (i.e. after closing operation). The smallest holes disappeared and the disk geometry seems to be proper. Whereas figure Fig. 5c shows places (holes) to be eliminated as a result of mutation applied to the disk. The effect of that operation can be seen in figure Fig. 5d. Morphology is a technique of image processing based on shapes. The value of each pixel in the output image is based on a comparison of the corresponding pixel in the input image with its neighbors. By choosing the size and shape of the neighborhood, you can construct a morphological operation that is sensitive to specific shapes in the input image. Morphologic operations are especially suited to the processing of binary images and greyscale images. Dilation and erosion are two fundamental morphological operations. Dilation adds pixels to the boundaries of objects in an image, while erosion removes pixels on object boundaries.

The image $D = I \oplus S$ is the dilation of image I by structuring element S .

$$D(\underline{x}) = \begin{cases} 1 & in \quad S\ hits\ I\ at\ \underline{x} \\ 0 & otherwise \end{cases} \tag{9}$$

$$D = \{\underline{x} : \underline{x} - \underline{s}, \underline{y} \in I\ and\ \underline{s} \in S\} \tag{10}$$

Erosion and dilation are dual operations:

$$(I \ominus S)^C = I^C \oplus S \tag{11}$$

Commutativity and associativity

$$I \oplus S = S \oplus I\ (I^C \oplus S) \oplus T = I \oplus (S \oplus T) \tag{12}$$

$$I \ominus S \neq S \ominus I\ (I^C \ominus S) \ominus T = I \ominus (S \oplus T) \tag{13}$$

Combining Dilation and Erosion Morphological Opening is defined as an erosion, followed by a dilation. The opening of I by S is

$$I \circ S = (I \ominus S) \oplus S \tag{14}$$

Morpholohical Closing is defined as a dilation, followed by on erosion.

$$I \cdot S = (I \oplus S) \ominus S \tag{15}$$

4 Conclusion

Despite promising results of genetic algorithms operation, one can consider improvement of the method efficiency. The improvement efforts can be made in two directions, the first of them shall remain in GA domain and will require using selection methods characterised by higher determinism. The second one them seems to be more interesting. Here the problem is solved by genetic algorithms and it is interesting how other AI optimising methods will cope with it. Special attention should be drawn to potential application of ACO (Ant Colony Optimization) as well as particle swarm optimisation. The following tasks can be carried out as part of future studies:

- description of the functional model of the mechanical grinder system,
- development of the optimising methodology for the structure of multi-edge grinders using ant colony optimisation and the variant of particle swarm optimisation,
- implementation the appropriate software designed to perform optimising calculations,
- integration of the developed software with the commercial CAD/CAM system,
- assessment of energy consumption per grinding unit,
- assessment of operating conditions for laboratory and industrial grinding,
- grinding product analysis.

References

1. Angryk, R., Czerniak, J.: Heuristic algorithm for interpretation of multi-valued attributes in similarity-based fuzzy relational databases. Int. J. Approximate Reasoning **51**(8), 895–911 (2010)
2. Apiecionek, L., Czerniak, J., Dobrosielski, W.: Quality of Services Method as a DDoS Protection Tool. Advances in Intelligent Systems and Computing. Springer, Berlin (2015)
3. Apiecionek, L., Czerniak, J., Zarzycki, H.: Protection tool for distributed denial of services attack. Commun. Comput. Inf. Sci. Springer **424**, 405–414 (2014)
4. Chan, F.T.S., Au, K.C., Chan, L.Y., Lau, T.L.: Using genetic algorithms to solve quality-related bin packing problem. Robot. Comput. Integr. Manuf. **23**, 71–81 (2007)
5. Ciesielska, D.: Ocena wpywu techniki rozdrabniania odpadw z tworzyw termoplastycznych na struktur i waciwoci otrzymywanych recyklatw. Archiwum Technologii Maszyn i Automatyzacji. **25**(1), 169–178 (2005)
6. Czerniak, J.: Evolutionary approach to data discretization for rough sets theory. Fundamenta Informaticae **92**(1–2), 43–61 (2009)
7. Czerniak, J., Dobrosielski, W., Zarzycki, H., Apiecionek, L.: A Proposal of the New Owlant Method for Determining the Distance Between Terms in On-tology. Advances in Intelligent Systems and Computing. Springer, Berlin (2015)
8. Czerniak, J.M. Apiecionek, Ł., Zarzycki, H.: Application of ordered fuzzy numbers in a new ofnant algorithm based on ant colony optimization. Commun. Comput. Inf. Sci. Springer **424**, 259–270 (2014)
9. De Jong, K., Spears, W.: Using genetic algorithms to solve np complete problems. In: Schaffer, J.D. (ed.) Proceeding of the 3th International Conference on Genetic Algorithms, pp. 124–132 (1989)

10. Ewald, D., Czerniak, J., Zarzycki, H.: Approach to Solve a Criteria Problem of the ABC algorithm used to the WBDP Multicriteria Optimization. Advances in Intelligent Systems and Computing. Springer, Berlin (2015)
11. Farzanegan, A., Vahidipour, S.: Optimization of comminution circuit simulations based on genetic algorithms search method. Miner. Eng. **22**, 719–726 (2009)
12. Flizikowski, J., Bieniaszewski, W., Macko, M.: Integron and innovation algorithm of the cereal grain grinder construction. In: Proceedings of the TMCE 2006 Ljubljana, Slovenia (2006)
13. Flizikowski, J., Kamyk, W.: Algorytmy genetyczne w konstrukcji rozdrabniaczy wielotarczowych ziarna kukurydzy. Inynieria Maszyn **22** (2005)
14. Flizikowski, J., Macko, M.: Method for estimating the efficiency of quasi cutting of recycled optical telecommunications pipes. Polimery **46**(1), 53 (2001)
15. Jekiel, J., Tam, E.K.L.: Plastics waste processing: comminution size distribution and prediction. Envir. Eng. **133**(2), 245–254 (2007)
16. Macko, M.: Aspekty poboru mocy w rozdrabniaczach wielokrawdziowych. Inynieria i Aparatura Chemiczna (4), 103–104 (2006)
17. Macko, M., Flizikowski, J., Zych, G.: Konstruowanie rozdrabniaczy w recyklingu z zastosowaniem systemw ekspertowych—re-cykling materiaw polimerowych. Nauka—Przemys (2003)
18. Marbac-Lourdelle, M.: Model-based clustering for categrocial and mixed data sets. Statitics, Universite de Lille **1** (2014)
19. Powell, M., Morrison, R.: The future of comminution modelling. Int. J. Miner. Process. **84**, 228–239 (2007)
20. Quagliarella, D.: Genetic Algorithms and Evolution Strategy in Engineering and Computer Science: Recent Advances and Industrial Applications. Wiley, New York (1998)
21. Rutkowska, D., Piliński, M., Rutkowski, L.: Sieci neuronowe, algorytmy genetyczne i systemy rozmyte. PWN, Warszawa (1999)
22. Sadrai, S., Meech, J., Ghomshei, M., Sassani, F., Tromans, D.: Influence of impact velocity on fragmentation and the energy efficiency of comminution. Int. J. Impact Eng. **33**, 723–734 (2006)
23. Sameon, D., Shamsuddin, S.M. and Sallehuddin, R., Zainal, A.: Compact classification of optimized boolean, reasoning with particle swarm optimization. Intelligent Data Analysis 16 IOS Press, pp. 915–931 (2012)
24. Shuiping, L., Hongzan, B., Zhichu, H., Jianzhong, W.: Nonlinear comminution process modeling based on ga-fnn in the computational comminution system. J. Mater. Process. Technol. **120**, 84–89 (2002)
25. Sikora, R.: Przetwrstwo tworzyw polimerowych—podstawy logiczne, formalne i terminologiczne. Wyd. Polit. Lubelskiej (2006)
26. Woldt, D., Schubert, G., Jäckel, H.G.: Size reduction by means of low-speed rotary shears. Int. J. Miner. Process. **74S**, 405–415 (2004)

Part VI
Applications and Implementations

A Proposal of a Fuzzy System for Hypertension Diagnosis

Juan Carlos Guzmán, Patricia Melin and German Prado-Arechiga

Abstract One of the most dangerous diseases for humans is the Arterial Hypertension, which this kind of disease that often leads to fatal outcomes, such as heart attack, stroke and renal failure. The hypertension seriously threats the health of the people worldwide. One of the dangerous aspects of the hypertension is that you may not know that you have it. In fact, nearly one-third of people who have high blood pressure don't know it. The only way to know if the blood pressure is high is through the regular checkups. The evaluation of a patient with Hypertension should (1) confirm the diagnosis of hypertension, (2) detect causes of secondary hypertension y (3) assess cardio vascular risk and organ damage. Therefore, is very important a correct measurement of the blood pressure (BP). Traditionally, office BP measurement has been performed using a sphygmomanometer and stethoscope. Recently, automated office and home BP measurements has been proposed as an alternative to traditional measurement. It has several advantages over manual BP, especially in routine clinical practice. Therefore, we have developed a Fuzzy System for the diagnosis of the Hypertension. Firstly, the input parameters include Systolic Blood Pressure and Diastolic Blood Pressure. Secondly, we have as an output parameter: Blood Pressure Levels (BPL). The input linguistic value includes Low, Low Normal, Normal, High Normal, High, Very High, Too High and Isolated Systolic Hypertension. Finally, we have 14 fuzzy rules to determine the diagnosis output.

Keywords Fuzzy system · Hypertension · Diagnosis

J.C. Guzmán (✉) · P. Melin
Tijuana Institute of Technology, Tijuana, BC, Mexico

G. Prado-Arechiga
Cardio-Diagnostico, Tijuana, BC, Mexico

K.T. Atanassov et al. (eds.), *Novel Developments in Uncertainty Representation and Processing*, Advances in Intelligent Systems and Computing 401,
DOI 10.1007/978-3-319-26211-6_29

1 Introduction

Nowadays different techniques of artificial intelligence, such as fuzzy systems are largely used in medical areas. As we know, the control of the hypertension is considered when the systolic blood pressure >140 mmHg and the diastolic blood pressure >90 mmHg. Thus, the use of an expert system that provides information to the user about the factors and dangers of high blood pressure is very important.

Fuzzy logic is used to model nonlinear systems, which are difficult to model mathematically. It is a logic system that is based on fuzzy set theory and continuous variables. Conclusions that are based on vague, imprecise, missing input information are simply provided by fuzzy logic (FL). Fuzzy logic uses different concepts, i.e. fuzzification, defuzzification, membership function, linguistic variables, domain, rules etc. In Boolean algebra or Boolean logic crisp sets are used, which have only two values 0 and 1, but in fuzzy logic, sets have an infinite number of values between 0 and 1. In Boolean logic an element is completely inclusive or exclusive membership is used, but in a FL completely inclusive, exclusive or between these two memberships is used. Also the fuzzy system is a system in which fuzzy rules are used with membership functions (MF) to find the conclusion or result. Fuzzy logic has been applied to many areas or fields of application, for example fuzzy logic has played an important role in the field of medicine [5, 6 and 8]. They are used in control, automobiles, household appliances and decision making systems.

Hypertension or high blood pressure, sometimes called arterial hypertension, is a chronic medical condition in which the blood pressure in the arteries is elevated [14]. The optimal level for blood pressure is below 120/80, where 120 represent the systolic measurement (peak pressure in the arteries) and 80 represents the diastolic measurement (minimum pressure in the arteries). Systolic BP between 120–129 and/or Diastolic BP between 80–84 is called Normal BP. And Systolic BP between 130–139 and/or Diastolic BP 85–89 is called High Normal. And a blood pressure of 140/90 or above is considered hypertension in three different grades and other like the Isolated systolic hypertension that is Systolic BP over 140 and Diastolic BP under 90 mm hg ESH/ESC guidelines [10].

Hypertension may be classified as essential or secondary. Essential hypertension is the term for high blood pressure with an unknown cause [15]. It accounts for about 95 % of cases. Secondary hypertension is the term for high blood pressure with a known direct cause, such as kidney disease, tumors or others.

The paper is organized as follows: in Sect. 2 a methodology of hypertension is presented, in Sect. 3 simulation and results of the prediction of the data that will be the input to the fuzzy system are presented, in Sect. 4 the design and development of the fuzzy logic system is described, and in Sect. 5 the conclusion obtained after tests the fuzzy system of diagnosis of hypertension.

2 Methodology

2.1 Type of Blood Pressure Diseases

Hypertension is the most common disease and it markedly increases both morbidity and mortality from cardiovascular and many other diseases [12]. Different types of hypertension are observed when the disease is sub-categorized. These types are shown in Table 1.

In Table 1 the blood pressure (BP) category is defined by the highest level of BP, whether systolic or diastolic. And should be graded 1, 2 or 3 according to the systolic or diastolic BP value. Isolated systolic hypertension it is according to the systolic BP value in the ranges indicated.

2.2 Risk Factors

Some of the primary risk factors for essential hypertension include the following [1]:

- Obesity
- Lack of exercise
- Smoking
- Consumption of salt
- Consumption of alcohol
- Stress level
- Age
- Sex
- Genetic factors

Table 1 Definitions and classification of the blood pressure levels (mmHg)[a]

Category	Systolic		Diastolic
Hypotension	<90	and/or	<60
Optimal	<120	and	<80
Normal	120–129	and/or	80–84
High normal	130–139	and/or	85–89
Grade 1 hypertension	140–159	and/or	90–99
Grade 2 hypertension	160–179	and/or	100–109
Grade 3 hypertension	≥180	and/or	≥110
Isolated systolic hypertension	≥140	and	<90

2.3 *Fuzzy Logic and Hypertension*

Nowadays we cannot be comfortable with the traditional medical analysis because the complexity of medical practices makes traditional quantitative approaches of analysis inappropriate [13]. Every trust worthy expert knows that his/her medical knowledge and resulting diagnosis are pervaded by uncertainty with imprecise formulations. Medical processes can be so complex and unpredictable that physicians sometimes must make decisions based on their experience or intuition or sometimes they might requires a specialization like Cardiology, Internal Medicine, etc. Computers are capable of making calculations at high and constant speed and of recalling large amounts of data and can, therefore, be used to manage decision networks of high complexity [11]. Fuzzy logic developed by Zadeh [16] makes it possible to define these inexact medical entities as fuzzy sets. Fuzzy logic together with the appropriate rules of inference provides a power framework for managing uncertainties pervaded in medical diagnosis [3, 4, 7, 9]. Fuzzy logic technology is adopted in this paper for the diagnosis of hypertension. This is because, fuzzy logic can adequately address the issue of uncertainty and lexical imprecision of knowledge [2], but fuzzy systems still requires human expert to discover rules about data relationship.

By applying fuzzy logic, a fuzzy rule base system for the diagnostic of hypertension was developed with the help of the domain expert.

3 Simulation and Results

The following graphic interface in Fig. 1 shows the information to be simulated and used for prediction of blood pressure following the result is selected.

The following graphic interface in Fig. 2 simulates the monitoring of blood pressure of a patient, which is based on the information given. A prediction of its next blood pressure is performed, and the result of the prediction is systolic and diastolic and this information is the input to the fuzzy system.

4 Design and Development of the Fuzzy Logic System

A fuzzy logic system is a collection of membership functions and fuzzy rules that are used to determine the diagnosis. This design has been divided into several steps. The steps are fuzzification, rule evaluation and finally defuzzification. To design the system, the FIS tool in MATLAB R2013a is used.

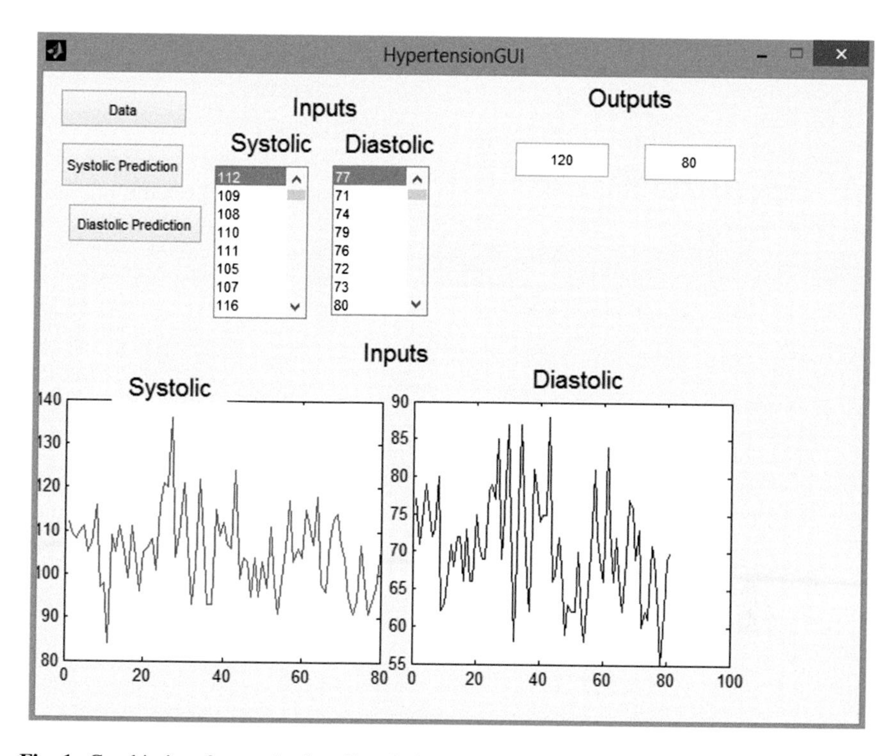

Fig. 1 Graphic interface and select file window

In this study, we propose a fuzzy system for the diagnosis of the hypertension. The fuzzy system has 2 inputs including the Systolic and Diastolic, and 1 output BP_level, in the inputs we have eight membership functions, such as Low, low normal, normal, high normal, high, very high, too high and isolated systolic Hypertension(ISH) and in the output there are eight member functions such as Hypotension, Optimal, Normal, High normal, Grade 1, Grade 2, Grade 3 and Isolated Systolic Hypertension (ISH) and Mamdani inference engine and centroid defuzzification.

The analysis focused on how to design a fuzzy logic system for diagnosis hypertension. This is performed by using a range of systolic and diastolic blood pressure. First, the linguistic values and corresponding membership functions have been determined in the next figures: Fig. 3 shows the fuzzy system of diagnosis of hypertension, Fig. 4 shows linguistic variable and membership function of "Systolic", Fig. 5 shows linguistic variable and membership function of the input "Diastolic", Fig. 6 shows the linguistic variable and membership function of the output "BP_level".

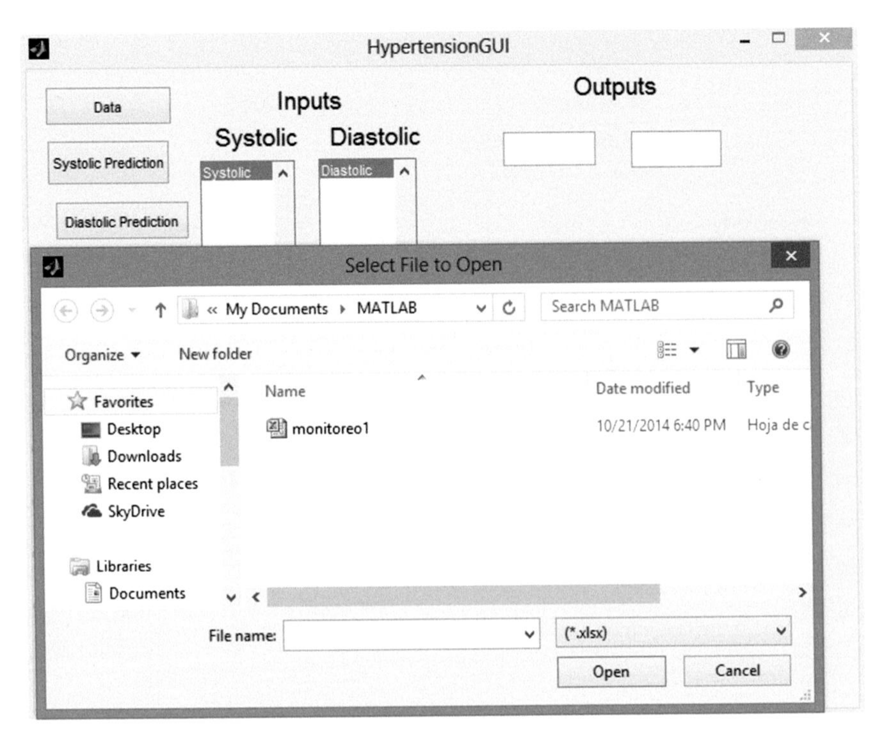

Fig. 2 Simulation and results of the graphic interface

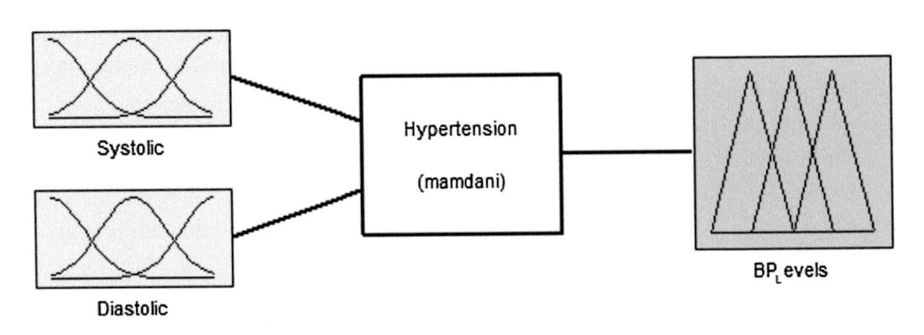

Fig. 3 Fuzzy system for diagnosis of hypertension

Now we will show in the following more details of the fuzzy system: Fig. 7 shows the rules of our fuzzy system of diagnosis of hypertension, Fig. 8 shows the result of rules of the fuzzy system of diagnosis of hypertension and finally Fig. 9 shows the surface view of the fuzzy system of diagnosis of hypertension.

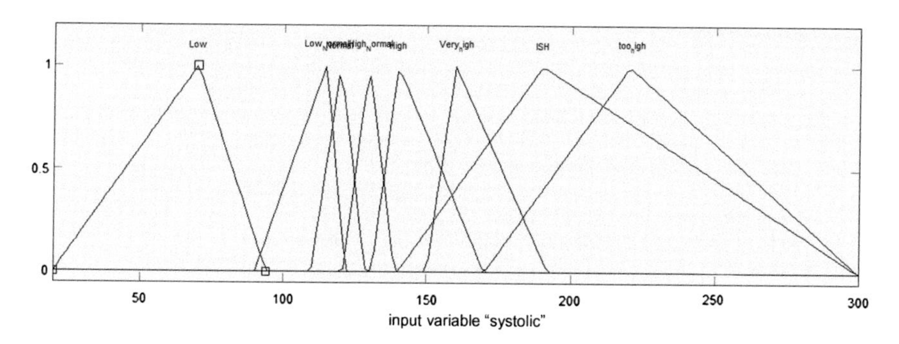

Fig. 4 Linguistic variable and membership functions of "Systolic"

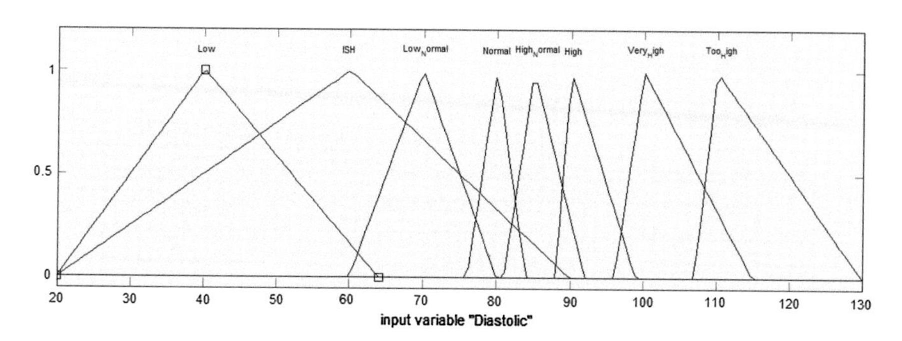

Fig. 5 Linguistic variable and membership functions of the input "Diastolic"

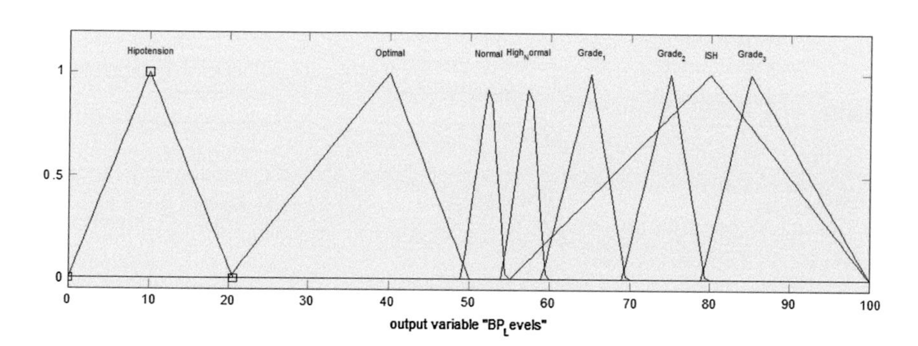

Fig. 6 Linguistic variable and membership functions of the output "BP_level"

1. If (Systolic is Low) and (Diastolic is Low) then (BP_Levels is Hipotension) (1)
2. If (Systolic is Low_Normal) and (Diastolic is Low_Normal) then (BP_Levels is Optimal) (1)
3. If (Systolic is Normal) and (Diastolic is Normal) then (BP_Levels is Normal) (1)
4. If (Systolic is High_Normal) and (Diastolic is High_Normal) then (BP_Levels is High_Normal) (1)
5. If (Systolic is High) and (Diastolic is High) then (BP_Levels is Grade_1) (1)
6. If (Systolic is Very_high) and (Diastolic is Very_High) then (BP_Levels is Grade_2) (1)
7. If (Systolic is too_high) and (Diastolic is Too_High) then (BP_Levels is Grade_3) (1)
8. If (Systolic is ISH) and (Diastolic is ISH) then (BP_Levels is ISH) (1)
9. If (Systolic is Very_high) and (Diastolic is High) then (BP_Levels is Grade_2) (1)
10. If (Systolic is too_high) and (Diastolic is Very_High) then (BP_Levels is Grade_3) (1)
11. If (Systolic is too_high) and (Diastolic is High) then (BP_Levels is Grade_3) (1)
12. If (Systolic is High) and (Diastolic is Very_High) then (BP_Levels is Grade_2) (1)
13. If (Systolic is High) and (Diastolic is Too_High) then (BP_Levels is Grade_3) (1)
14. If (Systolic is Very_high) and (Diastolic is Too_High) then (BP_Levels is Grade_3) (1)

Fig. 7 Fuzzy rules of the fuzzy system for diagnosis of hypertension

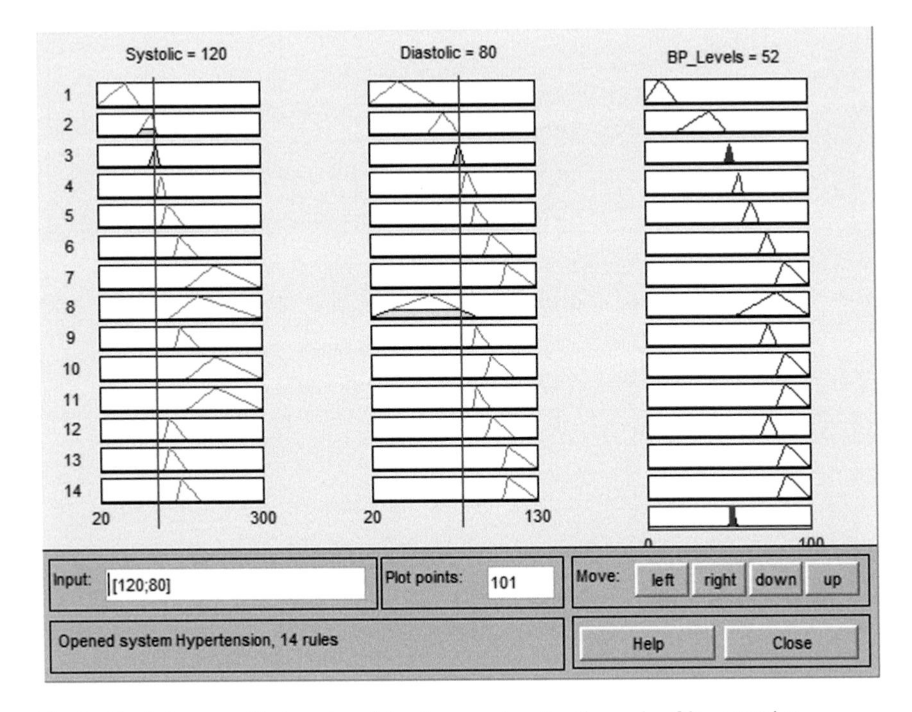

Fig. 8 The inference with the rules of the fuzzy system for diagnosis of hypertension

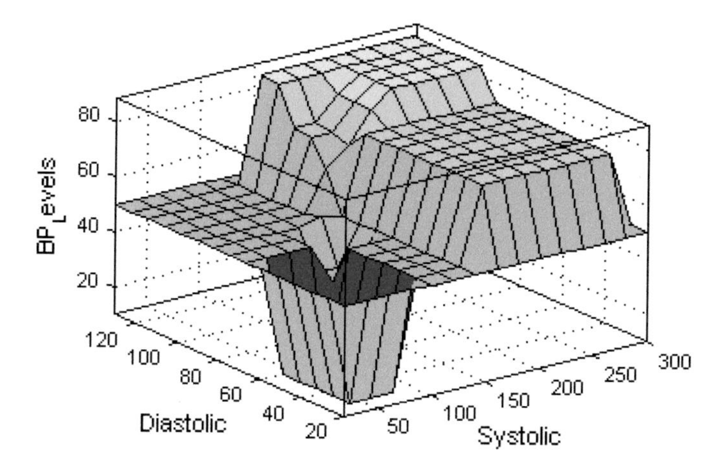

Fig. 9 Surface view of the fuzzy system for diagnosis of hypertension

5 Conclusions

This type of fuzzy systems actually implements the human intelligence and reasoning. Using a set of decision rules, provide different suggestions for diagnosing diseases, in this case hypertension. This is a very efficient, less time consuming and more accurate method to diagnose the risk and the grade of hypertension. Finally we can see that is a very effective method for an early and accurate diagnostic of hypertension, which can help a physician to get a better medical treatment when giving a diagnosis to the patient.

Acknowledgment We would like to express our gratitude to the CONACYT and Tijuana Institute of Technology for the facilities and resources granted for the development of this research.

References

1. Abrishami, Z., Tabatabaee, H.: Design of a fuzzy expert system and a multi-layer neural network system for diagnosis of hypertension. MAGNT Res. Rep. **2**(5), 913–926. ISSN: 1444-8939 (2014)
2. Akinyokun, O.C., Adeniji, O.A.: Experimental study of intelligent computer aided diagnostic and therapy. AMSE J. Model. Simul. Control **27**(3), 9–20 (1991)
3. Azamimi, A.A., Zulkarnay, Z., Nur Farahiyah, M.: Design and development of fuzzy expert system for diagnosis of hypertension. In: IEEE International Conference on Intelligent Systems, Modelling and Simulation (2011)
4. Das, S., Ghosh, P.K., Kar, S.: Hypertension diagnosis: a comparative study using fuzzy expert system and neuro fuzzy system. In: Fuzzy Systems, IEEE International Conference on. IEEE (2013)

5. Djam, X.Y., Kimbi, Y.H.: Fuzzy expert system for the management of hypertension. Pac. J. Sci. Technol. **12**(1) (2011) (Spring)
6. Fuller, R., Giove, S.: A Neuro-fuzzy approach to FMOLP problems. In: Proceedings of CIFT'94, pp. 97–101. Trento, Italy (1994)
7. Kaur, A., Bhardwaj, A.: Genetic neuro fuzzy system for hypertension diagnosis. Int. J. Comput. Sci. Inf. Technol. **5**(4), 4986–4989 (2014)
8. Kaur, R., Kaur, A.: Hypertension diagnosis using fuzzy expert system. In: International Journal of Engineering Research and Applications (IJERA) National Conference on Advances in Engineering and Technology, AET-29th March (2014). ISSN: 2248-9622
9. Ludmila, I.K., Steimann, F.: Fuzzy Medical Diagnosis. School of Mathematics, University of Wales, Bangor (2008)
10. Mancia, G., Fagard, R., Narkiewicz, K., Redon, J.: 2013 ESH/ESC guidelines for the management of arterial hypertension. J. Hypertens. **31**, 1281–1357 (2013)
11. Merouani, M., Guignard, B., Vincent, F., Borron, S.W., Karoubi, P., Fosse, J.P., Cohen, Y., Clec'h, C., Vicaut, E., Marbeuf-Gueye, C., Lapostolle, F., Adnet, F.: Can fuzzy logic make things more clear? Crit. Care **13**, 116 (2009)
12. O'Brien, E., Parati, G., Stergiou, G.: European society of hypertension position paper on ambulatory blood pressure monitoring. J. Hypertens. **31**, 1731–1768 (2013)
13. Rahim, F., Deshpande, A., Hosseini, A.: Fuzzy expert system for fluid management in general anesthesia. J. Clin. Diagn. Res. 256–267 (2007)
14. Srivastava, P:. A note on hypertension classification scheme and soft computing decision making system. ISRN Biomathematics (2013)
15. Sumathi, B. Santhakumaran, A.: Pre-diagnosis of hypertension using artificial neural network. Glob. J. Comput. Sci. Technol. **11**(2) Version 1.0 (2011)
16. Zadeh, L.A.: Fuzzy sets and systems. In Proceedings Symposium on System Theory. Fox, J. (ed.) Polytechnic Institute of Brooklyn, pp. 29–37. New York, NY. April (1965)

Using Intercriteria Analysis for Assessment of the Pollution Indexes of the Struma River

Tatiana Ilkova and Mitko Petrov

Abstract In this paper we are presenting the recently proposed approach Intercriteria Analysis (ICrA) for assessment of the pollution index of the Struma River in Bulgaria. The approach is based on the apparatus of the index matrices and the intuitionistic fuzzy sets. At the first we have investigated all indexes at the all measurement point with ICrA and we have searched the dependences between points. Results show the measurement points are dependent criteria and we have ignored some over others. At the second we have applied the ICrA to establish the pollution relations and the model structure based on different criteria involved in the Struma River. The investigations show that there are three positive consonances and dissonances between criteria. Using of a Modification of the Time Series Analysis (MTSA) method we have developed an adequate mathematical model of the pollution dynamic as function of time.

Keywords Intercriteria analysis · Index matrices · Intuitionistic fuzzy sets · Pollution index · Modelling · Modification times series analysis · Struma river

1 Introduction

The transboundary Struma River is located in the western part of Bulgaria and Greece. Its spring is near to the peak Cherni vrah in the Vitosha mountain. The river flows into Strimonikos bay of the Aegean Sea with the. The Struma River catchment area is the second largest catchment area in Bulgaria. It is shown in Fig. 1.

T. Ilkova (✉) · M. Petrov
Institute of Biophysics and Biomedical Engineering, Bulgarian Academy of Sciences, 105 Acad. G. Bonchev, 1113 Sofia, Bulgaria
e-mail: tanja@biomed.bas.bg

M. Petrov
e-mail: mpetrov@biomed.bas.bg

K.T. Atanassov et al. (eds.), *Novel Developments in Uncertainty Representation and Processing*, Advances in Intelligent Systems and Computing 401,
DOI 10.1007/978-3-319-26211-6_30

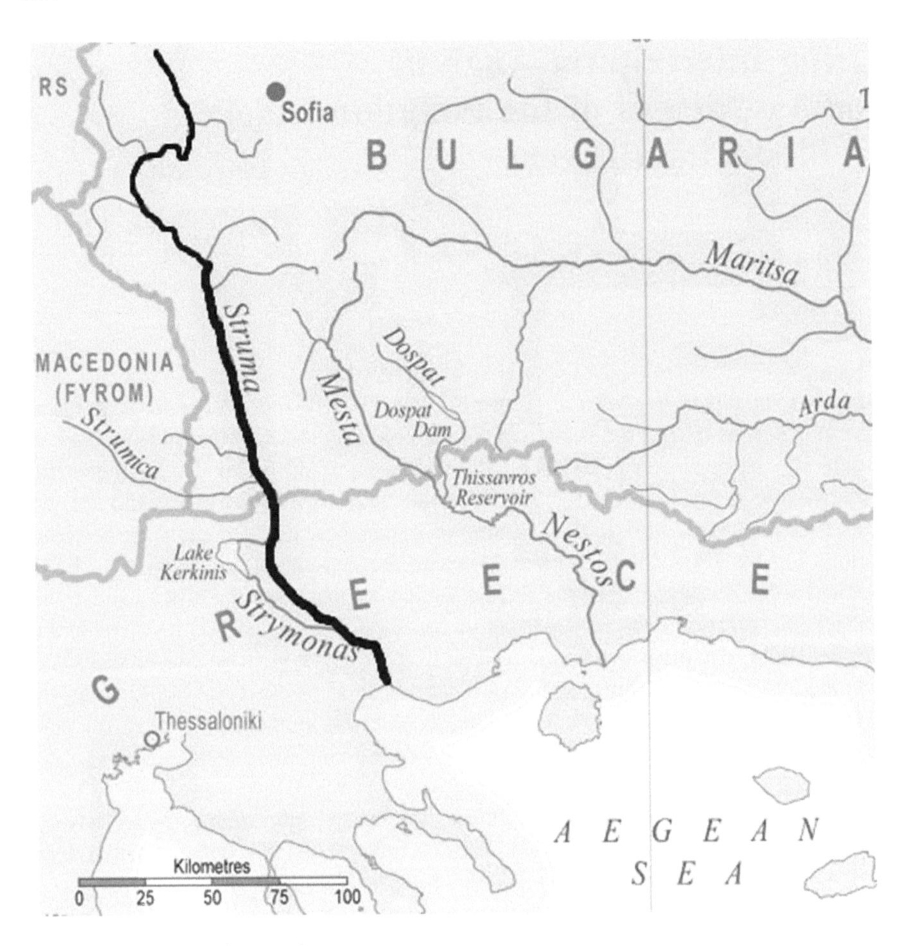

Fig. 1 The Struma River catchment area

In this paper we will investigate the river pollution dynamic. We have modelled the organic and biogenic water pollution using the following indices—Biological oxygen demand (BOD_5), Permanganate oxidation (KMn_3O_4-oxidation) (these are the indices for organic pollution), and the indices for biogenic pollution—Ammonia nitrogen ($N\text{-}NH_4$), Nitrate nitrogen ($N\text{-}NO_3$), Nitrite nitrogen (NN), Phosphates--general (PO_4), Orthophosphates (OPh), Dissolved substances (DS) and Unsolved substances (US).

In order to model the pollution through these and other indices of rivers in Bulgaria, we have used a modification of the Time Series Analysis (MTSA) method and a Modification of the Regression Analysis (MRA) method. We have obtained a series of adequate models [1–3].

The assessment of the river pollution indexes is based in a multicriteria analysis method called Intercriteria Analysis (ICrA) [4]. The ICrA is based on the apparatus of the index matrices (IMs) and the intuitionistic fuzzy sets (IFs). The approach employs the concept of IMs, making particular use of some of the operations introduced over them, and the concept of the intuitionistic fuzzy sets fuzziness, giving us the tools, to construct the IMs of intuitionistic fuzzy pairs (IFPs), defining the presence or absence of dependency/correlation between any pair of criteria within the set. The role of both concepts for the ICrA approach has been presented in details in [5–9].

Atanassova and co-authors [10, 11] use ICrA for an analysis of EU member states competitiveness. They investigated [12, 13] the possibility of defining intuitionistic fuzzy threshold values which enable the definition of the positive consonance, negative consonance and dissonance among the criteria. Ilkova et al. [13] used ICrA for modelling the Mesta River pollution.

In this paper we have used the ICrA for assessment of the pollution indexes of the Struma River.

2 The Intercriteria Analysis Method

2.1 *Short Remarks on Intuitionistic Fuzzy Pairs*

The IFPs [9] is an object in the form of an ordered pair $\langle a+b \rangle \leq 1$, where $a, b \in [0, 1]$ and $a+b \leq 1$. It is used as an evaluation of an object or a process and its components (a and b) are interpreted respectively as degrees of membership and non-membership to a given set, degrees of validity and non validity, degrees of correctness and non-correctness, etc. Let us have two IFPs $x = \langle a, b \rangle$ and $y = \langle c, d \rangle$. Atanassov et al. defined the relations [6]:

$$x < y \quad \text{iff } a < c \text{ and } b < d$$
$$x \leq y \quad \text{iff } a \leq c \text{ and } b \geq d$$
$$x = y \quad \text{iff } a = c \text{ and } b = d$$
$$x \geq y \quad \text{iff } a \geq c \text{ and } b \leq d$$
$$x > y \quad \text{iff } a > c \text{ and } b < d$$

2.2 *Short Remarks on Index Matrix (IMs)*

Atanassov [5–7] has presented the concept of IMs and has given the basic definitions and properties. Let I be a fixed set of indices and $\mathfrak{R}$ be the set of all numbers. By IMs with index sets K and L ($K, L \subset I$), we mean the following object:

$$[K, L, \{a_{k_i, l_j}\}] \equiv \begin{array}{c|cccc} & l_1 & l_2 & \dots & l_n \\ \hline k_1 & a_{k_1, l_1} & a_{k_1, l_2} & \dots & a_{k_1, l_n} \\ k_2 & a_{k_2, l_1} & a_{k_2, l_2} & \dots & a_{k_2, l_n} \\ \vdots & \vdots & \vdots & \vdots & \vdots \\ k_m & a_{k_m, l_1} & a_{k_m, l_2} & \dots & a_{k_m, l_n} \end{array}$$

where $K = \{k_1, k_2, \dots, k_m\}$, $L = \{l_1, l_2, \dots, l_n\}$ and for $1 \le i \le m$, and $1 \le j \le n$: $a_{k_i, l_j} \in \Re$.

On the basis of the above definition, Atanassov [4] has introduced the new object —the Intuitionistic Fuzzy Index Matrix (IFIM) in the form:

$$[K, L\{\langle \mu_{k_i l_j}, \nu_{k_i l_j} \rangle\}] \equiv \begin{array}{c|cccc} & l_1 & l_2 & \dots & l_n \\ \hline k_1 & \langle \mu_{k_1, l_1}, \nu_{k_1, l_1} \rangle & \langle \mu_{k_1, l_2}, \nu_{k_1, l_2} \rangle & \dots & \langle \mu_{k_1, l_n}, \nu_{k_1, l_n} \rangle \\ \vdots & \vdots & \vdots & \vdots & \vdots \\ k_m & \langle \mu_{k_m, l_1}, \nu_{k_m, l_1} \rangle & \langle \mu_{k_m, l_2}, \nu_{k_m, l_2} \rangle & \dots & \langle \mu_{k_m, l_n}, \nu_{k_m, l_n} \rangle \end{array}$$

where $1 \le i \le m$, $1 \le j \le n$, $0 \le \mu_{k_i, l_j}, \nu_{k_i, l_j}$, $\mu_{k_i, l_j} + \nu_{k_i, l_j} \le 1$, i.e. $\langle \mu_{k_i, l_j}, \nu_{k_i, l_j} \rangle$ is an IFPs.

2.3 The Proposed Intercriteria Decision Making Method

Let us have an IM:

$$A = \begin{array}{c|ccccccc} & O_1 & \dots & O_k & \dots & O_l & \dots & O_n \\ \hline C_1 & a_{C_1, O_1} & \dots & a_{C_1, O_k} & \dots & a_{C_1, O_l} & \dots & a_{C_1, O_n} \\ \vdots & \vdots & \vdots & \vdots & \vdots & \vdots & \vdots & \vdots \\ C_i & a_{C_i, O_1} & \dots & a_{C_i, O_k} & \vdots & a_{C_i, O_l} & \vdots & a_{C_i, O_n} \\ \vdots & \vdots & \vdots & \vdots & \vdots & \vdots & \vdots & \vdots \\ C_j & a_{C_j, O_1} & \dots & a_{C_j, O_k} & \dots & a_{C_k, O_l} & \dots & a_{C_k, O_n} \\ \vdots & \vdots & \vdots & \vdots & \vdots & \vdots & \vdots & \vdots \\ C_m & a_{C_m, O_1} & \dots & a_{C_m, O_k} & \dots & a_{C_m, O_l} & \dots & a_{C_m, O_n} \end{array}$$

where for every $p, q (1 \le p \le m, 1 \le q \le n)$:

- C_p is a criterion, taking part in the evaluation,
- O_q is an object, being evaluated.

- a_{C_p,O_q} is a real number or another object, that is comparable to relation R with other *a-object*, so that for each i,j,k: $R(a_{C_k,O_i}, a_{C_kO_j})$ is defined.

Let $\bar{R}$ be the dual relation of R in the sense that if R is satisfied, then $\bar{R}$ is not satisfied and vice versa. For example, if "R" is the relation "<", then $\bar{R}$ is the relation ">", and vice versa.

Let $S^{\mu}_{k,l}$ be the number of cases in which $R(a_{C_k,O_i}, a_{C_kO_j})$ and $R(a_{C_l,O_i}, a_{C_lO_j})$ are simultaneously satisfied. Let $S^{\nu}_{k,l}$ be the number of cases is which $R(a_{C_k,O_i}, a_{C_kO_j})$ and $\bar{R}(a_{C_l,O_i}, a_{C_lO_j})$ are simultaneously satisfied.

Obviously,

$$S^{\mu}_{k,l} + S^{\nu}_{k,l} \leq \frac{n(n-1)}{2}$$

Now, for every k, l such that $1 \leq k < l \leq m$, and for $n \geq 2$, it can be defined:

$$\mu_{C_k,C_l} = \frac{2S^{\mu}_{k,l}}{n(n-1)}, \qquad \nu_{C_k,C_l} = \frac{2S^{\nu}_{k,l}}{n(n-1)}$$

Therefore, $\langle \mu_{C_k,C_l}, \nu_{C_k,C_l} \rangle$ is an IFPs. The IMs can construct:

	C_1	…	C_m
C_1	$\langle \mu_{C_1,C_1}, \nu_{C_1,C_1} \rangle$	…	$\langle \mu_{C_1,C_m}, \nu_{C_1,C_m} \rangle$
⋮	⋮	…	⋮
C_m	$\langle \mu_{C_m,C_1}, \nu_{C_m,C_1} \rangle$	…	$\langle \mu_{C_m,C_m}, \nu_{C_m,C_m} \rangle$

that determined the degrees of correspondence between criteria C_1, …, C_m.

Let $\alpha, \beta \in [0,\ 1]$ be given, so that $\alpha + \beta \leq 1$. We say that criteria C_k and C_l are in:

- (α, β)—*positive consonance*, if $(\mu_{C_k,C_l} > \alpha)$ and $(\nu_{C_k,C_l} < \beta)$;
- (α, β)—*negative consonance*, if $(\mu_{C_k,C_l} < \beta)$ and $(\nu_{C_k,C_l} > \alpha)$;
- (α, β)—*dissonance, otherwise*.

The method is used for assessment of pollution index of the Struma River.

3 Results and Discussion

The investigations for the water quality were conducted along the entire river using data from all measuring points. The information used was provided by the West Aegean Water Basin Directorate, Ministry of Environmental and Water of Bulgaria. We have investigated the following measurement point in the Struma River catchment area—Point 50 (P50) it is on Waste water purification station in the city of Batanovtsi, Point 80 (P80)—at the village of Nevestino, Point 90 (P90)—at the

Dzherman River, Point 105 (P105)—before the city of Blagoevgrad, Point 120 (P120)—before the village of Krupnik, Point 123 (P123)—at the village of Lebnitsa, Point 140 (P140)—after the city of Petrich and Point 150 (P150)—it is on the Bulgarian-Greek border.

The program has been developed that realizes ICrA and calculates membership function (μ), and non membership function (ν). We have investigated all indexes at the all measurement point with ICrA. We have searched if there were measurement errors also if the pollution in the different points have had materially divergences, if there were the dependences between points. The calculated IFPs for the investigated measurement Points are shown in Table 1.

Figure 2 shows the relations between μ and ν for the different measurement points.

Let us see the Table 1 and Fig. 2. The membership function (μ) is changed in the interval $\mu \in [0.89, 0.99]$, the non-membership function (ν) is changed in the interval $\nu \in [0.01, 0.09]$, and we have a bit of uncertainty (π) which is changed in the interval $\pi \in [0.00, 0.03]$. The values of the membership function μ are high comparatively, the values of the non membership function—low high.

Since the experimental data of the river pollution are random values for determination of the positive consonance, negative consonance and dissonance the following minimal values are assumed $(\alpha, \beta) = (0.85, 0.15)$.

The obtained results (Table 1) show we have positive consonance for all investigated pairs of pollution indexes. That shows the measurement points are dependent criteria and we can ignore some over others. Thus we have investigated the P150.

We have modelled the following indexes for the water quality in the P150: BOD (C_1), KMn_3O_4 (C_2), N-NH_4 (C_3), NN (C_4), N-NO_3 (C_5), PO_4 (C_6), OPh (C_7), DS (C_8), and US (C_9). With the help of the program that realizes ICrA we have obtained the calculates membership function (μ), and non membership function (ν) for all pollution criteria. The calculated IFPs are shown in Table 2.

Figure 3 shows the relations between μ and ν for the different criteria (pollution indexes).

Let us see the Table 2 and Fig. 3. The membership function (μ) is changed in the interval $\mu \in [0.39, 0.90]$, the non-membership function (ν) is changed in the interval $\nu \in [0.04, 0.59]$, and we have a bit of uncertainty (π) which is changed in the interval $\pi \in [0.00, 0.07]$. The obtained results (Table 2) show we have 3 positive consonances and dissonances the other investigated pairs of pollution indexes. We have positive consonances of $\langle C_3\text{–}C_4 \rangle$, $\langle C_4\text{–}C_5 \rangle$ and $\langle C_6\text{–}C_7 \rangle$. That shows we can ignore NN over N-NH_4 and N-NO_3, also OPh over PO_4. The rest the pollution indexes have dissonance—they are independent criteria and we cannot ignore some over others.

At the end in this paper we have modelled the C_1, C_2, C_3, C_5, and C_6 using a MTSA method. We have not modelled DS and US because in this case [2, 3] they are not significant for the water river pollution. The authors have developed this modification [1]. The main difference between the TSA and the MTSA is that intend of using polynomial third degree for the trend function a polynomial of

Table 1 The calculated values of IFPs $\langle \mu, \nu \rangle$ for the investigated measurement Points

	P50	P80	P90	P105	P120	P123	P140	P150
P50		⟨0.94,0.06⟩	⟨0.92,0.08⟩	⟨0.94,0.04⟩	⟨0.93,0.04⟩	⟨0.99,0.01⟩	⟨0.91,0.09⟩	⟨0.99,0.01⟩
P80	⟨0.94,0.06⟩		⟨0.96,0.02⟩	⟨0.97,0.00⟩	⟨0.94,0.05⟩	⟨0.92,0.06⟩	⟨0.92,0.07⟩	⟨0.92,0.05⟩
P90	⟨0.92,0.08⟩	⟨0.96,0.02⟩		⟨0.97,0.02⟩	⟨0.92,0.07⟩	⟨0.92,0.08⟩	⟨0.96,0.04⟩	⟨0.92,0.01⟩
P105	⟨0.94,0.04⟩	⟨0.97,0.00⟩	⟨0.97,0.02⟩		⟨0.94,0.06⟩	⟨0.90,0.09⟩	⟨0.92,0.07⟩	⟨0.97,0.01⟩
P120	⟨0.93,0.04⟩	⟨0.94,0.05⟩	⟨0.92,0.07⟩	⟨0.94,0.06⟩		⟨0.90,0.08⟩	⟨0.96,0.01⟩	⟨0.89,0.08⟩
P123	⟨0.99,0.01⟩	⟨0.92,0.06⟩	⟨0.92,0.08⟩	⟨0.90,0.09⟩	⟨0.90,0.08⟩		⟨0.94,0.06⟩	⟨0.95,0.04⟩
P140	⟨0.91,0.09⟩	⟨0.92,0.07⟩	⟨0.96,0.04⟩	⟨0.92,0.07⟩	⟨0.96,0.01⟩	⟨0.94,0.06⟩		⟨0.93,0.06⟩
P150	**⟨0.99,0.01⟩**	**⟨0.92,0.05⟩**	**⟨0.92,0.08⟩**	**⟨0.97,0.01⟩**	**⟨0.89,0.08⟩**	**⟨0.95,0.04⟩**	**⟨0.93,0.06⟩**	

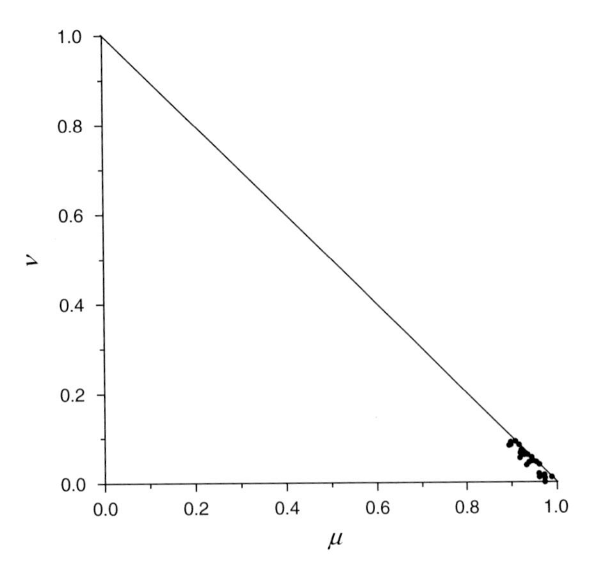

Fig. 2 Relations between μ and ν for the measurement Points

higher degree is used. Furthermore, instead of using Fourier components for the periodical component, periodical functions are used which have been defined in the process of modelling.

3.1 Short Remark of MTSA

For modelling water pollution, a Time Series Analysis (TSA) method is a determined component—$\mathbf{x}_{\mathrm{T}}$ describing the regularity of the development of the examined phenomenon, periodical component—$\mathbf{x}_{\mathrm{P}}$ and stochastic variable—ε_t [15, 16]:

$$\mathbf{x} = \mathbf{x}_{\mathrm{T}} + \mathbf{x}_{\mathrm{P}} + \varepsilon_t \quad (1)$$

where $\mathbf{x}$ is the vector of the pollution indexes, $\mathbf{x} = \mathbf{x}[\mathrm{C}_1, \mathrm{C}_2, \mathrm{C}_3, \mathrm{C}_5, \mathrm{C}_6]^{\mathrm{T}}$.

The determined component—$\mathbf{x}_{\mathrm{T}}$ is a polynomial of 1st to 3rd degrees and the periodical component—$\mathbf{x}_{\mathrm{P}}$ is described by the order of Fourier.

In contrast to the conventional method of TSA the determined component is a polynomial with a high degree [1–3]:

$$\mathbf{x}_{\mathrm{T}} = \sum_{j=0}^{r} a_j t^j \quad (2)$$

where a_j—coefficients of polynomial, j = 0, …, r—a degree of the polynomial, $r \leq 5$.

Table 2 The calculated values of IFPs $\langle \mu, \nu \rangle$

	C_1	C_2	C_3	C_4	C_5	C_6	C_7	C_8	C_9
C_1		⟨0.63,0.33⟩	⟨0.72,0.23⟩	⟨0.72,0.23⟩	⟨0.69,0.29⟩	⟨0.74,0.24⟩	⟨0.66,0.31⟩	⟨0.72,0.27⟩	⟨0.48,0.49⟩
C_2	⟨0.63,0.33⟩		⟨0.60,0.34⟩	⟨0.67,0.26⟩	⟨0.62,0.35⟩	⟨0.62,0.34⟩	⟨0.63,0.33⟩	⟨0.56,0.41⟩	⟨0.62,0.34⟩
C_3	⟨0.72,0.23⟩	⟨0.60,0.34⟩		**⟨0.90,0.04⟩**	⟨0.69,0.27⟩	⟨0.72,0.23⟩	⟨0.59,0.36⟩	⟨0.61,0.35⟩	⟨0.58,0.37⟩
C_4	⟨0.69,0.29⟩	⟨0.67,0.26⟩	⟨0.90,0.04⟩		**⟨0.87,0.12⟩**	⟨0.76,0.19⟩	⟨0.68,0.27⟩	⟨0.68,0.27⟩	⟨0.51,0.43⟩
C_5	⟨0.74,0.24⟩	⟨0.62,0.35⟩	⟨0.69,0.27⟩	⟨0.87,0.12⟩		⟨0.74,0.24⟩	⟨0.69,0.29⟩	⟨0.71,0.28⟩	⟨0.48,0.49⟩
C_6	⟨0.66,0.31⟩	⟨0.62,0.34⟩	⟨0.72,0.23⟩	⟨0.76,0.19⟩	⟨0.74,0.24⟩		**⟨0.88,0.12⟩**	⟨0.66,0.32⟩	⟨0.51,0.45⟩
C_7	⟨0.66,0.24⟩	⟨0.63,0.33⟩	⟨0.59,0.36⟩	⟨0.68,0.27⟩	⟨0.69,0.29⟩	⟨0.88,0.12⟩		⟨0.69,0.30⟩	⟨0.48,0.49⟩
C_8	⟨0.72,0.27⟩	⟨0.56,0.41⟩	⟨0.61,0.35⟩	⟨0.68,0.27⟩	⟨0.71,0.28⟩	⟨0.66,0.32⟩	⟨0.69,0.30⟩		⟨0.39,0.59⟩
C_9	⟨0.48,0.49⟩	⟨0.62,0.34⟩	⟨0.58,0.37⟩	⟨0.51,0.43⟩	⟨0.48,0.49⟩	⟨0.51,0.45⟩	⟨0.48,0.49⟩	⟨0.39,0.59⟩	

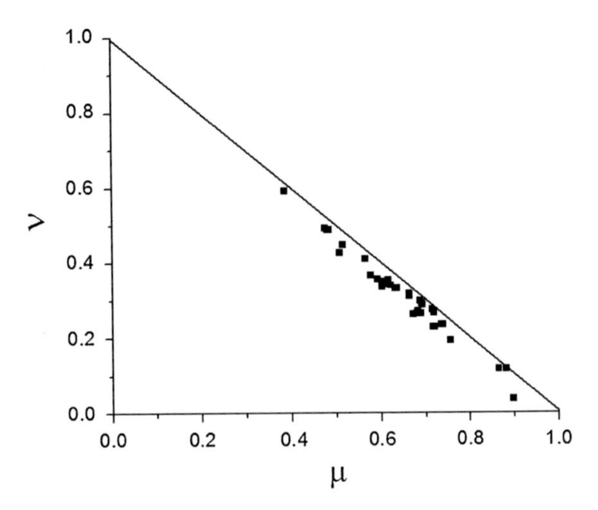

Fig. 3 Relations between μ and ν for the different criteria

The main trend shows the main tendencies in the alteration of the studied indices, and it is a straight line:

$$\mathbf{x}_{\mathrm{T}}^{\mathrm{M}} = A_0 + A_1 t \tag{3}$$

In contrast to the classical method for analyzing temporary series, where the Fourier series are used, the present research proposes the uses of the periodical functions of the type:

$$\mathbf{x}_{\mathrm{P}} = \sum_{k=0}^{p} b_k \sin(2\pi t/c_k + d_k) \tag{4}$$

where: p—number of the periodical functions; b_k, c_k and d_k—coefficients in the periodical functions, $k = 0,\ldots, p$.

The polynomial degree (2), and the number of the periodical functions p in (4) are determined on basis of the statistical criteria—experimental Fisher coefficient (F_E) and experimental correlation coefficient (R_E^2).

Then the model (1) for analysis and prognosis has the following form ($\varepsilon_t = 0$):

$$\mathbf{x} = \sum_{j=0}^{r} a_j t^j + \sum_{k=0}^{p} b_k \sin(2\pi t/c_k + d_k) \tag{5}$$

The established statistical model for water quality change can be used for modelling the water ecosystem condition. The results obtained by using a statistical water pollution model a together with the data of the National Monitoring Network can be used for water strategy management.

We have developed a validation of the models by the experimental correlation coefficient (R_E^2). The experimental correlation coefficients for the models are from

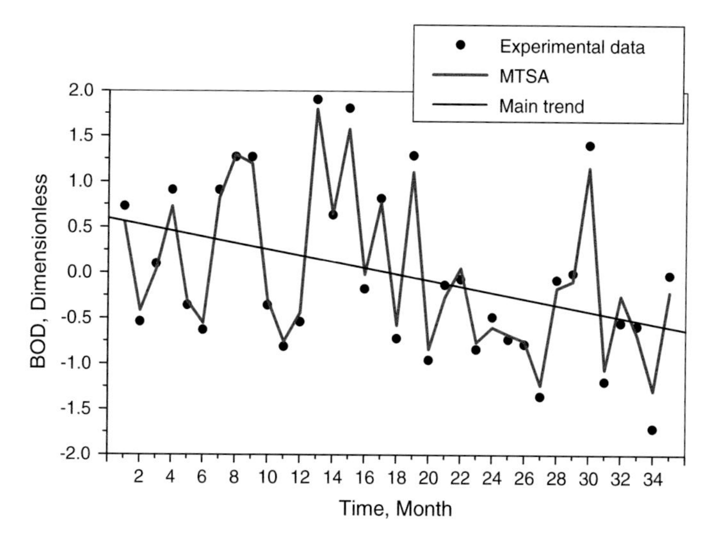

Fig. 4 Experimental and model data for BOD

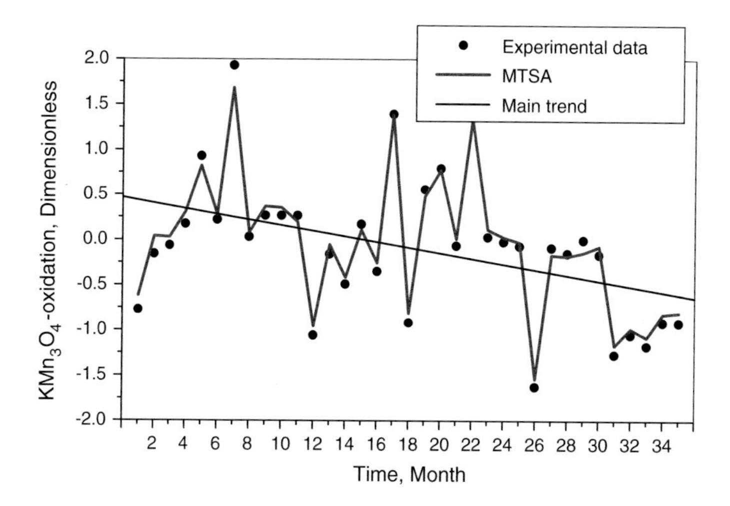

Fig. 5 Experimental and model data for KMn_3O_4-oxydation

$R_E^2 = 0.88$ to $R_E^2 = 0.95$. The tabular correlation coefficient is $R_T^2 = 0.35$ [17]. The results have shown that all models are adequate.

The simulation with MTSA method (5) and experimental data are shown in Figs. 4, 5, 6, 7 and 8.

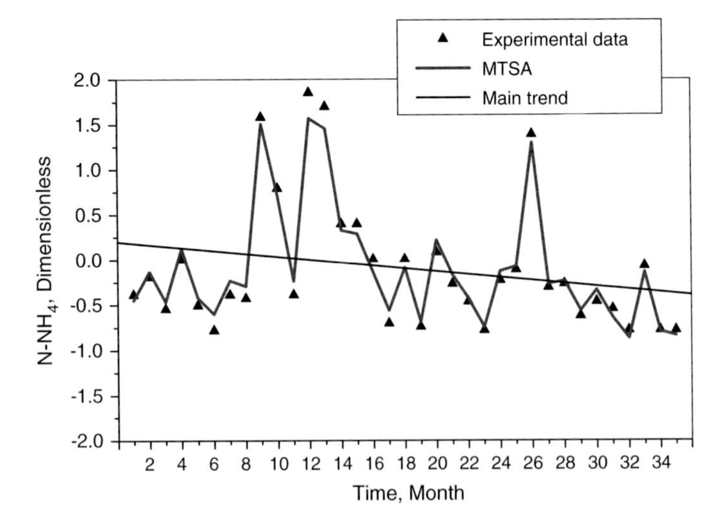

Fig. 6 Experimental and model data for $N\text{-}NH_4$

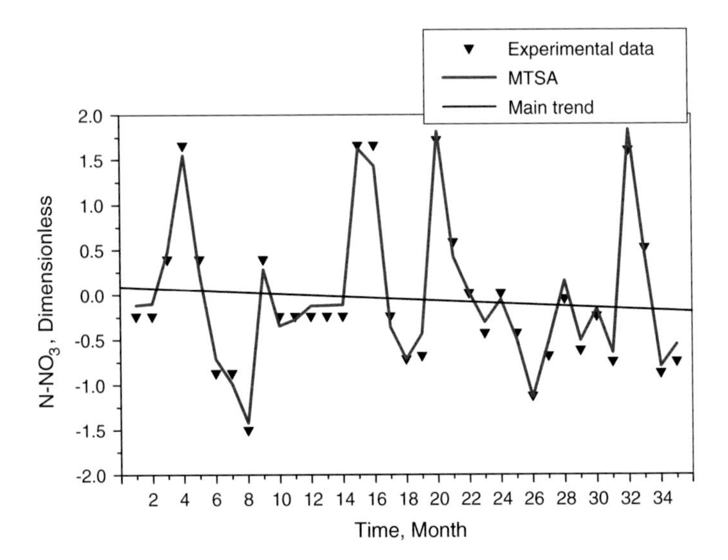

Fig. 7 Experimental and model data for $N\text{-}NO_3$

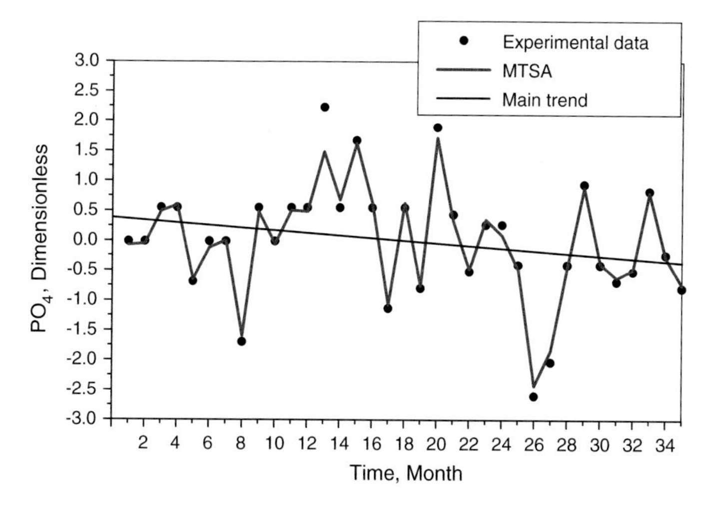

Fig. 8 Experimental and model data for PO_4

4 Conclusion

In this paper we have used a method named ICrA for assessment of index pollution at the different points of the Struma River in the Bulgarian area. The method has shown that there are positive consonances between different points, that means the measurement points are dependent criteria. The method also has presented are three positive consonances and dissonances between criteria. For these indexes that have dissonances and we could not develop a pollution model in which the criteria depend on each other. The ICrA proves that there are no correlations between criteria and each of these criteria are a time functions. The investigations are completed with in a part of stage of ICrA. In the following steps we will develop a commparise analysis with popular methods for multicriteria analysis.

We have modelled pollution by a MTSA method. The obtained models are adequate. Using of MTSA method for modeling of river ecosystems is completely enough. The integration of the determinate and statistical methods is necessary for the retrospective analysis and prognosis of water quality and the ecological processes in river flows.

Acknowledgements The authors are thankful for the support provided by the project DFNI-I-02-5/2014 "InterCriteria Analysis—New Approach for Decision Making", funded by the National Science Fund, Bulgarian Ministry of Education and Science.

References

1. Ilkova, T.C., Petrov, M.M: Investigation of the pollution of the Struma River by application of a modified times series analysis method. Modelling and Prognosis. Ecology and Safety. International Scientific Publication **2**, 495–506 (2008)
2. Ilkova, T.C., Petrov, M.M., Atanasova, M.P., Diadovski, I.K.: Modeling of the water pollution of the Struma River at the end of the Bulgarian part. J. Balkan Ecol. **9**, 435–441 (2007)
3. Ilkova, T.S., Petrov M.M.: Choice of Adequate Model for Modelling, Analysis and As-sessment of Water River Ecosystems. Int. J. Bioautomation **9**, 100–107 (2008)
4. Atanassov, K.T., Mavrov, D., Atanassova, V.K: Intercriteria decision making: a new approach for multicriteria decision making, based on index matrices and intuitionistic fuzzy sets. Issues Intuitionistic Fuzzy Sets Generalized Nets **11**, 1–8 (2014)
5. Atanassov, K.T.: Generalized Index Matrices: Comptesrendus de l'AcademieBulgare des Sciences **11**, 15–18 (1987)
6. Atanassov, K.T.: On index matrices, part 1: standard cases. Adv. Studies Contemp. Math. **20** (2), 291–302 (2010)
7. Atanassov, K.T.: On index matrices, part 2: intuitionistic fuzzy case. Proc. Jangjeon Math. Soc. **13**, 121–126 (2010)
8. Atanassov, K.T.: On Intuitionistic Fuzzy Sets Theory. Springer, Berlin (2012)
9. Atanassov, K.T., Szmidt, E., Kacprzyk, J.: On intuitionistic fuzzy pairs. Notes Intuitionistic Fuzzy Sets **19**, 1–13 (2013)
10. Atanassova, V.K., Doukovska, L., Mavrov, D., Atanassov, K.T.: InterCriteria decision making approach to eu member states competitiveness analysis: temporal and threshold analysis. Proc. IEEE Int. Syst. **1**, 97–106 (2014)
11. Atanassova, V.K., Doukovska, L., Atanassov, K.T., Mavrov, D.: Intercriteria decision making approach to EU member states competitiveness analysis. Proc. Int. Symp. Bus. Model. Soft. Des. **1**, 289–294 (2014)
12. Atanassova, V.K., Vardeva, I: Sum- and average-based approach to criteria shortlisting in the InterCriteria analysis. Notes Intuitionistic Fuzzy Sets **20**(4), 41–46 (2014)
13. Atanassova, V.K., Mavrov, D., Doukovska, L., Atanassov, K.T.: Discussion on the threshold values in the InterCriteria decision making approach. Notes Intuitionistic Fuzzy Sets **20**(2), 94–99 (2014)
14. Ilkova, T.S., Petrov, M.M.: Application of InterCriteria analysis to the Mesta River pollution modelling. Notes Intuitionistic Fuzzy Sets **21**(2), 118–125 (2015)
15. Lange, H.: Time-series Analysis in Ecology. Wiley (2001)
16. Nicklow, J.: Discrete-time optimal control for water resources engineering and management. Water Int. **25**(1), 89–95 (2000)
17. Stoyanov, S.: Optimization Methods and Algorithms. Technique, Sofia (1990) (in Bulgarian)

Application of the InterCriteria Decision Making Method to Universities Ranking

Maciej Krawczak, Veselina Bureva, Evdokia Sotirova and Eulalia Szmidt

Abstract In this paper we present an application of the InterCriteria Decision Making (ICDM) approach to real data extracted from the Polish University Ranking System [13] in the years 2012–2014. The aim is to analyze the correlations between the indicators used by the Ranking System.

Keywords Intuitionistic fuzzy intercriteria analysis method • Intuitionistic fuzzy sets • Index matrix • University ratings • Multicriteria decision making

1 Introduction

In this paper we present the second application of the ICDM method for the ratings of universities. The purpose of this development is to identify the most correlated indicators in the Ranking System for the Polish universities. By applying the ICDM approach over extracted data for ratings of the universities, we can find the indicators that have the highest dependencies. In this way we can observe the behavior of them in time (several years). Analogously we can receive the opposite indicators or indicators that frequently are independent from each other. In the current

M. Krawczak (✉) · E. Szmidt
Systems Research Institute, Polish Academy of Sciences, Newelska 6, 01-447 Warsaw, Poland
e-mail: krawczak@ibspan.waw.pl

E. Szmidt
e-mail: eulalia.szmidt@ibspan.waw.pl

V. Bureva · E. Sotirova
Laboratory of Intelligent Systems, University Professor Dr. Assen Zlatarov Burgas, 8010 Burgas, Bulgaria
e-mail: vbureva@btu.bg

E. Sotirova
e-mail: esotirova@btu.bg

K.T. Atanassov et al. (eds.), *Novel Developments in Uncertainty Representation and Processing*, Advances in Intelligent Systems and Computing 401,
DOI 10.1007/978-3-319-26211-6_31

investigation we analyze the data over the period 2012–2014. This application of the ICDM method using Ratings of Polish University Ranking System can help to determine the precision and confirm the current weights of the indicators [13].

We explore real data extracted from Universities Ranking System, i.e., from the sites of a relevant rating system which provide free access to data. Using InterCriteria Decision Making (ICDM) approach the behavior of the objects or criteria can be monitoring and optimized. For illustration we use data sets from overall ranking of the academic institutions (universities) of Perspektywy University Ranking in Poland [13].

2 Presentation of the ICDM Analysis

The ICDM method helps to discover the relationships and examine the correlations between the indicators used in the Bulgarian university ratings. The ICDM method is introduced by Atanassov et al. [1]. Several applications of the method have been already published [5–9]. The method is based on the theory of the intuitionistic fuzzy sets and the index matrices. The intuitionistic fuzzy sets are defined by Atanassov [2, 4]. They are an extension of the concept of fuzzy sets defined by Zadeh [11]. The theory of index matrices is introduced in [3].

The objects can be estimated on the base of several criteria. The number of the criteria can be reduced by taking into account the correlations of each pair of criteria presented in the form of intuitionistic fuzzy pairs of values [2]. The intuitionistic fuzzy pairs of values are the intuitionistic fuzzy evaluations in the interval [0, 1]. The relations can be established between any two group of indicators C_w and C_t.

Let us have a number of C_q group of indicators, $q = 1, \ldots, n$, and a number of O_p universities, $p = 1, \ldots, m$. So we use the following sets: a set of group of indicators $C_q = \{C_1, \ldots, C_n\}$ and a set of universities $O_p = \{O_1, \ldots, O_m\}$.

We will evaluate 13 universities (objects) using 6 groups of criteria. We obtain an index matrix M that contains two sets of indeces, one for rows and another for columns. For for every p, q $(1 \le p \le m,\ 1 \le q \le n)$, O_p in an evaluated object, C_q is an evaluation criterion, and $a_{Op,Cq}$ is the evaluation of the p-th object against the q-th criterion, defined as a real number or another object that is comparable according to relation R with all the rest elements of the index matrix M.

$$M = \begin{array}{c|cccccccc} & C_1 & \cdots & C_k & \cdots & C_l & \cdots & C_n \\ \hline O_1 & a_{O_1,C_1} & \cdots & a_{O_1,C_k} & \cdots & a_{O_1,C_l} & \cdots & a_{O_1,C_n} \\ \cdots & \cdots & \cdots & \cdots & \cdots & \cdots & \cdots & \cdots \\ O_i & a_{O_i,C_1} & \cdots & a_{O_i,C_k} & \cdots & a_{O_i,C_l} & \cdots & a_{O_i,C_n} \\ \cdots & \cdots & \cdots & \cdots & \cdots & \cdots & \cdots & \cdots \\ O_j & a_{O_j,C_1} & \cdots & a_{O_j,C_k} & \cdots & a_{O_j,C_l} & \cdots & a_{O_j,C_n} \\ \cdots & \cdots & \cdots & \cdots & \cdots & \cdots & \cdots & \cdots \\ O_m & a_{O_m,C_1} & & a_{O_m,C_k} & & a_{O_m,C_l} & \cdots & a_{O_m,C_n} \end{array},$$

The next step is applying the InterCriteria Analysis for calculating evaluations.

From the requirement for comparability above, it follows that for each i, j, k, l the relation $R(a_{Oi,Ck}, a_{Oj,Ck})$ holds. The relation R has dual relation $\bar{R}$, which is true in the case when relation R is false, and vice versa.

The pairwise comparisons between every two different criteria are made along all evaluated objects. During the comparison, it is maintained one counter of the number of times when the relation R holds, and another counter for the dual relation $\bar{R}$.

Let $S^{\mu}_{k,l}$ be the number of cases in which the relations $R(a_{Oi,Ck}, a_{Oj,Ck})$ and $R(a_{Oi,Cl}, a_{Oj,Cl})$ are simultaneously satisfied. Let also $S^{\nu}_{k,l}$ be the number of cases in which the relations $R(a_{Oi,Ck}, a_{Oj,Ck})$ and its dual $\bar{R}(a_{Oi,Cl}, a_{Oj,Cl})$ are simultaneously satisfied. As the total number of pairwise comparisons between the objects is $\frac{m(m-1)}{2}$, then the following inequality is held:

$$0 \le S^{\mu}_{k,l} + S^{\nu}_{k,l} \le \frac{m(m-1)}{2}.$$

For every k, l, such that $1 \le k \le l \le n$, and for $m \ge 2$ two numbers are defined:

$$\mu_{C_k,C_l} = 2\frac{S^{\mu}_{k,l}}{m(m-1)}, \nu_{C_k,C_l} = 2\frac{S^{\nu}_{k,l}}{m(m-1)}.$$

The pair constructed from these two numbers plays the role of the intuitionistic fuzzy evaluation of the relations that can be established between any two criteria C_k and C_l. In this way the index matrix M that relates evaluated objects with evaluating criteria can be transformed to another index matrix M^* that gives the relations among the criteria:

The pair constructed from these two numbers plays the role of the intuitionistic fuzzy evaluation of the relations that can be established between any two criteria C_k and C_l. In this way the index matrix M that relates evaluated objects with evaluating criteria can be transformed to another index matrix M^* that gives the relations among the criteria:

$$M^* = \begin{array}{c|ccc} & C_1 & \dots & C_n \\ \hline C_1 & \langle \mu_{C_1,C_1}, \nu_{C_1,C_1} \rangle & \dots & \langle \mu_{C_1,C_n}, \nu_{C_1,C_n} \rangle \\ \dots & \dots & \dots & \dots \\ C_n & \langle \mu_{C_q,C_1}, \nu_{C_q,C_1} \rangle & \dots & \langle \mu_{C_n,C_n}, \nu_{C_n,C_n} \rangle \end{array}$$

The last step of the algorithm is to determine the degrees of correlation between the indicators depending of the chosen threshold for μ and ν from the user. The correlations between the criteria are called "positive consonance", "negative consonance" or "dissonance".

The practical considerations have shown that it is more flexible to work with two index matrices M^{μ} and M^{ν}, rather than with the index matrix M^* of IF pairs.

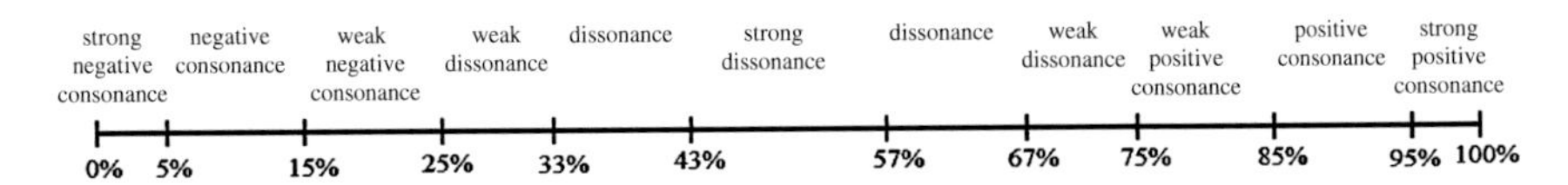

Fig. 1 Scale for determination of the type of the correlations between the criteria

The final step of the algorithm is to determine the degrees of correlation between the criteria, depending on the user's choice of μ and ν. We call these correlations between the criteria as: 'positive consonance', 'negative consonance' or 'dissonance'. Let $\alpha, \beta \in [0; 1]$ be the threshold values, against which we compare the values of $\mu_{Ck,Cl}$ and $\nu_{Ck,Cl}$. We call that criteria C_k and C_l are in:

- (α, β)-positive consonance, if $\mu_{Ck,Cl} > \alpha$ and $\nu_{Ck,Cl} < \beta$;
- (α, β)-negative consonance, if $\mu_{Ck,Cl} < \beta$ and $\nu_{Ck,Cl} > \alpha$;
- (α, β)-dissonance, otherwise.

Certainly, the larger α and/or the smaller β, the less number of criteria may be simultaneously connected with the relation of (α, β)-positive consonance. For practical purposes, the most information is carried when either the positive or the negative consonance is as large as possible, while the cases of dissonance are less informative and are skipped.

Here we use the scale that is shown in Fig. 1 [10].

3 Application of the ICDM to the Polish University Ranking System

The Polish University Ranking System contains information on 83 accredited universities in Poland, which offer education in a variety of majors [12, 13].

The ranking system contains information and data expressed by more than 33 indicators, which measure different aspects of university activities including prestige, innovation, academic potential, academic effectiveness, teaching and learning internationalisation. The final assessment is provided in the range from 0 to 100 [13].

The Perspektywy Ranking consists in fact of several rankings: the main and most important ranking is the overall ranking of academic institutions (universities), but it also includes: ranking of master level private institutions and ranking of state vocational schools. Rankings of 39 disciplines (fields ranking) have also been published.

Perspektywy University Ranking (Poland) is the first national university ranking in the World to pass the rigorous audit and to receive the "IREG Approved" awarded by the IREG Observatory on Academic Ranking and Excellence [13].

In the current paper the ICDM method is applied over Polish ratings of universities in the years 2012–2014. In the further text the indicators are named by numbers. The numbering of the indicators is following:

1 employer reputation
2 academic reputation (teaching)
3 international recognition
4 talented students application
5 patents and licenses
6 EU funding
7 infrastructure for innovation
8 parametric evaluation
9 right to confer PhD with habilitation degree
10 rights to confer PhD degrees
11 staff with highest qualifications
12 accreditations
13 faculty development
14 academic titles awarded
15 external funding for research and development
16 publications
17 citations
18 h-index
19 EU programmes
20 PhD students
21 students—teaching staff
22 e-holdings
23 printed library holdings
24 library facilities
25 support for students' scientific interests
26 sports achievements
27 programs in foreign languages
28 students studying in foreign language
29 student exchange (outbound)
30 student exchange (inbound)
31 international students
32 foreign teaching staff
33 multicultural composition of students body

In Table 1 the number of pairs of criteria for years 2012, 2013 and 2014th for the ratings of universities obtained by applying the ICDM method are shown.

In the case of the ICDM method we are interested in indicators that are in positive consonance. The pairs of indicators in positive consonance and weak positive consonance in 2012, 2013 and 2014 for the ratings of universities are shown below.

Table 1 Number of pairs of criteria

Type of correlations	Number of pairs of criteria for a year		
	2012	2013	2014
Positive consonance [0,85; 0,95)	3	0	1
Weak positive consonance [0,75; 0,85)	12	7	12
Weak dissonance [0,67; 0,75)	55	69	93
Dissonance [0,57; 0,67)	209	229	185
Strong dissonance [0,43; 0,57)	215	190	145
Dissonance [0,33; 0,43)	34	33	58
Weak dissonance [0,25; 0,33)	0	0	26
Weak negative consonance [0,15; 0,25)	0	0	27

3.1 Pair of Criteria in Positive Consonance [0,85; 0,95)

- for 2012 year: 16–17, 16–18, 17–18;
- for 2013 year: -;
- for 2014 year: 18–19;

3.2 Pair of Criteria in Weak Positive Consonance [0,75; 0,85)

- for 2012 year: 1–2, 1–10, 2-10, 2–17, 2–18, 6–10, 6–17, 10–17, 10–18, 14–20, 27–28, 29–30;
- for 2013 year: 1–2, 2–14, 14–20, 16–18, 17–18, 27–28, 29–30;
- for 2014 year: 1–3, 2–3, 2–13, 3–12, 3–13, 3–15, 12–15, 17–18, 15–21, 17–22, 28–29, 32–33.

Via the comparison of the results over the period of research (2012–2014) the following outcomes are obtained:

- According to the scale for determination of the type of the correlations from Fig. 1 the indicators are previously in weak dissonance, dissonance or strong dissonance. The is no strong dependences, i.e. the indicators are well chosen.
- The indicators "publications", "citations" and "h-index" become more independent. The correlation between indicator "publications" and "citations" changes from positive consonance in 2012 to weak dissonance in 2013 and 2014. The correlation between indicator "publications" and "h-index" changes from positive consonance in 2012, weak dissonance in 2013 to dissonance in 2014. The correlation between indicator "citations" and "h-index" changes from positive consonance in 2012 to dissonance in 2013 and 2014.

- For the pair of indicators “h-index” and “EU programmes” the correlations increase, namely, from dissonance in 2012 and 2013 to positive consonance in 2014.
- For the pairs of indicators “employer reputation”—“academic reputation (teaching)”, “EU funding” and “student exchange (outbound)”—“student exchange (inbound)” the correlations decrease from weak positive consonance to weak dissonance. The correlations between indicator “academic reputation (teaching)” and indicators “rights to confer PhD degrees”, “citations” and “h-index”, and between indicator ”rights to confer PhD degrees” and indicators “employer reputation” and “citations” change from weak positive consonance to dissonance. A similar trend, i.e., a change from weak positive consonance to strong dissonance can be observed for the correlation in the pairs of criteria: “EU funding”—“rights to confer PhD degrees”, “rights to confer PhD degrees”—“h-index”, “academic titles awarded”—“PhD students” and “programs in foreign languages”—“students studying in foreign language”.
- For the pairs of indicators “international recognition”—“faculty development”, “accreditations”—“external funding for research and development”, “external funding for research and development”—“students—teaching staff”, “foreign teaching staff—“multicultural composition of students body”, “employer reputation—“international recognition”, “academic reputation (teaching)—“international recognition”, “international recognition—“accreditations”, “academic reputation (teaching)—“faculty development”, “international recognition —“external funding for research and development”, “students studying in foreign language—“student exchange (outbound)”, “citations—“e-holdings” the correlations increase from strong dissonance or dissonance to weak positive consonance.

Having the above observations in mind we can conclude that ICDM method makes it possible to observe which indicator is strongly dependent with others and whether the correlation appears periodically. In order to determine the behavior of each indicator over time we should observe the results of the application of InterCriteria Decision Making (ICDM) method for several years. If such a criterion has a strong correlation, again, in the next step we can try to ignore it. Therefore InterCriteria Decision Making (ICDM) method is helpful for determining the behavior of the indicators. When comparing the results of applying InterCriteria Decision Making (ICDM) approach over the data from Polish university rankings over the years we can observe the possible differences or changes between them.

4 Conclusions

We used the ICDM method to find some hidden patterns in the data from Perspektywy University Ranking. Ratings of Polish Universities are compiled over three years. We analyzed the data to identify the best correlations between the indicators, to discover dependent and independent indicators and the relationships between them. The comparison can help to describe the behavior of the used indicators and their assessment. In the next research the authors will analyze the indicators individually—it will make possible to compare a single indicator with all the rest ones.

The increase of the coefficient of consonance and the entry in the zone of strong positive consonance means strong correlation between the respective pair of criteria, which may justify the removal of one of the criteria in the pair on the basis that its informational values is lesser. Removal of indicators leads to simplification of the process of evaluation.

Acknowledgments The authors are thankful for the support provided by the Bulgarian National Science Fund under Grant Ref. No. DFNI-I-02-5 "InterCriteria Analysis: A New Approach to Decision Making".

References

1. Atanassov, K., Mavrov, D., Atanassova, V.: Intercriteria decision making: a new approach for multicriteria decision making, based on index matrices and intuitionistic fuzzy sets. Issues Intuitionistic Fuzzy Sets Generalized Nets **11**, 1–8 (2014)
2. Atanassov, K., Szmidt, E., Kacprzyk, J.: On intuitionistic fuzzy pairs. Notes Intuitionistic Fuzzy Sets **19**(3), 1–13 (2013)
3. Atanassov, K.: Index Matrices: Towards an Augmented Matrix Calculus. Studies in computational intelligence series, vol. 573, Springer, Cham (2014)
4. Atanassov, K.: On Intuitionistic Fuzzy Sets Theory. Springer, Berlin (2012)
5. Atanassova, V., Mavrov, D., Doukovska, L., Atanassov, K.: Discussion on the threshold values in the intercriteria decision making approach. Notes on Intuitionistic Fuzzy Sets **20**(2), 94–99 (2014)
6. Atanassova, V., Doukovska, L., Atanassov, K. Mavrov, D.: InterCriteria decision making approach to eu member states competitiveness analysis. In: Proceedings of the International Symposium on Business Modeling and Software Design—BMSD' 14, 24–26 June 2014, Luxembourg, Grand Duchy of Luxembourg, pp. 289–294 (2014)
7. Atanassova, V., Vardeva, I.: Sum- and average-based approach to criteria short-listing in the intercriteria analysis. Notes Intuitionistic Fuzzy Sets **20**(4), 41–46 (2014)
8. Atanassova, V., Doukovska, L., Karastoyanov, D. & Capkovic, F.: InterCriteria decision making approach to eu member states competitiveness analysis: trend analysis. In: Angelov P. et al. (eds.) Intelligent Systems'2014, Advances in Intelligent Systems and Computing, vol. 322, pp. 107–115 (2014)
9. Atanassova, V., Doukovska, L., Mavrov, D., Atanassov, K.: InterCriteria decision making approach to eu member states competitiveness analysis: temporal and threshold analysis. In: Angelov P. et al. (eds.) Intelligent Systems'2014, Advances in Intelligent Systems and Computing, vol. 322, pp. 95–106 (2014)
10. Atanassov, K., Atanassova, V., Gluhchev, G.: Inter criteria analysis: ideas and problems. Notes Intuitionistic Fuzzy Sets **21**(1), 81–88 (2015)
11. Zadeh, L.A.: Fuzzy sets. Inf. Control **8**, 333–353 (1965)
12. http://www.ranking.perspektywy.org/2013/
13. http://www.ranking.perspektywy.org/2014/

The Algorithms of Automation of the Process of Creating Acoustic Units Databases in the Polish Speech Synthesis

Janusz Rafałko

Abstract This paper presents the new approach of creating the database of acoustic units in concatenative TTS synthesis. Nowadays databases like this are created manually, which is very time-consuming and takes at least several months of work. Creation such base in automatic way shortens this time to hours. One of the next problem in the concatenative synthesis is the problem of reproduction any text using a voice and a way of speaking of particular man. Presented algorithms allow to create the allophone units database of particular man after recciving a sample of his voice and as a result synthesizer speaking with exactly this voice.

1 Introduction

The main goal in speech technology is creation a speech which is almost as real as a voice of living person. In this point arises the task of full keeping the personal and acoustic voice properties, phonetic articulation, accent properties and prosodic individuality of speech. Within this task, this study was made and it shows algorithms which allow to create a database of particular speakers voices (acoustic units databases) in an automatic way.

One of the methods of speech synthesis based on the text (Text to Speech), which allows to reproduce the human personal speech characteristics is a concatenation method, which uses small and natural acoustic units, from which the speech is synthesised. These can be allophones, diphones or syllables. That type of system synthesizes the speech by joining the acoustic units in accordance of appropriate phonetic rules. The individual features of human voice are not included in this rules, but only in natural acoustic units and individual, prosodic voice characteristics, such as intonation. In order to synthesize the voice of particular man, there must be acoustic units database created.

J. Rafałko (✉)
Faculty of Mathematics and Information Science, Warsaw University of Technology, Warszawa, Poland
e-mail: j.rafalko@mini.pw.edu.pl

K.T. Atanassov et al. (eds.), *Novel Developments in Uncertainty Representation and Processing*, Advances in Intelligent Systems and Computing 401,
DOI 10.1007/978-3-319-26211-6_32

2 Acoustic Units Database

Different approaches to the speech synthesis from the text are described in details in [1, 2]. The basic feature of compilatory approach of speech synthesis is the use of elementary pieces of natural speech [3]. In this way, the necessity of modelling the complexed acoustic processes of synthesized speech is excluded. In the synthesizer, the signal compiled from natural speech segments is a subject of further modification which change the prosodic parameters of the signal.

In the [4] study are shown the basic assumptions of concatenative TTS system for Polish language, based on allophones in the context of multilingual synthesis. The natural elements from which the speech is synthesized may be allophones, diphones, multiphones as well as syllables. The sequence of phonetic elements is given to the signal processing block, which selects the appropriate sound realizations from the natural speech segments database and joins them into continuous speech signal. The generated signal is ginen to the acoustic processing block. This block performs the appropriate modifications of speech signal prosodic parameters in accordance to the input prosodic markings. In compilatory method of speech synthesis the type of basic speech units have very influence on the obtaining individual speech characteristics. This paper refers to the databases of acoustic units, which include several context groups of particular phoneme, which may be identified with acoustic allophone, described in the study of Victor Jassem [5]. The advantages of the choice of allophones as a basic units [6–9] base on the fact, that first of all, speech units remain the effects of sounds interference, and secondly, the number of basic units is relatively low holds in the range of 400–2000 in different systems. The difficulty of this approach is a necessity of precise allophones marking during the segmentation of natural speech signal. In most cases, determining the exact limit is not a difficult task [8], because the beginning and the end of particular allophone is clear. However, there are some cases, in which determining the exact limit is difficult. This is due to the co-articulation and assimilation phonemes, i.e. the influence of some phones one to another.

3 The Technology of Automatic Segmentation and Creation the Acoustic Units Databases

If in the speech synthesis the compilation elements contain only the phonetical and acoustical characteristics, the segmentation task is about to “cutting” the basic segments from the speech stream and placing them into the database.

The general scheme of acoustic units databases creation technology is shown on the Fig. 1. The main stages of this algorithms include:

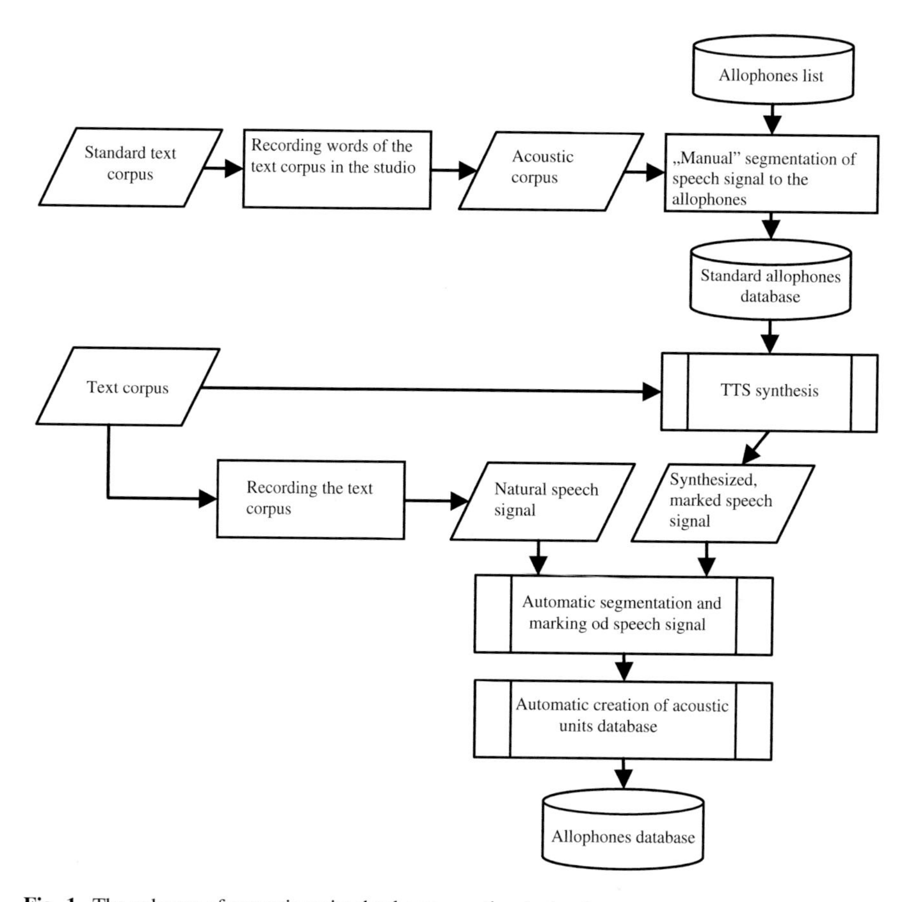

Fig. 1 The scheme of acoustic units databases creation technology

1. Selection and preparation of text and acoustic corpuses.
2. Manual segmentation of speech signal to the allophones.
3. The automation of creation of acoustic units databases of the particular speaker voice.

There are about 1800 words in the prepared standard text corpus. The acoustic corpus that is used in this system is created on the base of professional speaker records. The standard allophones database also include about 1800 allophones.

To create the new allophone database of particular voice, we can use the same standard text corpus and on its basis create new record. In this system, the automatic segmentation of speech signal is based on the DTW algorithm, however, in contrast to the classical DTW, it is not based on the signal in the time domain but on the frequency domain. Standard (synthesized) and natural speech signal is divided into frames, which may overlap and the Fourier transform (FFT) is being calculated in each frame. The first step in the DTW method is calculating the local distances matrix. This matrix are calculated for the spectral features vectors in every frame:

$$c(n,m) = \|S(n), E(m)\| = \sum_{k=1}^{K} |S(n,k) - E(m,k)| \tag{1}$$

where:

S(n)—the synthesized signal spectral features vector in the "n" frame
E(m)—the natural signal spectral features vector in the "m" frame
k—the length of spectral feature vector

The synthesized signal frame is correlated with natural signal frame and on the base of signal spectrum in the frame, the distance between vectors is calculated. The next step is calculation of global distances matrix and alignment path. It is shown on the Fig. 2. This figure shows the speech signals, which uses frames with the length of 256 samples, the Hamming window and overlapping of 200 samples. The used metrics is a Manhattan metrics. The blue colour means the small distances values between signal frames, and the red one means a long distance, that is signals with different frequencies.

We can also see the natural speech signal allophones borders determined on the basis of alignment path and limits of allophones from the synthesized signal. This is

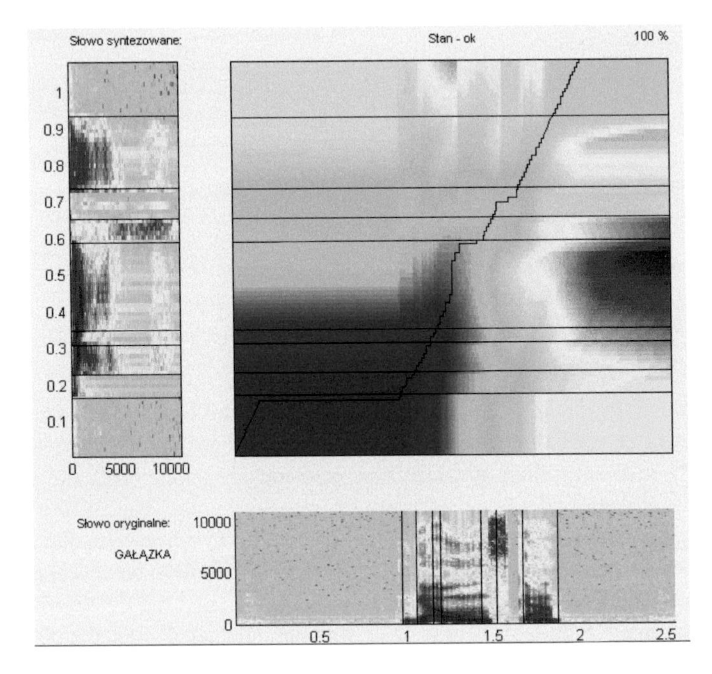

Fig. 2 Global distances matrix

the essence of units marking and cutting automation. We do not mould the signal to the standard one, but using the alignment path we determine the allophones boundaries.

4 Quality Control Algorithms

As a result of this system work, we get many of the same units. This is because the corpus of the speech is surplus, it contains whole words, sentences and texts in which the particular unit is presented many times. On the other hand, we need only one piece of each element to our base. In order to create an acoustic database it is necessary to analyze received units in details, and delete those phonetic units, in which the acceptable error during reading or automatic segmentation was exceeded and save only the best of them—operation called “rejection”. If there are more than one identical allophones, choose the best of—operation called “selection”. And finally we must perform the control of quality of each element that left and mark deviations and perform the correction of segments parameters with noticed deviations—“correction”.

The “rejection” operation consists of correlation of synthesized acoustic characteristics and natural speech segments obtained from the segmentation process. When the differences between them are higher than the threshold quantity, it means, that such segment will not be able to provide the minimum level of necessary quality of synthesized speech and should be rejected. This operation is performed by testing the time and acoustically-phonetical parameters:

- Duration
- Cost of matching

The T_T duration of tested unit, obtained in the segmentation process is compared with the T_W duration of the standard unit used in the speech synthesis. Since the allophone units differs in length, we cannot take the absolute error as a measure, we only can take the relative error in the respect of the standard allophone length, according to the formula (2).

$$\delta_t = \frac{|T_T - T_W|}{T_W} > \alpha \tag{2}$$

If the units duration relative error is greater than α threshold quantity, the element is rejected. It was determined during experiments, that if the error exceeds 80 % the allophone is not suitable for primary base. There is about 9 % of this kind of allophones in the exemplary set.

The second parameter in this operation is the cost of both units, standard and tested, matching in the DTW algorithm. This cost is the sum of local distances within the alignment path determined for those units. If we mark on the alignment path the number of points for particular element by *P*, then the cost of such matching will be:

$$C_P = \sum_{p=1}^{P} c(x_p, y_p) \tag{3}$$

This cost, similarly as a unit duration, may differ for particular units. In the tested units sets, the cost contained within the range of 0.05 to approx 50. Just as before, the relative coefficient ought to be done. The 4 formula shows this coefficient:

$$\delta_C = \frac{|C_P - C_{\mathrm{Psr}}|}{C_{\mathrm{Psr}}} > \beta \tag{4}$$

C_P—the allophone unit matching cost
C_{Psr}—the medium matching cost for all instances for particular unit

In this case, for some sets, also was experimentally determined that, the error greater than 50 % disqualifies the unit for use in the primary base. This is about 3.5 % of total number of units. As a result of the "rejection" operation, all pieces for which $\delta_t > \alpha$ lub $\delta_C > \beta$ are excluded from further treatment. As the experiments shown, the best results of "rejection" operation are obtained when $\alpha = 0.8$ and $\beta = 0.5$.

For other pieces that left after "rejection" operation, the "selection" operation is used, as a result of we get the best representative of any unit. The most typical unit due to the value of prosodic characteristics must be chosen: the F_0 frequency of basic tone, amplitude A and duration T. Since "selection" operation, the unit with characteristics the most approximate to the medium values of this parameters are selected. If after the "rejection" operation the number of pieces of particular unit equals n, then the duration T_i, average amplitude value A_i^{AVE} and the average value of basic tone F_i^{ave} is calculated. For the soundless units, the basic tone frequency value equals zero. Then, the average values T^{ave}, A^{ave} an F_0^{ave} for all sets of units of one kind are calculated.

The selection coefficient, normalised in the scale [0…1] is set as:

$$D_i = \frac{1}{3}\left(\frac{|T_i - T^{AVE}|}{\max_{j=1}^{n}|T_j - T^{AVE}|} + \frac{|A_i^{ave} - A^{AVE}|}{\max_{j=1}^{n}|A_j^{ave} - A^{AVE}|} + \frac{|F_{0i}^{ave} - F_0^{AVE}|}{\max_{j=1}^{n}|F_{0j}^{ave} - F_0^{AVE}|}\right) \tag{5}$$

As a result of "selection" operation, the allophone piece, which meets this condition, is selected:

$$k = \arg \min_{i=1}^{n}(D_i) \tag{6}$$

The "correction" operation is performed for those segments, which are obtained in accordance to above mentioned parameters. The segments with inaccurate defined limits are subject to correction by use of proper proceedings, involving removing the inaccurate periods of basic tone and inserting the missing terminal periods of basic tone. The diagnosis of determining the limits of units is performed by determining the level of time and acoustic signal characteristics similarities on the first and the second period of element basic tone, as well as on the penultimate and the last one. The "correction" is performed only in the case of voiced units.

To determine the level of similarity of terminal and pre-terminal period of acoustic characteristics, the (formula 7a) is used. The distance between time characteristics is calculated as a ratio of duration of these periods (formula 7b).

$$\text{(a)}\ L_A = 1 - \frac{\sum_{i=1}^{N} \left| s_i^G - s_i^P \right|}{\sum_{i=1}^{N} \left(\left| s_i^G \right| + \left| s_i^P \right| \right)} \qquad \text{(b)}\ L_T = 1 - \frac{\min(T^G, T^P)}{\max(T^G, T^P)} \tag{7}$$

S_i^G—value of the signal at the "i" segment of terminal period
S_i^P—value of the signal at the "i" segment of pre-terminal period
T^G, T^P—duration of periods, both the terminal and the pre-terminal

The "correction" deletes the terminal period and doubles the pre-terminal one. As a result, the number of basic tone periods doesn't change. The second "correction" case is a complete rejection of terminal period. However, in the case of proper marking of the unit limit, such period remains. The analysis of all this cases for different databases had led to the selection of appropriate values of this coefficients and determination, when the particular allophone should be revised and how. If:

- $L_T < 0.2$ and $L_A > 0.6$—terminal period doesn't change
- $0.2 < L_T < 0.7$ and $L_A > 0.4$—terminal period replaced by the pre-terminal one
- $L_T > 0.7$ or $L_A < 0.4$—terminal period deleted

5 The Experiments

In the case of synthesized speech, the most important assessments include articulation, transparency and effortlessness, the best methods to assess are subjective ones. In the case of this study, acoustic units databases are created, and they will be use to synthesize speech, therefore, the studies were about the comparison of the speech obtained from the automatic databases derived from the system and the test databases obtained in manual way.

The most commonly used subjective speech testing methods are [10–12]:

- ACR (Absolute Category Rating)—the method of absolute speech quality assessment.
- DCR (Degradation Category Rating)—the method describing the level of speech quality degradation.

- CCR (Comparison Category Rating)—speech quality assessment comparison method.
- The logatom articulation research method.

The ACR and DCR are described and recommended by the International Telecommunication Unit (ITU) to assess the quality of speech signal transmission in the analog and digital telecommunications channels and speech encoding systems [13]. The logatom method is described in the Polish Standard PN-90/T-05100 [14]. The studies were performed on two databases obtained in automatic way. The database of professional radio speaker—Speaker, and the voice of this study author—Janusz. The study involved 73 people.

5.1 ACR Method

In the ACR method the absolute quality of presented voice samples is determined, without the use of a reference signal. Then the MOS parameter is calculated (Mean Opinion Score)—the averaged listeners opinion, which characterizes the sound quality [10, 12]. The participants listen to the recorded speech, and then they assesses it in the scale from 1 to 5. In this case, two scales recommended by the ITU were used:

- Listening quality scale
- Auditory effort scale

The assessment of automatically obtained databases were performed in comparison with the standard databases obtained manually. This means that, the listeners, who assessed the speaker synthesized voice, assessed the samples created on the basis of the automatic database, as well as the standard database samples. The order of the samples was random. Based on the results, the MOS parameters for particular databases were obtained. The results are shown in the Table 1:

As the table shows, in any case, the assessment was a little below 4, what means, that the synthesized speech quality was good and allowed to understand the speech without any difficulties with a light intensity of attention. The standard deviation

Table 1 The results of the ACR method tests

Databases	Speaker standard	Speaker automatic	Janusz standard	Janusz automatic
Absolute assessment—speech quality				
MOS	3.96	3.85	3.72	3.69
Standard deviation	0.68	0.58	0.53	0.46
Auditory effort				
MOS	3.96	3.9	3.81	3.81
Standard deviation	0.65	0.59	0.54	0.53

calculated from the sample keeps the level a little above 0.5 what means, that were not huge differences between listeners. The standard databases looked better in this assessment, what should be understandable. The "Janusz" database was assessed a little lower than, the "Speaker" database. Important is, that the "Janusz" database was created on the basis of another speaker voice, and yet it practically doesn't differ from the standard database.

5.2 *DCR Method*

The another method is DCR [10, 12]. This method is used for examining so-called speech degradation degree. The measurement is based on the comparison of the standard (natural) signal with the examining one, determining its degradation in the five-point scale (from "imperceptible" to "very clear"). On the basis of obtained results, the DMOS coefficient (degradation mean opinion score) was determined. Results are shown in the Table 2. In the case of this coefficient, we can see that, the "Janusz" database, both standard and automatic, got higher score.

5.3 *CCR Method*

In this method, similar as in the previous method, two samples are presented, one of them is the standard—natural one. This time, the order of samples is random. On the basis of the results, the CMOS coefficient (Comparison Mean Opinion Score) is calculated—comparative, averaged listeners opinion. The sound samples are random, which means that, the examined sound may be assessed better than the standard one. The sample that is compared may be better or worse quality, which means that, the assessment can be either positive or negative. The results are shown in the Table 3.

In the case of this coefficient, the standard speaker voice database was assessed very high. At this time the automatic "Speaker" database was worse than the

Table 2 The results of the DCR method tests

Databases	Speaker standard	Speaker automatic	Janusz standard	Janusz automatic
DMOS	3.73	3.63	3.91	3.87
Standard deviation	0.53	0.56	0.45	0.44

Table 3 The results of the CCR method tests

Databases	Speaker standard	Speaker automatic	Janusz standard	Janusz automatic
CMOS	−0.47	−0.83	−0.87	−0.75
Standard deviation	0.28	0.65	0.43	0.5

Table 4 The results of the logatom method tests

Databases	Speaker standard	Speaker automatic	Janusz standard	Janusz automatic
Logatom sharpness	52.86 %	57.63 %	44.93 %	61.83 %
Standard deviation	0.1	0.11	0.15	0.12

standard one, but the quality of this database is acceptable. The "Janusz" standard and automatic databases were assessed in very similar way, close to the speaker's voice automatic database. Received grades indicate the acceptable databases quality, only slightly worse than the natural speech.

5.4 Logatom Articulation Method

The logatom articulation method determines the percent of logatoms correctly received by the listeners in relation to the total number of logatoms presented [11–13]. The logatom recognition is the result of hearing all of its component phonemes and not an association with the known word. The results of averaged logatom articulation for all study are shown in the Table 4. During analyzing, we can see that, the logatom sharpness is better for both voices in automatic databases. In case of the speaker voice, the average difference is approx 5 %, and in case of "Janusz" databases this difference is about 15 %. In any case and any type of logatoms, their recognition is better when they are created using automatic databases.

6 Summary

The performed studies showed, that the quality of databases created automatically with the use of presented algorithms is comparable with the quality of standard databases created manually. This means that they can be used in the concatenative speech synthesisers. The presented system was developed for creation databases intended for synthesiser which uses allophones as a basic units. However, after some modifications, it may be adapted to other types of basic units, such as diphones or syllables.

References

1. Dutoit, T.: An Introduction to text-to-speech synthesis. Kluwer Academic Publishers, Dordrecht (1997)
2. Taylor, P.: Text-to-speech synthesis. Cambridge University Press, Cambridge (2009)

3. Van Santen, J., Sproat, R., Olive, J., Hirshberg, J.: Progress in speech synthesis. Springer, New York (1997)
4. Szpilewski, E., Piórkowska, B., Rafałko, J., Lobanov, B., Kiselov, V., Tsirulnik, L.: Polish TTS in multi-voice slavonic languages speech synthesis system. In: SPECOM'2004 Proceedings, 9th International Conference Speech and Computer, pp. 565–570. Saint-Petersburg, Russia (2004)
5. Jassem, W.: Podstawy fonetyki akustycznej, wyd. PWN, Warszawa (1973)
6. Lobanov, B., Piórkowska, B., Rafałko, J., Cyrulnik, L.: Реализация межъязыковых различий интонации завиершённости и незавиершённости в синтезаторе русской и полской речи по тексту. In: Computational Linguistics and Intellectual Technologies, International Conference Dialogue'2005 Proceedings, pp. 356–362. Zvenigorod, Russia (2005)
7. Matoušek, J.: Building a new czech text-to-speech system using triphonebased speech units. In: Text, Speech and Dialog, Proceedings of the 3rd International Workshop TSD'2000, pp. 223–228. Czech Republic, Brno (2000)
8. Piórkowska, B., Popowski, K., Rafałko, J., Szpilewski, E.: Polish language speech synthesis basis on text information. New trends in audio and video, vol. I, pp. 507–526. Rozprawy Naukowe Nr 134, Białystok (2006)
9. Skrelin, P.: Allophone-based concatenative speech synthesis system for Russian. In: Text, Speech and Dialog, Proceedings of the 2nd International Workshop TSD'99, pp. 156–159. Czech Republic, Pilsen (1999)
10. Brachmański, S.: VoIP—ocena jakości transmisji mowy metodą ACR i DCR. Przegląd Telekomunikacyjny i Wiadomości Telekomunikacyjne, nr **8–9**, 424–427 (2003)
11. Janicki, A., Księżak, B., Kijewski, J., Kula, S.: Badanie jakości sygnału mowy w telefonii internetowej z wykorzystaniem zdań nieprzewidywalnych semantycznie. KSTiT 2006, Bydgoszcz (2006)
12. Trzaskowska, M.J., Mucha, B.: Metody obiektywnej oceny jakości usługi głosowej QoS w sieciach łączności elektronicznej. w Metody obiektywnej oceny jakości usługi głosowej QoS w sieciach łączności elektronicznej oraz urządzenia do takiej oceny i do badania dostępności usług poprzez numery alarmowe—etap 1, załącznik "X", Instytut Łączności, Państwowy Instytut Badawczy, Warszawa (2006)
13. ITU-T Recommendation P.800.: Method for subjective determination of transmission quality (1996)
14. PN-90/T-05100.: Analogowe łańcuchy telefoniczne. Wymagania i metody pomiaru wyrazistości logatomowej. Warszawa (1993)

InterCriteria Analysis Approach to Parameter Identification of a Fermentation Process Model

Tania Pencheva, Maria Angelova, Peter Vassilev and Olympia Roeva

Abstract In this investigation recently developed InterCriteria Analysis (ICA) is applied aiming at examination of the influence of a genetic algorithm (GA) parameter in the procedure of a parameter identification of a fermentation process model. Proven as the most sensitive GA parameter, generation gap is in the focus of this investigation. The apparatuses of index matrices and intuitionistic fuzzy sets, laid in the ICA core, are implemented to establish the relations between investigated here generation gap, from one side, and model parameters of fed-batch fermentation process of *Saccharomyces cerevisiae*, from the other side. The obtained results after ICA application are analysed towards convergence time and model accuracy and some conclusions about observed interactions are derived.

Keywords InterCriteria analysis · Genetic algorithms · Generation gap · Parameter identification · *S. cerevisiae*

1 Introduction

InterCriteria Analysis (ICA), given in details in [2] is a contemporary approach for multicriteria decision making. ICA implements the apparatuses of index matrices (IM) and intuitionistic fuzzy sets (IFS) in order to compare some criteria or estimated by them objects. In [16] ICA has been applied for the first time in the field

T. Pencheva (✉) · M. Angelova · P. Vassilev · O. Roeva
Institute of Biophysics and Biomedical Engineering, Bulgarian Academy of Sciences, 105 Acad. G. Bonchev Str., 1113 Sofia, Bulgaria
e-mail: tania.pencheva@biomed.bas.bg

M. Angelova
e-mail: maria.angelova@biomed.bas.bg

P. Vassilev
e-mail: peter.vassilev@gmail.com

O. Roeva
e-mail: olympia@biomed.bas.bg

K.T. Atanassov et al. (eds.), *Novel Developments in Uncertainty Representation and Processing*, Advances in Intelligent Systems and Computing 401,
DOI 10.1007/978-3-319-26211-6_33

of parameter identification of fermentation processes (FP) models. The ICA implementation allowed to authors to establish relations and dependencies between two of the main genetic algorithms (GA) parameters—numbers of individuals and number of generations, on the one hand, and convergence time, model accuracy and model parameters on the other hand. The reported results confirmed some existing dependencies that result from the physical meaning of the model parameters and from the stochastic nature of GA. The current investigation is a successive attempt ICA to be applied in the establishment of correlations between GA parameters and FP model parameters.

Fermentation processes (FP) are objects of increase research interest because of their widespread used in different branches of industry. Due to the fact that FP models have a complex structures based on systems of non-linear differential equations with several specific growth rates [8] their modeling and optimization are a real challenge for the investigators. Thus the choice of appropriate model parameter identification procedure is the most important problem for their adequate modeling. Among others biologically inspired optimization techniques, GA [9, 13] has been proved as a global search methods for solving many engineering and optimization problems [17] and especially for parameter identification of fermentation processes [1, 15, 18]. GA workability are quided mainly by different operators, functions, parameters, and settings that can be implemented specifically in different problems. Current research is focused on the investigation of the influence of proven as the most sensitive GA operator [1], namely generation gap (*ggap*). Simple GA (SGA) is applied for the purposes of model parameter identification of *S. cerevisiae* fed-batch fermentation process. This representative of yeast has numerous applications in food and pharmaceutical industry. Also it is widely used model organism in genetic engineering and cell biology due to their well known metabolic pathways [12].

ICA could be appropriate approach for establishing the correlations between model and optimization algorithm parameters, when given parameters are considered as criteria. This may lead to additional exploring of the model or the relation between model and optimization algorithm, which will be valuable especially in the case of modelling of living systems, such as FP. (Due to that reason in this investigation ICA is applied to identify the influences and relations between model parameters, on the one hand, and generation gap on the other hand based on results of parameters identification procedures of yeast.)

In this paper aiming to derived additional knowledge for existing correlations that will be useful in identification procedures when modelling FP, presented in [16] investigations are expanded and further developed. Since the focus of the research is on the relations between FP model parameters, here ICA is applied based on results of parameters identification procedures applying SGA with different values of *ggap*. (Current investigation is carried out based on the parameter identification *S. cerevisiae* fed-batch fermentation processes.)

The paper is organized as follows: the problem formulation is given in Sect. 2, while Sect. 3 presents the background of ICA. Numerical results and discussion are presented in Sect. 4 and conclusion remarks are given in Sect. 5.

2 Problem Formulation

2.1 *Mathematical Model of S. cerevisiae Fed-Batch Fermentation Process*

The mathematical model of *S. cerevisiae* fed-batch process is presented by the following non-linear differential equations system [15]:

$$\frac{dX}{dt} = (\mu_{2S}\frac{S}{S + k_S} + \mu_{2E}\frac{E}{E + k_E})X - \frac{F_{in}}{V}X \quad (1)$$

$$\frac{dS}{dt} = -\frac{\mu_{2S}}{Y_{S/X}}\frac{S}{S + k_S}X + \frac{F_{in}}{V}(S_{in} - S) \quad (2)$$

$$\frac{dE}{dt} = -\frac{\mu_{2E}}{Y_{E/X}}\frac{E}{E + k_E}X + \frac{F_{in}}{V}(S_{in} - S) \quad (3)$$

$$\frac{dV}{dt} = F_{in} \quad (4)$$

where X is the biomass concentration, [g/l]; S—substrate concentration, [g/l]; E—ethanol concentration, [g/l]; F_{in}—feeding rate, [l/h]; V—bioreactor volume, [l]; S_{in}—substrate concentration in the feeding solution, [g/l]; μ_{2S}, μ_{2E}—the maximum values of the specific growth rates, [1/h]; k_S, k_E—saturation constants, [g/l]; $Y_{S/X}$, $Y_{E/X}$—yield coefficients, [-].

For the considered here model (Eqs. (1–4)), the following parameter vector should be identified: $p_1 = [\mu_{2S}\ \mu_{2E}\ k_S\ k_E\ Y_{S/X}\ Y_{E/X}]$.

Model parameters identification of a *S. cerevisiae* fed-batch fermentation process is performed based on experimental data for biomass, glucose and ethanol concentrations. The detailed description of the process conditions and experimental data can be found in [15].

2.2 *Optimization Criterion*

The objective function is considered as a mean square deviation between the experimental data trajectories and model predicted ones, defined as:

$$J = \sum_{i=1}^{m} \left(X_{\exp}(i) - X_{\mathrm{mod}}(i)\right)^2 + \sum_{i=1}^{n} \left(S_{\exp}(i) - S_{\mathrm{mod}}(i)\right)^2 \rightarrow \min \quad (5)$$

where m and n are the experimental data dimensions; $X_{\exp}$ and $S_{\exp}$—available experimental data for biomass and substrate; X_{mod} and S_{mod}—model predictions for biomass and substrate with a given model parameter vector.

2.3 Generation Gap

Simple genetic algorithm initially presented in Goldberg [13] searches a global optimal solution using three main genetic operators in a sequence selection, crossover and mutation. SGA starts with a creation of a randomly generated initial population. Each solution is then evaluated and assigned a fitness value. According to the fitness function, the most suitable solutions are selected. After that, crossover proceeds to form a new offspring. Mutation is then applied with determinate probability aiming to prevent falling of all solutions into a local optimum. The execution of the GA has been repeated until the termination criterion (i.e. reached number of populations, or found solution with a specified tolerance, etc.) is satisfied.

Parameter generation gap ($ggap$) defines the percentage of the population that is replaced in each generation of the GA. A $ggap$ of 0 means none of the population is replaced; conversely, a generation gap of 100 means that the entire population is replaced in each generation.

De Jong [11] introduces $ggap$ as a GA parameter for the first time in 1975. He evaluated empirically the performance of GAs with overlapping populations and found that when the value of $ggap$ is low the algorithm had loss of alleles. As a result searching performance of GA becomes poor [10]. Very big $ggap$ value does not lead to improve GA performance too, especially regarding how fast the solution will be found [14].

In the previous investigation [1] is shown that $ggap$ is the most sensitive GA parameter towards the convergence time when GA is applied in the field of parameter identification of fermentation process models. With appropriate tuned $ggap$ almost 40 % of the algorithms calculation time can be saved, while keeping the model accuracy.

For the considered here parameter identification of *S. cerevisiae* fed-batch fermentation process, the GA operators and parameters are tuned as proposed in [1], namely: crossover operators are double point; mutation operators—bit inversion; selection operators—roulette wheel selection; number of generations—$maxgen = 100$; crossover rate—$xovr = 0.95$; mutation rate—$mutr = 0.05$, number of individuals—$nind = 20$. Here the following values of $ggap$ are used $ggap = 0.5 : 0.01 : 1$.

3 InterCriteria Analysis

Here the idea proposed in [2] is expanded. Following [2, 4] an Intuitionistic Fuzzy Pair (IFP) [3] as an estimation of the degrees of "agreement" and "disagreement" between two criteria applied to different objects is obtained. Recall that an IFP is an ordered pair of real non-negative numbers $\langle a, b \rangle$ such that: $a + b \leq 1$.

Consider an Index Matrix (IM) [5, 6] whose index sets consist of the criteria (for rows) and objects (for columns). The elements of this IM are further assumed to be real numbers. An IM with index sets consisting of the criteria (for rows and for columns) with elements IFPs corresponding to the "agreement" and "disagreement" between the respective criteria is then constructed.

Let O denotes the set of all objects $O_1, O_2, \ldots, O_n$ being evaluated, and $C(O)$ be the set of values assigned by a given criteria C to the objects, i.e.

$$O \stackrel{\text{def}}{=} \{O_1, O_2, \ldots, O_n\};\; C(O) \stackrel{\text{def}}{=} \{C(O_1), C(O_2), \ldots, C(O_n)\}$$

Then, let:

$$C^*(O) \stackrel{\text{def}}{=} \{\langle x, y\rangle \mid x \neq y \;\&\; \langle x, y\rangle \in C(O) \times C(O)\}$$

In order to find the "agreement" of two criteria the vector of all internal comparisons of each criteria, which fulfill exactly one of three relations R, $\overline{R}$ and $\tilde{R}$, is constructed. In other words, it is required that for a fixed criterion C and any ordered pair $\langle x, y\rangle \in C^*(O)$ it is true:

$$\langle x, y\rangle \in R \Leftrightarrow \langle y, x\rangle \in \overline{R} \tag{6}$$

$$\langle x, y\rangle \in \tilde{R} \Leftrightarrow \langle x, y\rangle \notin (R \cup \overline{R}) \tag{7}$$

$$R \cup \overline{R} \cup \tilde{R} = C^*(O) \tag{8}$$

From the above it is seen that only a subset of $C(O) \times C(O)$ needs to be considered for the effective calculation of the vector of internal comparisons (further denoted by $V(C)$) since from Eqs. (6–8) it follows that if the relation between x and y is known, then so is the relation between y and x. Thus of interest are only the lexicographically ordered pairs $\langle x, y\rangle$. Denote for brevity:

$$C_{i,j} = \langle C(O_i), C(O_j)\rangle$$

Then for a fixed criterion C, the vector with $\frac{n(n-1)}{2}$ elements is obtained:

$$V(C) = \{C_{1,2}, C_{1,3}, \ldots, C_{1,n}, C_{2,3}, C_{2,4}, \ldots, C_{2,n}, C_{3,4}, \ldots, C_{3,n}, \ldots, C_{n-1,n}\}$$

Let $V(C)$ be replaced by $\hat{V}(C)$, where for the kth component $(1 \leq k \leq \frac{n(n-1)}{2})$:

$$\hat{V}_k(C) = \begin{cases} 1, & \text{iff } V_k(C) \in R \\ -1, & \text{iff } V_k(C) \in \overline{R} \\ 0, & \text{otherwise} \end{cases}$$

When comparing two criteria the degree of "agreement" is determined as the number of matching components of the respective vectors (divided by the length of the vector for normalization purposes). This can be done in several ways, e.g. by counting the matches or by taking the complement of the Hamming distance. The degree of "disagreement" is the number of components of opposing signs in the two vectors (again normalized by the length). An example pseudocode for two criteria C and C' is presented below.

Algorithm 1 Calculating "agreement" and "disagreement" between two criteria

Require: Vectors $\hat{V}(C)$ and $\hat{V}(C')$

```
 1: function DEGREES OF AGREEMENT AND DISAGREEMENT(V̂(C), V̂(C'))
 2:     V ← V̂(C) − V̂(C')
 3:     μ ← 0
 4:     ν ← 0
 5:     for i ← 1 to n(n−1)/2 do
 6:         if V_i = 0 then
 7:             μ ← μ + 1
 8:         else if abs(V_i) = 2 then          ▷ abs(V_i): the absolute value of V_i
 9:             ν ← ν + 1
10:         end if
11:     end for
12:     μ ← (2 / n(n−1)) μ
13:     ν ← (2 / n(n−1)) ν
14:     return μ, ν
15: end function
```

If the respective degrees of "agreement" and "disagreement" are denoted by $\mu_{C,C'}$ and $\nu_{C,C'}$, it is obvious (from the way of computation) that $\mu_{C,C'} = \mu_{C',C}$ and $\nu_{C,C'} = \nu_{C',C}$. Also it is true that $\langle \mu_{C,C'}, \nu_{C,C'} \rangle$ is an IFP.

4 Numerical Results and Discussion

In order to obtain reliable results for computational time, optimization criterion and model parameters estimations, thirty independent runs of SGA have been performed for each value of $ggap$. Obtained results have been averaged and IM $A_{(ggap)}$ (Eq. 9) is constructed. Full IM is available at http://intercriteria.net/studies/gengap/s-cerev/

$$A_{(ggap)} = \begin{array}{c|ccccccccc} & C_1 & C_2 & C_3 & C_4 & C_5 & C_6 & C_7 & C_8 & C_9 \\ \hline O1 & 0.02214 & 167.14 & 0.5 & 0.9550 & 0.1283 & 0.1264 & 0.7993 & 0.4047 & 1.6917 \\ \ldots & \ldots & \ldots & \ldots & \ldots & \ldots & \ldots & \ldots & \ldots & \ldots \\ \ldots & \ldots & \ldots & \ldots & \ldots & \ldots & \ldots & \ldots & \ldots & \ldots \\ O29 & 0.02207 & 252.63 & 0.78 & 0.9889 & 0.1493 & 0.1427 & 0.7995 & 0.3974 & 2.0102 \\ \ldots & \ldots & \ldots & \ldots & \ldots & \ldots & \ldots & \ldots & \ldots & \ldots \\ O51 & 0.02364 & 355.54 & 1 & 0.9614 & 0.1084 & 0.1162 & 0.7182 & 0.4007 & 1.4754 \end{array} \quad (9)$$

IM $A_{(ggap)}$ presents averaged values for the objective function value J (C_1), computational time T (C_2), as well as the averaged estimates of model parameters, as follows: $\mu_{2S}(C_4)$, $\mu_{2E}(C_5)$, $k_S(C_6)$, $k_E(C_7)$, $Y_{S/X}(C_8)$, and $Y_{E/X}(C_9)$. The parameter *ggap* takes 51 different values from 0.5 to 1, with step 0.01. All mentioned criteria are listed in Table 1 for all considered here *ggap* values.

ICA algorithm calculate the IFP $\langle \mu, \nu \rangle$ for every two pairs of considered criteria using obtained IM $A_{(ggap)}$. To evaluate the variation in $\langle \mu, \nu \rangle$-values 6 different groups are selected.

The first group G_1 has 6 objects, namely for the following values of $ggap : G_1 = \{0.5,\ 0.6,\ 0.7,\ 0.8,\ 0.9,\ 1\}$, i.e. in the input IM only the objects $O_1, O_{11}, O_{21}, O_{31}, O_{41}$ and O_{51} are included for the corresponding six criteria. The second group consist of 11 objects, which are constructed by adding the some new values of *ggap* to the first group, namely $G_2 = \{G_1 \cup \{x + 0.05 | x \in G_1 \setminus \{1\}\}\}$. In the same manner the other groups may be represented as $G_i = \{G_{i-1} \cup G'_i,\ i = 3, 4, 5, 6\}$, where

$$G'_i = \{x + \frac{2i-4}{100} | x \in G_1 \setminus \{1\}\} \cup \{x + \frac{2i - 4 + (-1)^{\max(0,2i-9)} * 5}{100} | x \in G_1 \setminus \{1\}\}.$$

Obtained results after the ICA application are presented in Table 1. Figure 1a, b present distribution of the obtained here altogether 36 criteria pairs towards the number of considered objects. The results are discussed according proposed in [7] scale (see Table 2).

As it could be seen from Table 1 and especially from the plotted results in Fig. 1a–d, there are six pairs of criteria with positive consonance, namely $J \leftrightarrow Y_{S/X}$ ($C_1 \leftrightarrow C_8$) (Fig. 1a), $T \leftrightarrow ggap$ (Fig. 1b) ($C_2 \leftrightarrow C_3$), and pairs $\mu_{2E} \leftrightarrow Y_{E/X}$ ($C_5 \leftrightarrow C_9$), $\mu_{2E} \leftrightarrow k_S$ ($C_5 \leftrightarrow C_6$), $k_S \leftrightarrow Y_{E/X}$ ($C_6 \leftrightarrow C_9$), $\mu_{2S} \leftrightarrow k_S$ ($C_4 \leftrightarrow C_6$) (Fig. 1d). All these six pairs start and finish in the corresponding ranges: pairs $J \leftrightarrow Y_{S/X}$, $\mu_{2E} \leftrightarrow k_S$, $k_S \leftrightarrow Y_{E/X}$, and $\mu_{2S} \leftrightarrow k_S$ are in weak positive consonance; the pair $T \leftrightarrow ggap$ is in the strong positive consonance; while the pair $\mu_{2E} \leftrightarrow Y_{E/X}$ is on the edge between positive consonance and strong positive consonance. It should be noted, that the pair $T \leftrightarrow ggap$ is with the highest recorded in this investigation degree of "agreement" $\mu = 1$. This is to prove the expected result that the convergence time is on a high dependence with *ggap* values.

On the other pole are the recorded results with degree of "agreement" μ, falling in the range of negative consonance. The most of pairs with such behaviour are towards the objective function value J. Looking at Fig. 1a, there are three such pairs—$J \leftrightarrow \mu_{2E}$ ($C_1 \leftrightarrow C_5$), $J \leftrightarrow k_S$ ($C_1 \leftrightarrow C_6$) and $J \leftrightarrow Y_{E/X}$ ($C_1 \leftrightarrow C_9$). The pairs $\mu_{2E} \leftrightarrow Y_{S/X}$ ($C_5 \leftrightarrow C_8$), $Y_{S/X} \leftrightarrow Y_{E/X}$ ($C_8 \leftrightarrow C_9$) and in the most of the time—$k_S \leftrightarrow Y_{S/X}$ ($C_6 \leftrightarrow C_8$) fall in the scale of weak negative consonance, going at the end in the scale of weak dissonance. It should be noted that the first two pairs start from weak dissonance at small number of objects, finishing with weak negative consonance at bigger number of objects, and vice versa for the third pair. Looking deeply insight, it can be concluded that the model parameters μ_{2E}, k_S and $Y_{E/X}$ are the parameters, causing the negative dissonance with J (Fig. 1a), and with $Y_{S/X}$ (Fig. 1d).

Table 1 Results from the ICA in case of *E. coli* fed-batch cultivation process

Criteria correlation	Number of objects					
	6	11	21	31	41	51
	$\langle\mu,\nu\rangle$	$\langle\mu,\nu\rangle$	$\langle\mu,\nu\rangle$	$\langle\mu,\nu\rangle$	$\langle\mu,\nu\rangle$	$\langle\mu,\nu\rangle$
$C_1 \leftrightarrow C_2$	⟨0.58, 0.42⟩	⟨0.55, 0.45⟩	⟨0.58, 0.42⟩	⟨0.59, 0.41⟩	⟨0.71, 0.29⟩	⟨0.87, 0.13⟩
$C_1 \leftrightarrow C_3$	⟨0.59, 0.41⟩	⟨0.56, 0.44⟩	⟨0.60, 0.40⟩	⟨0.59, 0.41⟩	⟨0.71, 0.29⟩	⟨0.87, 0.13⟩
$C_1 \leftrightarrow C_4$	⟨0.36, 0.64⟩	⟨0.34, 0.66⟩	⟨0.32, 0.68⟩	⟨0.29, 0.71⟩	⟨0.25, 0.75⟩	⟨0.27, 0.73⟩
$C_1 \leftrightarrow C_5$	⟨0.09, 0.91⟩	⟨0.10, 0.90⟩	⟨0.10, 0.90⟩	⟨0.13, 0.87⟩	⟨0.15, 0.85⟩	⟨0.07, 0.93⟩
$C_1 \leftrightarrow C_6$	⟨0.21, 0.79⟩	⟨0.20, 0.80⟩	⟨0.19, 0.81⟩	⟨0.15, 0.85⟩	⟨0.16, 0.84⟩	⟨0.20, 0.80⟩
$C_1 \leftrightarrow C_7$	⟨0.33, 0.67⟩	⟨0.33, 0.67⟩	⟨0.37, 0.63⟩	⟨0.31, 0.69⟩	⟨0.31, 0.69⟩	⟨0.47, 0.53⟩
$C_1 \leftrightarrow C_8$	⟨0.83, 0.17⟩	⟨0.83, 0.17⟩	⟨0.81, 0.19⟩	⟨0.77, 0.23⟩	⟨0.78, 0.22⟩	⟨0.80, 0.20⟩
$C_1 \leftrightarrow C_9$	⟨0.10, 0.90⟩	⟨0.11, 0.89⟩	⟨0.11, 0.89⟩	⟨0.12, 0.88⟩	⟨0.15, 0.85⟩	⟨0.13, 0.87⟩
$C_2 \leftrightarrow C_3$	⟨0.96, 0.04⟩	⟨0.95, 0.05⟩	⟨0.95, 0.05⟩	⟨0.98, 0.02⟩	⟨1.00, 0.00⟩	⟨1.00, 0.00⟩
$C_2 \leftrightarrow C_4$	⟨0.47, 0.53⟩	⟨0.46, 0.54⟩	⟨0.41, 0.59⟩	⟨0.42, 0.58⟩	⟨0.40, 0.60⟩	⟨0.40, 0.60⟩
$C_2 \leftrightarrow C_5$	⟨0.45, 0.55⟩	⟨0.49, 0.51⟩	⟨0.47, 0.53⟩	⟨0.50, 0.50⟩	⟨0.40, 0.60⟩	⟨0.20, 0.80⟩
$C_2 \leftrightarrow C_6$	⟨0.45, 0.55⟩	⟨0.46, 0.54⟩	⟨0.40, 0.60⟩	⟨0.42, 0.58⟩	⟨0.35, 0.65⟩	⟨0.33, 0.67⟩
$C_2 \leftrightarrow C_7$	⟨0.42, 0.58⟩	⟨0.40, 0.60⟩	⟨0.40, 0.60⟩	⟨0.36, 0.64⟩	⟨0.35, 0.65⟩	⟨0.33, 0.67⟩
$C_2 \leftrightarrow C_8$	⟨0.50, 0.50⟩	⟨0.45, 0.55⟩	⟨0.49, 0.51⟩	⟨0.46, 0.54⟩	⟨0.53, 0.47⟩	⟨0.67, 0.33⟩
$C_2 \leftrightarrow C_9$	⟨0.45, 0.55⟩	⟨0.49, 0.51⟩	⟨0.46, 0.54⟩	⟨0.50, 0.50⟩	⟨0.40, 0.60⟩	⟨0.27, 0.73⟩
$C_3 \leftrightarrow C_4$	⟨0.45, 0.55⟩	⟨0.44, 0.56⟩	⟨0.38, 0.62⟩	⟨0.42, 0.58⟩	⟨0.40, 0.60⟩	⟨0.40, 0.60⟩
$C_3 \leftrightarrow C_5$	⟨0.44, 0.56⟩	⟨0.48, 0.52⟩	⟨0.45, 0.55⟩	⟨0.50, 0.50⟩	⟨0.40, 0.60⟩	⟨0.20, 0.80⟩
$C_3 \leftrightarrow C_6$	⟨0.44, 0.56⟩	⟨0.45, 0.55⟩	⟨0.39, 0.61⟩	⟨0.43, 0.57⟩	⟨0.35, 0.65⟩	⟨0.33, 0.67⟩
$C_3 \leftrightarrow C_7$	⟨0.41, 0.59⟩	⟨0.40, 0.60⟩	⟨0.39, 0.61⟩	⟨0.36, 0.64⟩	⟨0.35, 0.65⟩	⟨0.33, 0.67⟩
$C_3 \leftrightarrow C_8$	⟨0.50, 0.50⟩	⟨0.46, 0.54⟩	⟨0.50, 0.50⟩	⟨0.47, 0.53⟩	⟨0.53, 0.47⟩	⟨0.67, 0.33⟩
$C_3 \leftrightarrow C_9$	⟨0.45, 0.55⟩	⟨0.48, 0.52⟩	⟨0.45, 0.55⟩	⟨0.50, 0.50⟩	⟨0.40, 0.60⟩	⟨0.27, 0.73⟩
$C_4 \leftrightarrow C_5$	⟨0.61, 0.39⟩	⟨0.65, 0.35⟩	⟨0.65, 0.35⟩	⟨0.65, 0.35⟩	⟨0.64, 0.36⟩	⟨0.67, 0.33⟩
$C_4 \leftrightarrow C_6$	⟨0.79, 0.21⟩	⟨0.80, 0.20⟩	⟨0.78, 0.22⟩	⟨0.77, 0.23⟩	⟨0.80, 0.20⟩	⟨0.80, 0.20⟩
$C_4 \leftrightarrow C_7$	⟨0.59, 0.41⟩	⟨0.58, 0.42⟩	⟨0.60, 0.40⟩	⟨0.60, 0.40⟩	⟨0.58, 0.42⟩	⟨0.40, 0.60⟩
$C_4 \leftrightarrow C_8$	⟨0.39, 0.61⟩	⟨0.36, 0.64⟩	⟨0.35, 0.65⟩	⟨0.32, 0.68⟩	⟨0.11, 0.89⟩	⟨0.07, 0.93⟩
$C_4 \leftrightarrow C_9$	⟨0.63, 0.37⟩	⟨0.65, 0.35⟩	⟨0.65, 0.35⟩	⟨0.65, 0.35⟩	⟨0.64, 0.36⟩	⟨0.60, 0.40⟩
$C_5 \leftrightarrow C_6$	⟨0.77, 0.23⟩	⟨0.79, 0.21⟩	⟨0.80, 0.20⟩	⟨0.83, 0.17⟩	⟨0.76, 0.24⟩	⟨0.73, 0.27⟩
$C_5 \leftrightarrow C_7$	⟨0.60, 0.40⟩	⟨0.61, 0.39⟩	⟨0.56, 0.44⟩	⟨0.58, 0.42⟩	⟨0.58, 0.42⟩	⟨0.47, 0.53⟩
$C_5 \leftrightarrow C_8$	⟨0.18, 0.82⟩	⟨0.19, 0.81⟩	⟨0.20, 0.80⟩	⟨0.22, 0.78⟩	⟨0.33, 0.67⟩	⟨0.27, 0.73⟩
$C_5 \leftrightarrow C_9$	⟨0.95, 0.05⟩	⟨0.96, 0.04⟩	⟨0.95, 0.05⟩	⟨0.95, 0.05⟩	⟨0.96, 0.04⟩	⟨0.93, 0.07⟩
$C_6 \leftrightarrow C_7$	⟨0.63, 0.37⟩	⟨0.62, 0.38⟩	⟨0.60, 0.40⟩	⟨0.59, 0.41⟩	⟨0.56, 0.44⟩	⟨0.33, 0.67⟩
$C_6 \leftrightarrow C_8$	⟨0.26, 0.74⟩	⟨0.24, 0.76⟩	⟨0.23, 0.77⟩	⟨0.22, 0.78⟩	⟨0.20, 0.80⟩	⟨0.13, 0.87⟩
$C_6 \leftrightarrow C_9$	⟨0.78, 0.22⟩	⟨0.78, 0.22⟩	⟨0.79, 0.21⟩	⟨0.82, 0.18⟩	⟨0.76, 0.24⟩	⟨0.80, 0.20⟩
$C_7 \leftrightarrow C_8$	⟨0.44, 0.56⟩	⟨0.44, 0.56⟩	⟨0.49, 0.51⟩	⟨0.48, 0.52⟩	⟨0.45, 0.55⟩	⟨0.67, 0.33⟩
$C_7 \leftrightarrow C_9$	⟨0.61, 0.39⟩	⟨0.60, 0.40⟩	⟨0.56, 0.44⟩	⟨0.60, 0.40⟩	⟨0.58, 0.42⟩	⟨0.40, 0.60⟩
$C_8 \leftrightarrow C_9$	⟨0.20, 0.80⟩	⟨0.20, 0.80⟩	⟨0.21, 0.79⟩	⟨0.24, 0.76⟩	⟨0.33, 0.67⟩	⟨0.33, 0.67⟩

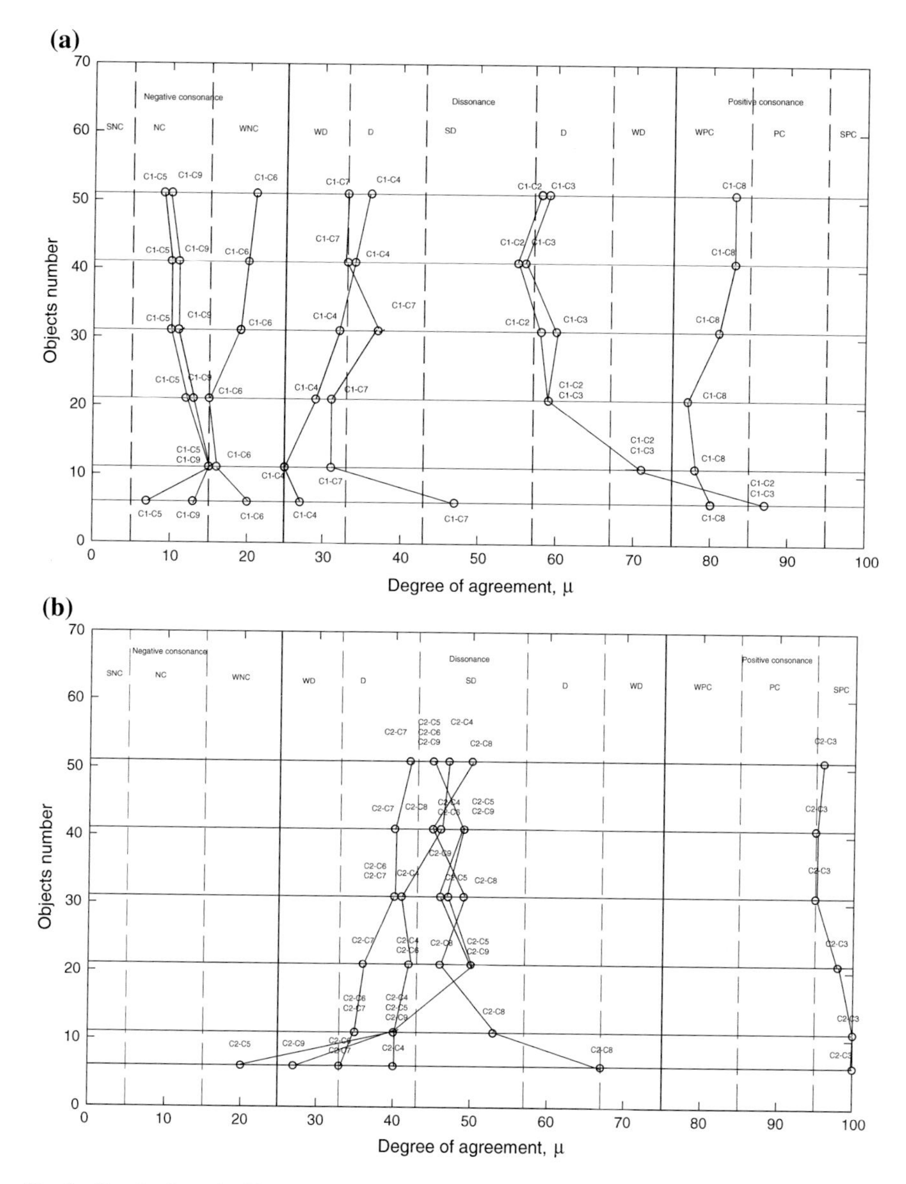

Fig. 1 Results from the ICA application towards the number of considered objects

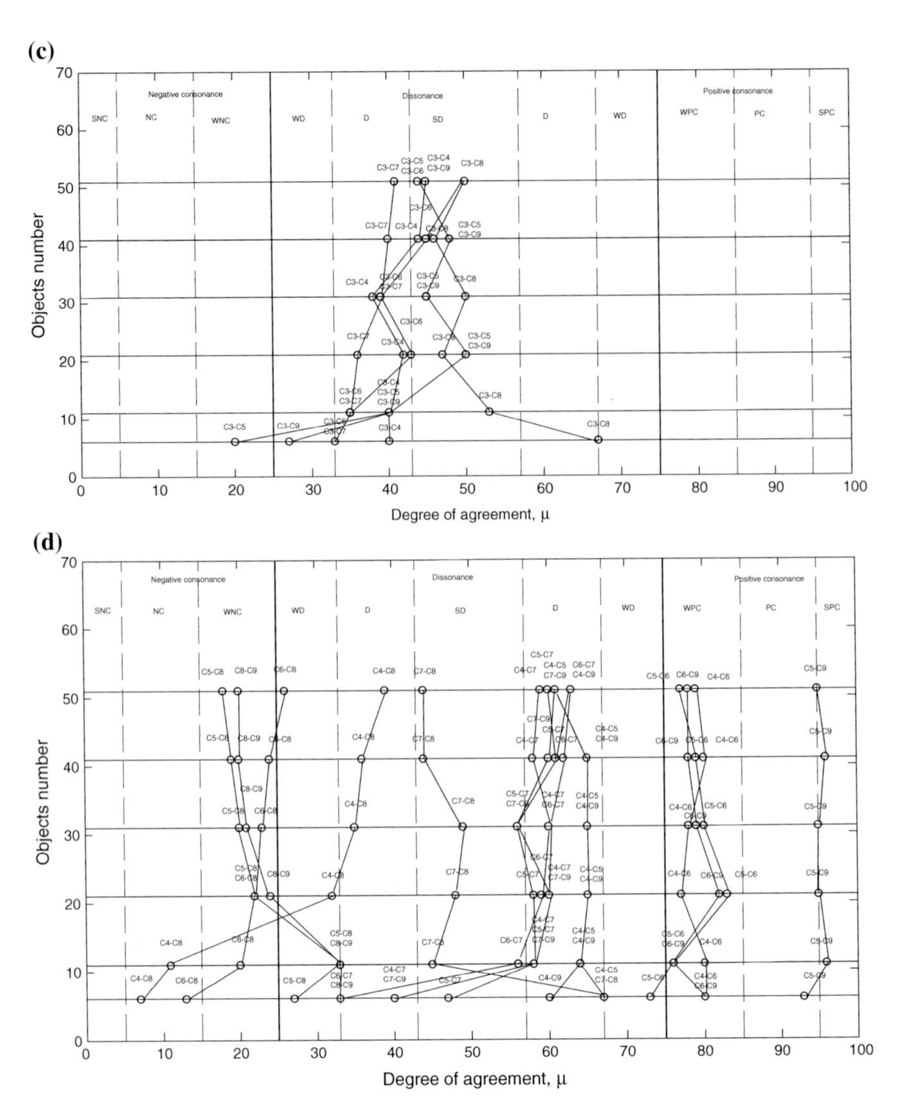

Fig. 1 (continued)

Looking further in Fig. 1a–d, the biggest part of the pairs are with the degree of "agreement" μ, falling in the range of dissonance—altogether 25 pairs of criteria. Those presented in Fig. 1a, b have a reasonable explanation due to the stochastic nature of GA. Because of that, it is quite difficult for some very strong relations between the optimization criterion J and the convergence time T on the one hand, and model parameters on the other hand, to be established. If the randomly chosen initial conditions are sufficiently good, the relation between J and T might be of higher agreement—already mentioned pair $T \leftrightarrow Y_{S/X}$ (Fig. 1a). Or, conversely, GA starting

Table 2 Consonance and dissonance scale [7]

Interval of $\mu_{C,C'}$	Meaning
[0–0.5]	Strong negative consonance (SNC)
(0.5–0.15]	Negative consonance (NC)
(0.15–0.25]	Weak negative consonance (WNC)
(0.25–0.33]	Weak dissonance (WD)
(0.33–0.43]	Dissonance (D)
(0.43–0.57]	Strong dissonance (SD)
(0.57–0.67]	Dissonance (D)
(0.67–0.75]	Weak dissonance (WD)
(0.75–0.85]	Weak positive consonance (WPC)
(0.85–0.95]	Positive consonance (PC)
(0.95–1.00]	Strong positive consonance (SPC)

from more "distanced" initial conditions may need more time to find a satisfactory solution, thus leading to a lower degree of "agreement" between J and T, that happen in the most of the criteria pairs considered here.

The following things are worth to be noted: criteria pairs of $T \leftrightarrow \mu_{2E}$ (Fig. 1b) and $ggap \leftrightarrow \mu_{2E}$ (Fig. 1c) start with a weak negative consonance at the small number of objects, while finish in the range of strong dissonance at the bigger number of objects. The pairs $\mu_{2S} \leftrightarrow Y_{S/X}$ and $k_S \leftrightarrow Y_{S/X}$ (mentioned also above) start in the range of negative consonance and finish respectively in the range of dissonance and weak dissonance at the bigger number of objects. There are many other pairs, showing such "improvement" in the degree of "agreement"—start with lower μ at the small number of objects and finish with higher μ at the bigger number of objects. Meanwhile, there are some pairs showing quite opposite behaviour—starting with bigger μ at the small number of objects and finishing with lower μ at the bigger number of objects. The most "representatives" among them are the pairs between $T \leftrightarrow J$ and $J \leftrightarrow ggap$ (Fig. 1a), crossing "the border" between positive consonance and dissonance. The similar is the pair behaiour $T \leftrightarrow Y_{S/X}$ (Fig. 1b) and $ggap \leftrightarrow Y_{S/X}$ (Fig. 1c) and $k_E \leftrightarrow Y_{S/X}$ (Fig. 1d).

When consider the influence of the generation gap $ggap$ and model parameters, in most of the cases it is observed mentioned above "improvement" in the degree of "agreement". All they start with lower μ at the small number of objects and finish with higher μ at the bigger number of objects, except one obvious "deviation"—the pair $ggap \leftrightarrow Y_{S/X}$. It should be noticed that all of connections between $ggap$ and model parameters show degree of "agreement" in the range of dissonance (except the pair $ggap \leftrightarrow \mu_{2E}$ at small number of objects), that lead to the suggestion that there are no significant correlations between them.

The last presented group of examined correlations is between model parameters themselves (Fig. 1d). The most interesting results have been mentioned yet. This figure is with the mostly dispersed degree of "agreement" μ. Although, the biggest part of the criteria pairs fall in the range of dissonance, meaning that there are no significant dependencies to be outlined. This fact has logically explanation—more difficult process model structure trouble the establishment of some very strong relations between the parameters themselves. Although this fact, there are four pairs with proven positive consonance and three with negative consonance (all they analysed above). Thus, new knowledge almost impossible without ICA application has been achieved during the investigation.

5 Conclusion

In this paper the idea of recently proposed InterCriteria Analysis is applied for establishing the relations and dependencies between GAs parameter generation gap on the one hand, and convergence time, model accuracy and model parameters on the other hand. Simple GA with different values of $ggap$ is used for parameter identification of fed-batch fermentation process of *S. cerevisiae*.

The obtained results from ICA implementation might be summarised as follows: there are (1) six pairs of criteria are in positive consonance; (2) five pairs of criteria are in negative consonance; (3) twenty five pairs of criteria are in dissonance. It should be noted, that the pair $J \leftrightarrow ggap$ is with the highest recorded in this investigation μ value. This is to prove the expected result that the value of optimization criterion is on a high dependence with proven as the most sensitive GA parameter $ggap$.

The obtained results from ICA show some existing relations and dependencies that result from the physical meaning of the model parameters on the one hand, and from stochastic nature of the considered metaheuristic on the other hand. Moreover, derived additional knowledge for existing correlations will be useful in further identification procedures of fermentation process models and, in general, for more accurately SGA application.

Acknowledgments The work is supported by the Bulgarian National Scientific Fund under the grant DFNI-I-02-5 "InterCriteria Analysis—A New Approach to Decision Making".

References

1. Angelova, M.: Modified Genetic Algorithms and Intuitionistic Fuzzy Logic for Parameter Identification of Fed-batch Cultivation Model. Ph.D. Thesis, Sofia (2014) (in Bulgarian)
2. Atanassov, K., Mavrov, D., Atanassova, V.: Intercriteria decision making: a new approach for multicriteria decision making, based on index matrices and intuitionistic fuzzy sets. Issues IFSs GNs **11**, 1–8 (2014)

3. Atanassov, K., Szmidt, E., Kacprzyk, J.: On intuitionistic fuzzy pairs. Notes IFS **19**(3), 1–13 (2013)
4. Atanassov, K.: On Intuitionistic Fuzzy Sets Theory. Springer, Berlin (2012)
5. Atanassov, K.: On index matrices, part 1: standard cases. Adv. Stud. Contemp. Math. **20**(2), 291–302 (2010)
6. Atanassov, K.: On index matrices, part 2: intuitionistic fuzzy case. Proc. Jangjeon Math. Soc. **13**(2), 121–126 (2010)
7. Atanassov, K., Atanassova, V., Gluhchev, G.: InterCriteria analysis: ideas and problems. Notes Intuitionistic Fuzzy Sets **21**(1), 81–88 (2015)
8. Bastin, G., Dochain, D.: On-line Estimation and Adaptive Control of Bioreactors. Elsevier Scientific Publications, Amsterdam (1991)
9. Boussaid, I., Lepagnot, J., Siarry, P.: A survey on optimization metaheuristics. Inf. Sci. **237**, 82–117 (2013)
10. Cantu-Paz, E.: Selection Intensity in Genetic Algorithms with Generation Gaps. In: Proceedings of the Genetic and Evolutionary Computation Conference, pp. 911–918. Morgan Kaufmann, Las Vegas (2000)
11. De Jong, K.A.: An Analysis of the Behavior of a Class of Genetic Adaptive Systems. Doctoral Dissertation, University of Michigan, Ann Arbor, University Microfilms, No. 76–9381 (1975)
12. Dickinson, R.J., Schweizer, M.: Metabolism and Molecular Physiology of Saccharomyces Cerevisiae, 2nd edn. CRC Press, Boca Raton (2004)
13. Goldberg, D.E.: Genetic Algorithms in Search, Optimization and Machine Learning. Addison Wesley Longman, London (2006)
14. Obitko, M.: Genetic Algorithms. http://www.obitko.com/tutorials/genetic-algorithms/
15. Pencheva, T., Roeva, O., Hristozov, I.: Functional State Approach to Fermentation Processes Modelling. Prof. Marin Drinov Academic Publishing House, Sofia (2006)
16. Pencheva, T., Angelova, M., Atanassova, V., Roeva, O.: InterCriteria analysis of genetic algorithm parameters in parameter identification. Notes Intuitionistic Fuzzy Sets **21**(2), 99–110 (2015)
17. Roeva, O. (Ed.): Real-world Application of Genetic Algorithms, InTech (2012)
18. Roeva, O., Fidanova, S., Paprzycki, M.: Population size influence on the genetic and ant algorithms performance in case of cultivation process modeling. Recent Adv. Comput. Optim. Stud. Comput. Intell. **580**, 107–120 (2015)

Author Index

K.T. Atanassov et al. (eds.), *Novel Developments in Uncertainty Representation and Processing*, Advances in Intelligent Systems and Computing 401,
DOI 10.1007/978-3-319-26211-6

Printed in the United States
By Bookmasters